PRESTRESSED CONCRETE DESIGN

PRESTRESSED CONCRETE DESIGN

Praveen Nagarajan

Department of Civil Engineering
NIT Calicut, Kerala

Pearson

Senior Editor—Acquisitions: Sandhya Jayadev
Editor—Production: M. Balakrishnan

ISBN 978-93-325-1375-4

First Impression

Published by Dorling Kindersley (India) Pvt. Ltd, licensees of Pearson Education in South Asia.

Head Office: 7th Floor, Knowledge Boulevard, A-8(A), Sector 62, Noida 201 309, UP, India.
Registered Office: 11 Community Centre, Panchsheel Park, New Delhi 110 017, India.

Compositor: White Lotus InfoTech Pvt. Ltd., Puducherry.
Printed in India at Repro India Limited

To my parents,
Mrs. Rani Nagarajan and Dr N. M. Nagarajan,
and my Teachers at
NIT Calicut and IIT Madras

To my parents,

Contents

Preface

This book deals with the analysis and design of prestressed concrete members as per the IS: 1343–1980 specification. It is intended to be an easy-to-read textbook for a first course in prestressed concrete design for senior undergraduate and postgraduate students of Civil Engineering. The book will also be useful for practising engineers and help them understand the basis behind the code recommendations.

Spread across 13 chapters, the key features of the book are:

- The concepts of prestressed concrete design are explained in a simple manner.
- The contents of each chapter are enhanced with new and better figures for improved understanding.
- Detailed derivations are presented for almost all equations.
- The book follows a step-by-step approach for problem solving, with each step explained from first principles.

Chapter 1 explains the behaviour of the prestressed concrete member and introduces the different methods of prestressing and types of prestressed concrete.

The need for high tensile steel and high strength concrete for prestressed concrete structure is brought out in Chapter 2, which also describes briefly some of the important properties of these materials.

The philosophy of limit state design is discussed in Chapter 3. The various sources and methods to evaluate the loss of prestressing force in both pre-tensioned and post-tensioned prestressed concrete members are dealt with in Chapter 4.

Chapter 5 analyzes the behaviour of prestressed concrete member in flexure at both serviceability and ultimate limit state. It elaborates on the different methods to calculate elastic stress distribution in concrete, namely, the combined load approach, the internal couple approach and the equivalent load approach. The chapter also examines the procedures to find the ultimate moment of resistance of both rectangular and flanged sections.

Chapters 6 and 7 examine the design for shear, torsion and anchorage zones. Methods to predict the short term and long term deflections of prestressed concrete beams are elucidated in Chapter 8.

Chapter 9 outlines the procedure to be followed for the design of statically determinate prestressed concrete beams (Type 1 and Type 2 members). The methods of analysis and design of composite members are revealed in Chapter 10.

Chapter 11 gives a brief introduction to the analysis of indeterminate prestressed concrete members. The design of one-way and two-way slabs is delineated in Chapter 12. Finally, Chapter 13 covers the design of prestressed concrete pipes, watertanks and ring beams.

I thank Ms. Sandhya Jayadev of Pearson Education, for her support and Mr. Beljith P., for typing the manuscript.

Comments, feedback and suggestions for the improvement of the book are welcome.

Praveen Nagarajan

About the Author

Dr Praveen Nagarajan is an alumnus of NIT Calicut and IIT Madras. An illustrious student of Civil Engineering throughout his college life, he worked briefly for L&T Ramboll, Chennai, before joining as faculty at NIT Calicut.

Included among Dr Praveen's many areas of interest are reinforced and prestressed concrete, steel structures, bridge engineering, structural reliability and structural dynamics – topics in which he has published numerous technical papers. A recipient of several awards for technical excellence such as the Valli Anantharamakrishnan Merit Prize (2003) from IIT Madras, E P Nicolaides Prize (2009) from the Institution of Engineers (India) and the Best Young Teacher award (for the academic year 2008–2009) from NIT Calicut, he has guided more than 25 M.Tech. projects. Currently Assistant Professor at the Department of Civil Engineering, NIT Calicut, he is also guiding three research scholars to their doctoral degrees.

chapter 1

Basic Principles

1.1 Introduction

Prestressed concrete can be defined as concrete in which internal compressive stresses are deliberately introduced so as to counteract the tensile stresses (and consequent cracking) resulting from external loads to a desired degree. In simple words, a precompressed concrete is known as prestressed concrete. In concrete members, the prestress is commonly introduced by tensioning high tensile steel reinforcements known as *tendons*.

The advantage of prestressing a structural member can be explained by considering a row of books shown in Fig. 1.1. Though each book is a discrete element, the row of books can be lifted as a single unit by applying axial compressive force C at the ends.

The row of books has zero tensile capacity. Hence the beam made of books cannot even carry its own self-weight. However, if the books are externally compressed (prestressed), the beam can resist the self-weight and even some additional loads kept at the top till the tensile stress developed by bending overcomes the compressive stress induced by prestress. The same resisting mechanism exists in a prestressed concrete member.

1.2 Need for Prestressed Concrete

Concrete is a building material which is strong in compression and at the same time very weak in tension. One way to overcome this limitation is to use steel reinforcement in concrete to take the tensile stresses.

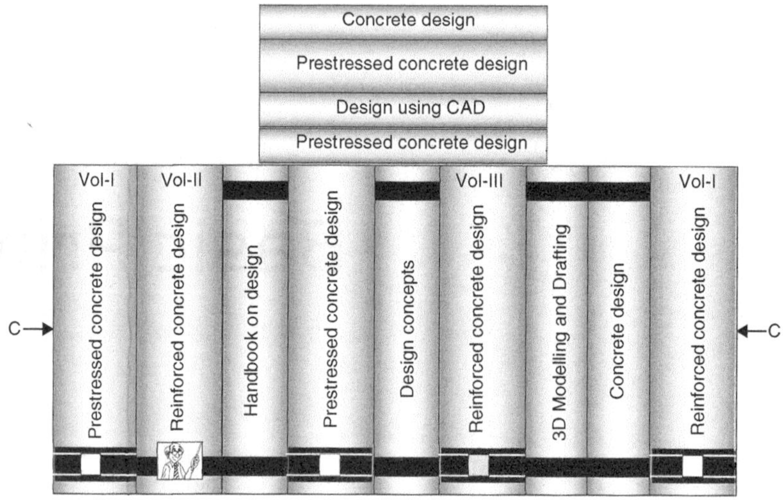

Fig. 1.1 Row of books lifted as a single unit

In reinforcement concrete, steel reinforcements are placed in the regions of tensile stresses. But the steel reinforcements do not prevent the development of tensile cracks and hence cannot eliminate the loss of stiffness due to cracking. This decrease in stiffness results in larger deflection of the member even when subjected to service loads.

To minimize the loss in stiffness by cracking, it is necessary to restrict the width of cracks which in turn sets a limit to the maximum allowable stress in steel reinforcement. Due to this, not much benefit is obtained by using high strength steel and high strength concrete in reinforced concrete members.

In a prestressed concrete, compressive stresses are introduced in order to reduce or eliminate tensile stresses and cracking of the member. Hence a prestressed concrete member is more stiffer and durable than a similar reinforced concrete member.

1.3 Brief History of Prestressed Concrete

The concept of prestressing, i.e., to preload a structure so as to reduce the effects of external loads has been used for a long time. The practice of heating the steel rim before fitting on wheel, winding ropes around wooden staves of barrel, tensioning the spokes in a bicycle wheel are some examples where the principle of prestressing are used.

The first attempt to prestress concrete was done by Jackson of USA. In 1886, he obtained a patent for constructing artificial stone and concrete arches by tightening steel rods. In 1888, C. E. W. Doehring of Germany used tensioned wires to prestress concrete slabs and small beams. These earlier attempts to prestress concrete were not successful due to the use of low strength steel. The long term effects like shrinkage and creep of concrete, stress relaxation in steel lead to loss in prestress within a few months of construction.

In 1925, R. H. Dill of USA used high strength unbonded steel rods for prestressing concrete. Eugene Freyssinet of France systematically investigated the time dependent prestress losses due to shrinkage and creep of concrete. He successfully developed the technique of practically prestressing concrete structures using high strength wires in 1928. He is in fact known as the 'Father of Prestressed Concrete'. After this development, prestressed concrete became a widely accepted method of construction.

1.4 Structural Behaviour of Prestressed Concrete Member

The behaviour of a prestressed concrete beam is explained by considering a simply supported plain concrete beam subjected to a uniformly distributed external load w (Fig. 1.2(a)). In this example, the self-weight of the beam is neglected. The beam will fail suddenly when the stress at the bottom fibre of the beam at section 1 – 1 (location of maximum bending moment) exceeds the tensile strength (modulus of rupture) of concrete. If it is assumed that concrete has no tensile strength, then the beam can carry no load. Therefore, the intensity of external load that this beam can carry (w_l) is equal to zero.

$$w_l = 0$$

As the next case, consider that the beam is subjected to an axial compressive or prestressing force P (Fig. 1.2(b)). The force P causes a uniform compression stress of intensity P/A at all sections in the beam. Where A is the

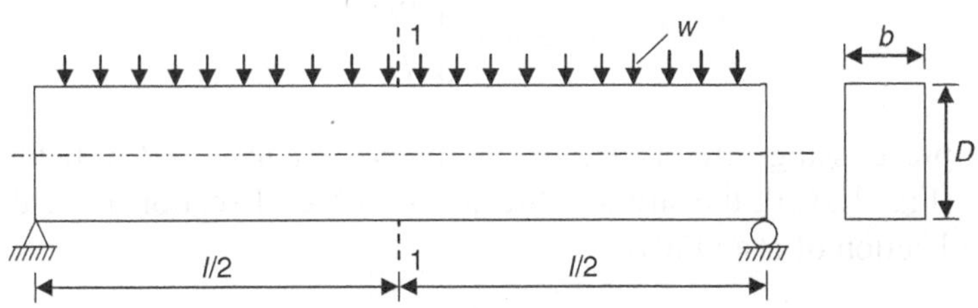

(a)

Fig. 1.2 (a) Plain concrete beam with no prestress

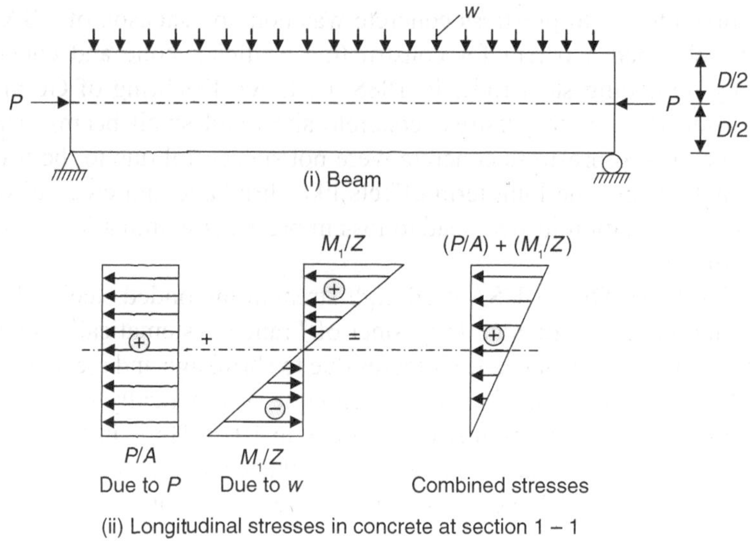

(b)

Fig. 1.2 (b) Plain concrete beam with axial prestress (e_s = 0)

cross-sectional area equal to bD. The uniformly distributed load w causes a tensile stress equal to M_1/Z at the bottom fibre of section 1 – 1. Where M_1 is the bending moment due to w at the section ($M_1 = wl^2/8$) and Z is the section modulus ($Z = bD^2/6$). The beam will fail when the tensile stress due to w exceeds the compressive stress due to P. Hence just before failure, stress at the bottom fibre (σ_b) of section 1 – 1 will be equal to zero.

Thus the uniformly distributed load causing failure of the beam (w_{II}) is:

At failure, $$\sigma_b = 0 \Rightarrow \frac{P}{A} - \frac{M_1}{Z} = 0$$

(*Note:* Compressive stress is taken as positive and tensile stress as negative.)

Substituting the values of M_1, A and Z in the equation gives:

$$w_{II} = w = \frac{4}{3}\frac{PD}{l^2}$$

If the prestressing force is applied at an eccentricity $e_s = D/6$ below the centroid (Fig. 1.2(c)), the stress at the bottom fibre of section 1 – 1 due to combined action of P and w is:

$$\sigma_b = \frac{P}{A} + \frac{Pe_s}{Z} - \frac{M_1}{Z}$$

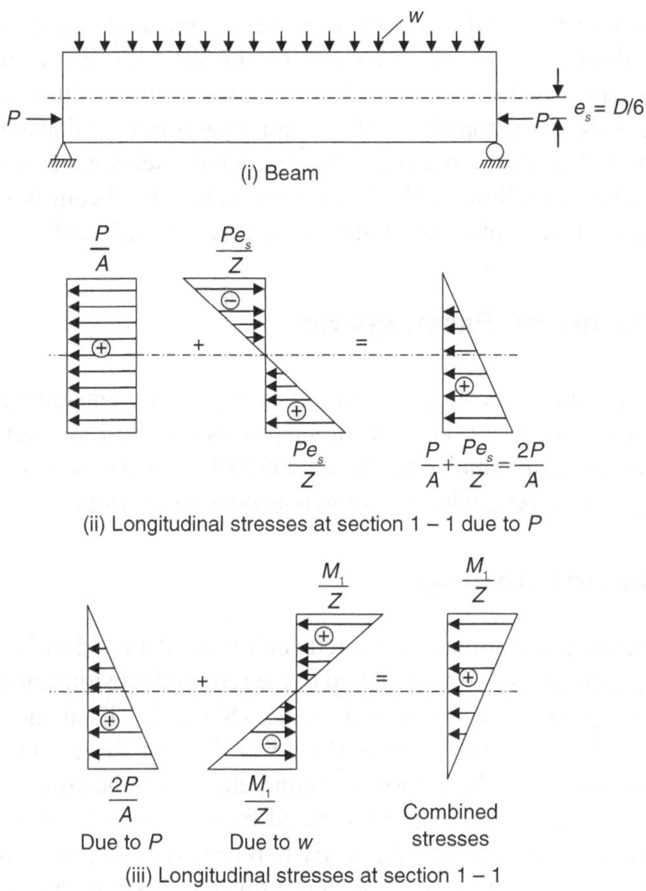

(c)

Fig. 1.2 (c) Plain concrete beam with eccentric prestress ($e_s = D/6$)

Substituting the values for e_s, A, Z and M_1, and equating σ_b to zero gives the distributed load at failure (w_{III}) as:

$$w_{III} = w = \frac{8}{3}\frac{PD}{l^2} = 2w_{II}$$

Thus the application of the prestressing force at an eccentricity of $D/6$ has doubled the load carrying capacity of the beam.

The eccentric prestressing force develops bending moment opposite in nature of the moment caused by external loads. By properly selecting P and e_s, it is possible to counteract the effects of external loading to any degree. However, the magnitude of P cannot be increased arbitrarily to compensate the tensile stress. Increase in P also increases the compressive stress P/A.

Hence P should be restricted such that the compressive stress does not exceed the allowable compressive stress of concrete. Similarly, increasing e_s beyond a critical value causes the failure of the beam. In this example, at $e_s = b/6$, the stress at the top fibre of the beam at section 1 – 1 due to prestress is zero (Fig. 1.2(c)(ii)). Increasing e_s beyond $b/6$ causes tensile stress at the top fibre leading to failure of the beam even before the beam is subjected to w. Hence the values of P and e_s should be carefully selected.

1.5 Methods of Prestressing

In concrete members, prestress is commonly applied by tensioning the tendons using hydraulic jacks. The tensioning process can be either done before casting of concrete, in which case the method is known as *pre-tensioning* or after casting of concrete, which is known as *post-tensioning*.

1.5.1 Pre-tensioning

The different stages involved in pre-tensioning are depicted in Fig. 1.3. In the first stage, the tendons are stretched to the required tension and anchored to the rigid supports known as *bulk heads* (Stage 1). Then the concrete member is cast around the tendons (Stage 2). During the curing process, bond develops between the tendon and concrete. When the concrete attains sufficient strength, the tendons are released from the bulk heads (Stage 3). Immediately after release, the tendons try to return to its original length. But the bond between the steel and concrete prevents this movement thereby forcing the concrete into compression. Since in pre-tensioning method, the prestressing force from the tendon is transferred to concrete by bond, pre-tensioned members usually have large numbers of prestressing steel of smaller diameter so as to increase the area of contact.

For members having significant self-weight, it is advantages to use tendons having curved profile. In pre-tensioning, it is very difficult to layout a tendon in a curved shape. The tendons can be only given a sharp change in direction at selected points using hold-down forces (Fig. 1.4).

The pre-tensioning method is ideally suited for factory production where casting yard and hold-down forces can be easily arranged. Several members can be cast at the same time. Hence this method is suitable for the manufacture of precast members of similar size. Since the members will be transported from the factory to the site, there will be constraints on the size and weight of members prestressed by pre-tensioning. This method is widely used for the mass production of short-span structures like railway sleepers, electric poles, fencing posts etc.

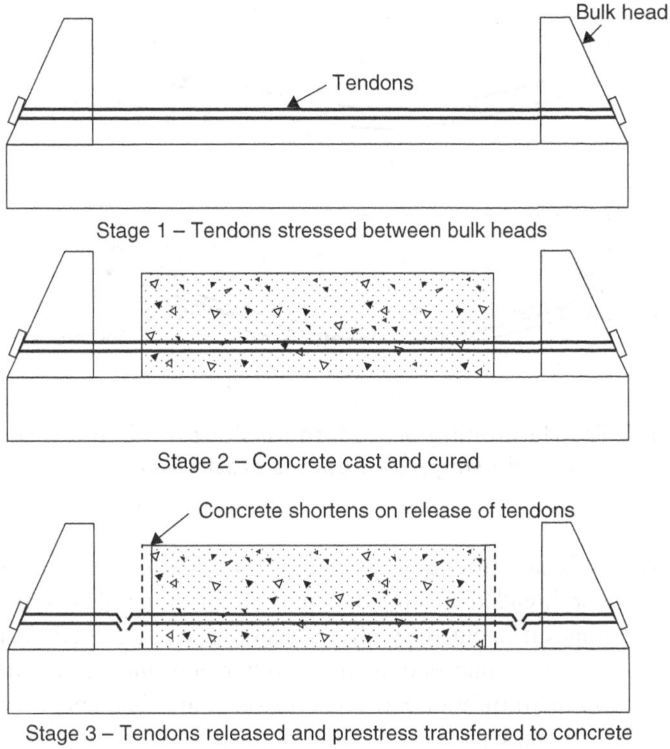

Fig. 1.3 Pre-tensioning procedure

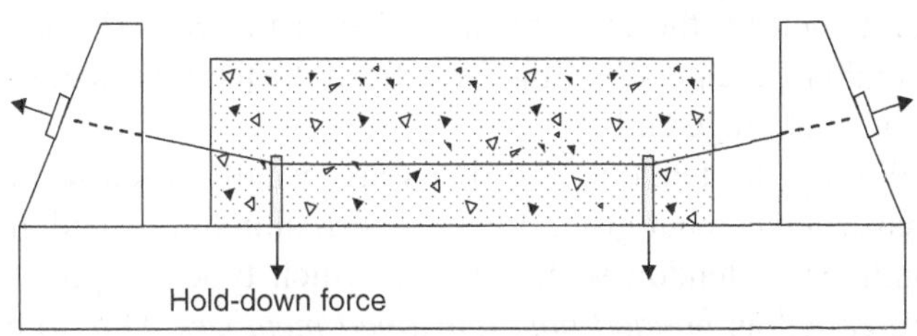

Fig. 1.4 Deflected pre-tensioning tendons

1.5.2 Post-tensioning

One of the important limitations of pre-tensioning is the difficulty in fixing the tendon in the required profile. This limitation can be overcome by post-tensioning the member. The different stages involved in post-tensioning process are shown in Fig. 1.5.

In the beginning, metal ducts or corrugated high density polyethylene ducts are placed in the required tendon profile by tying them to the reinforcement

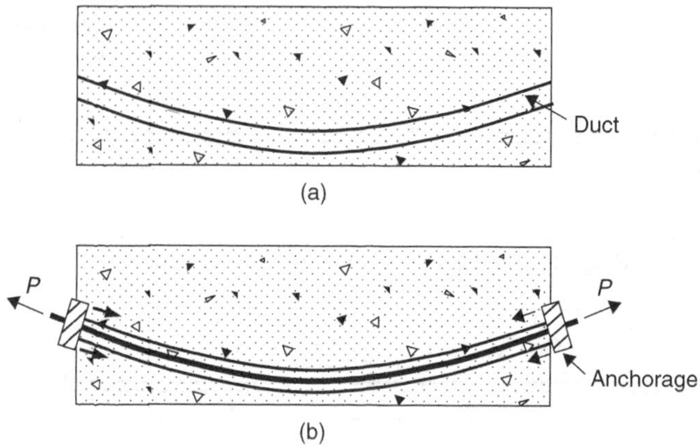

Fig. 1.5 Post-tensioning procedure (a) stage 1 – beam is cast and (b) stage 2 – beam is prestressed

cage and concrete is cast around the duct (Stage 1). The tendons can be either kept in the ducts or threaded through the ducts after casting of concrete. Once the concrete attains sufficient strength, the tendons are stretched against the hardened concrete from one end with the other end anchored (known as *dead end anchor*) or from both ends. The tendons are then anchored at the stressing ends (also known as *live anchor*) using external anchorage systems (Stage 2).

During the stretching process, the concrete gets compressed and this prestress is maintained by the tendons anchored at the ends. Hence in post-tensioned members, the prestressing force from the tendons is transferred to concrete at the anchorage ends.

After anchoring the tendons, the ducts are usually filled using cement grout. The hardened cement grout prevents corrosion of the tendon and also helps in bonding the tendon with concrete. Such types of post-tensioned members are called as *bonded post-tensioned members*. Due to economic and practical difficulties, sometimes the ducts may be left as such without grouting. In these members, there will be no bond between the tendon and concrete, and the members are known as *unbonded members*.

Almost all in-situ prestressed concrete members are post-tensioned.

External Prestressing

In conventional post-tensioning, it is not possible to replace the corroded tendons. This difficulty can be overcome by tensioning the tendon kept outside the concrete. This type of prestressing is known as *external prestressing*. There will be no bond between the tendon and concrete, and the tendon can be replaced or added whenever required. This method can be also used for retrofitting and strengthening existing structures.

Post-tensioning Anchorage Systems

The prestressing force in the tendons is transferred to concrete through anchorages (Fig. 1.5(b)). Though there are different types of anchorages used in practice, they can be grouped under three categories:

1. Wedge anchorages
2. Buttonhead anchorages
3. Threaded bar anchorages

Wedge Anchorages: Wedge anchorages are the most widely used form of anchorages. In this type, the tendons are anchored by the friction forces developed between the tendons and wedges bearing against the sides of a conical hole in a steel plate.

The common form of a wedge anchorage system consists of a steel plate (*anchor head*) with number of conical holes, conical wedges and a bearing plate (Fig. 1.6). The bearing plate will be embedded in concrete and they are used to transfer the prestressing force to concrete. The anchor head consists of number of tapered holes for gripping the tendon using wedges.

The tendons (wires or strands) are passed through the wedges. When the tendons are tensioned to the required force, the jack is released and the tendons try to move towards the anchor head. Since the wedges cannot move in, they get locked into the anchor head along with the tendons. The anchor head bears against the bearing plate which in turn transfers the force to concrete by bearing.

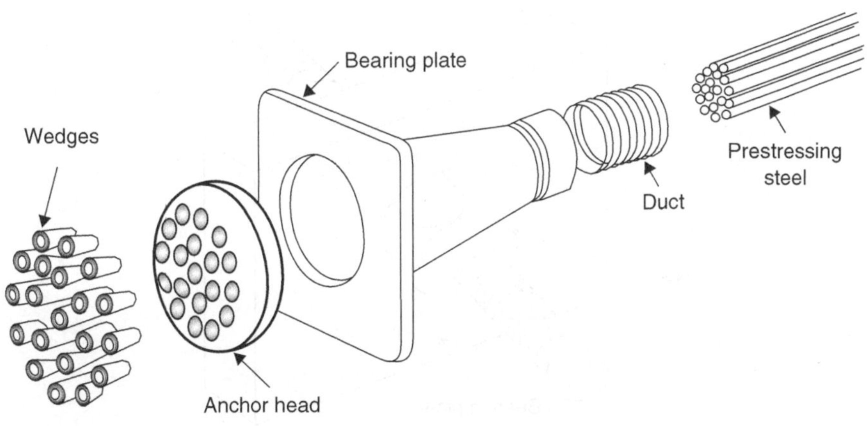

Fig. 1.6 Wedge anchorages

Buttonhead Anchorages: In these types of anchorage systems (Fig. 1.7) the tendons are cut into the required lengths and buttonheads are formed at the ends. The buttonheads bears against a threaded anchorhead which can be fitted inside a jack so that all tendons can be stretched together. After tensioning to the required force, steel shims are inserted between the anchor head and bearing plate.

Threaded Bar Anchorages: This anchorage system consists of steel bars as tendons threaded at the ends with anchor nut bearing against the steel bearing plate (Fig. 1.8). During the tensioning process, as the steel bar is pulled by the jack, the nut is continuously screwed down, and it transfers the force to the bearing plate after the jack is released.

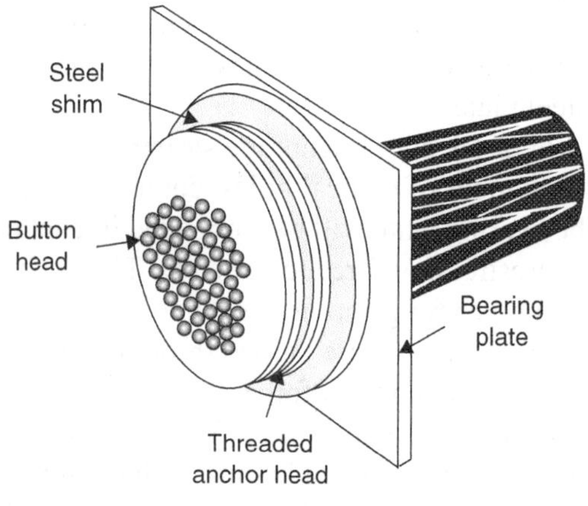

Fig. 1.7 Buttonhead anchorage

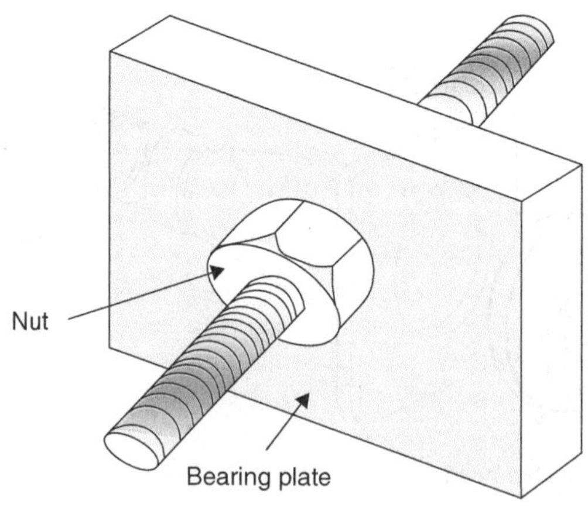

Fig. 1.8 Bar anchorage

1.6 Types of Prestressed Concrete

Prestressed concrete can be classified in various ways as discussed below.

1.6.1 Classification as per IS:1343–1980[1]

The Indian standard IS: 1343–1980 classifies all prestressed concrete members into three types as:

- *Type 1:* Members in which no tensile stress is permitted under working conditions are known as type 1 or *fully prestressed members*.
- *Type 2:* Members in which tensile stress is allowed but cracking is not permitted within working conditions are type 2 or *limited prestressed members*.
- *Type 3:* Members in which cracking is permitted under working conditions are known as type 3 or *partially prestressed members*. In these members, the width of cracks should be within the allowable limits.

1.6.2 Other Classifications

- *Bonded prestressed concrete:* In these members, the tendon is attached to surrounding concrete through bond. All pre-tensioned members and post-tensioned members having grouted ducts are bonded prestressed concrete members.
- *Unbonded prestressed concrete:* In this case, there is no bond between the tendon and concrete. Externally prestresed members and post-tensioned members with ungrouted ducts are examples for unbonded prestressed concrete members.
- *Concentrically prestressed members:* In these members, the centroid of the tendon coincides with the centroid of the section. These members are subjected to a uniform compressive stress.
- *Eccentrically prestressed members:* When the centroid of the tendon does not coincide with the centroid of the section, then the member is said to be eccentrically prestressed.
- *Uniaxial, biaxial or triaxial prestressing:* Uniaxial, biaxial or triaxial prestressing refers to the situation in which the member is prestressed in one, two or three mutually perpendicular directions.

---

[1] IS: 1343–1980 (Reaffirmed 2004), *Code of Practice for Prestressed Concrete* (First Revision), Bureau of Indian Standards, Manak Bhavan, New Delhi.

- *Linear prestressing and circular prestressing:* When the member is straight in the direction of prestressing, then the prestressing is known as linear prestressing. Prestressing in curved members are called circular prestressing.

1.7 Comparison with Reinforced Concrete

One of the main advantages of prestressed concrete over reinforced concrete is that by prestressing, the structure can be made crack-free under full working loads. Hence the full section is effective in resisting the loads. On the other hand, in reinforced concrete, due to low tensile strength, the concrete located in the flexural tensile zone is ineffective. Since there is no crack in prestressed concrete (type 1 and type 2 members), they are more durable, and can be used in aggressive atmosphere and for making liquid-retaining structures.

Due to pre-compression and reduction in shear due to component of prestressing force in curved cable, the principle tensile stress in prestressed concrete will be lesser than in reinforced concrete. Hence shear resistance in prestressed concrete can be provided by using thin webs and low amount of shear reinforcements.

In prestressed concrete, only high tensile steel can be used as longitudinal reinforcement. Hence the quantity of steel required in prestressed concrete is very less when compared with reinforced concrete. Thus concrete can be better placed in prestressed concrete due to low percentage of longitudinal and shear reinforcements.

Eccentrically prestressing the member causes deflections opposite to that caused due to applied loads. Hence by properly choosing the prestress, the deflection of the member can be reduced or even completely eliminated.

Due to the above reasons, when compared with reinforced concrete, prestressed concrete requires smaller sections. Hence, prestressed concrete members are lighter and have more aesthetic appeal. Prestressed concrete not only saves material, it also reduces dead load. This aspect makes it the most preferred construction method for long-span structures like bridges where dead load is a significant part of the total design load.

In both reinforced and prestressed concrete members, the external moment is resisted by the internal couple formed by steel in tension and concrete in compression. In a reinforced concrete section, the lever arm remains constant and the stresses in concrete and steel increases with increase in external moment. But in the case of uncracked prestressed concrete section, the stresses in steel almost remain constant and the increase in moment is resisted by increasing the lever arm. This means that the stress in prestressing steel remains a constant even if there is a fluctuation in the applied loading. Hence uncracked prestressed members have high fatigue resistance, since the fatigue failure depends on the amplitude of stresses induced by fluctuations in loads.

A prestressed concrete structure can be considered to be a pre-tested structure. This is due to the fact that both steel and concrete are subjected to very high stresses at the time of prestressing. If the structure behaves satisfactory at this stage, then it can be expected to perform well at all other loading stages. This implies that a prestressed concrete structure is more reliable than a reinforced concrete structure.

In spite of these advantages, there are some limitations of prestressed concrete which needs to be considered. At all loading stages, the prestressed concrete is subjected to compression. Hence the problems due to creep will be more in these structures.

The construction of prestressed concrete requires a high level of quality control and inspection. Even if there is saving in materials, the cost of high strength materials and prestressing equipments makes prestressed concrete uneconomical for ordinary structures. Prestressed concrete are ideally suited for long-span structures. Sometimes the combination of prestressed and reinforced concrete construction known as *composite construction* results in economic structures.

1.8 Applications of Prestressed Concrete

Construction of structural members with large values of span to depth ratio using reinforced concrete is almost impractical. In such cases, prestressed concrete can be used to achieve economical, durable and elegant looking structures. Therefore prestressed concrete is used for long-span beams and bridges. Prestressed concrete are used in the construction of floor slabs in buildings where large column free space are required. They are also widely used in the construction of huge structures like containment walls of nuclear reactors, large capacity liquid retaining structures, cement silos, chimneys, towers etc.

1.9 Design Code

The analysis and design procedures discussed in this book are based on the Indian Code:

IS: 1343–1980 (Reaffirmed 2004) – *Code of Practice for Prestressed Concrete* (First Revision).

This Code will be referred as 'the Code' in the remaining chapters of the book.

chapter 2

Materials

2.1 Introduction

In order to design prestressed concrete structures, it is necessary that the designer should have a proper understanding of the properties of constituent materials. The principal materials used in prestressed concrete are high tensile steel and high strength concrete. In addition to these materials, plain or deformed reinforcing bars are used as shear reinforcement, longitudinal reinforcement and anchorage zone reinforcement. For bonded post-tensioned members, grout in the form of cement-sand slurry is used to fill the duct.

This chapter describes briefly some of the important properties of high tensile steel and high strength concrete, as they significantly affect the behaviour of prestressed concrete structures.

2.2 Prestressing Steel

2.2.1 High Tensile Steel

As per Cl. 2.3 of IS:2090–1983[1], alloy steels having tensile strength f_p not less than 980 N/mm^2 are known as *high tensile steel*. The high strength

[1] IS:2090–1983 (Reaffirmed 2004), *Specification for High Tensile Steel Bars Used in Prestressed Concrete*, Bureau of Indian Standards, Manak Bhavan, New Delhi.

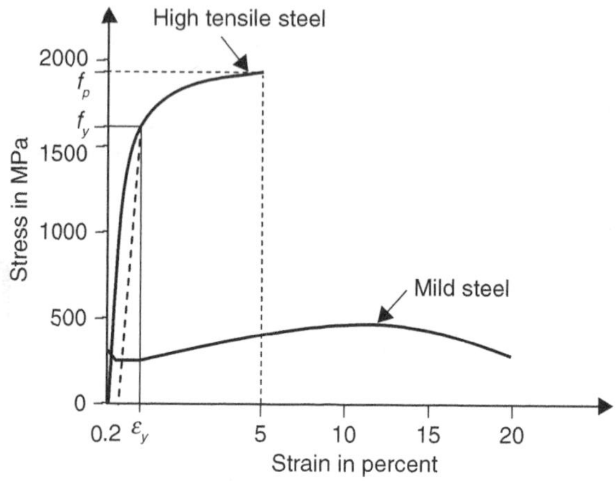

Fig. 2.1 Stress-strain curves of steel

in these types of steels is due to the presence of high carbon content. The carbon content in high tensile steel varies from 0.60 to 0.85 percent. But mild steel has carbon content in the range of 0.2 to 0.3 percent. Due to high carbon content, high tensile steel are harder and brittle in nature. The elongation at failure will be only in the order of 2.5 to 10 percent compared to 30 percent elongation in mild steel.

Fig. 2.1 shows typical stress-strain curves for both high tensile steel and mild steel. From the figure, it can be seen that high tensile steel does not have a well defined yield point, plastic and strain hardening range. In the absence of distinct yield point, the *proof stress*, i.e., the stress which produces a residual strain of 0.2 percent of the original gauge length is taken as the yield stress (f_y) for high tensile steel.

2.2.2 Need for High Tensile Steel

The initial attempts to prestress concrete structures failed due to the use of prestressing steel having low strength. The effect of prestress was completely lost due to prestress losses caused by shrinkage and creep of concrete.

The need for high strength prestressing steel is explained by considering the case of a concentrically prestressed concrete beam shown in Fig. 2.2. The length of the concrete beam and steel before prestressing are l_c and l_s respectively. After tensioning of steel and transfer of prestress, the stretched length of steel is l_{s1}. Due to application of compressive

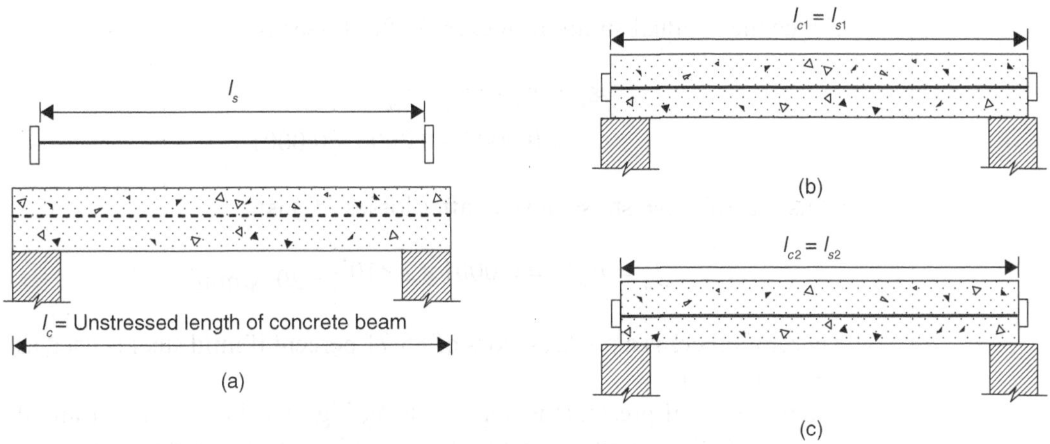

Fig. 2.2 Variation in length of prestressed concrete beam (a) before applying prestress, (b) at transfer of prestress (reduction in length of concrete due to elastic deformation) and (c) shortening due to shrinkage and creep of concrete

force, the length of the beam reduces to l_{c1} (due to elastic deformation of concrete). But $l_{c1} = l_{s1}$ (Fig. 2.2(b)). With the passage of time, shrinkage and creep in concrete reduces the length of beam and steel to l_{c2} and l_{s2} respectively (Fig. 2.2(c)).

The reduction in length of steel (Δl_s) due to shrinkage strain (ε_{sh}) and creep strain (ε_{cr}) in concrete is:

$$\Delta l_s = (\varepsilon_{sh} + \varepsilon_{cr}) l_s$$

The sum of shrinkage strain and creep strain is normally in the order of 0.0006. The reduction in stress in steel (Δf_s) i.e., loss in prestress due to these effects is:

$$\Delta f_s = (\varepsilon_{sh} + \varepsilon_{cr}) E_s$$

where E_s is the modulus of elasticity of steel which can be taken as 2×10^5 N/mm^2. Hence $\Delta f_s = 120$ N/mm^2.

If the beam is prestressed using mild steel at an initial stress of 140 N/mm^2, then the initial strain in steel ε_{si} is:

$$\varepsilon_{si} = \frac{f_s}{E_s} = \frac{140}{2 \times 10^5} = 0.0007$$

Hence the residual strain in steel (ε_{sr}) after losses is:

$$\varepsilon_{sr} = \varepsilon_{si} - (\varepsilon_{sh} + \varepsilon_{cr})$$
$$= 0.0007 - 0.0006 = 0.0001$$

Thus the effective stress in steel after losses (f_{se}) is:

$$f_{se} = \varepsilon_{sr} E_s = 0.0001 \times 2 \times 10^5 = 20 \text{ N/mm}^2$$

Therefore there is a prestress loss of 85.71 percent if mild steel is used as prestressing steel.

Alternatively, if prestress is applied using high tensile steel at an initial stress of 1000 N/mm^2, then the initial and residual strains in steel are:

$$\varepsilon_{si} = \frac{1000}{2 \times 10^5} = 0.005$$

$$\varepsilon_{sr} = 0.005 - 0.0006 = 0.0044$$

The effective stress in steel is:

$$f_{se} = 0.0044 \times 2 \times 10^5 = 880 \text{ N/mm}^2$$

In this case, there is only 12 percent loss in prestress.

The initial and effective prestress in mild steel and high tensile steel are shown in Fig. 2.3. From the figure, it can be concluded that in order to have a significant effective prestress (i.e., prestress after losses), only high tensile steel should be used.

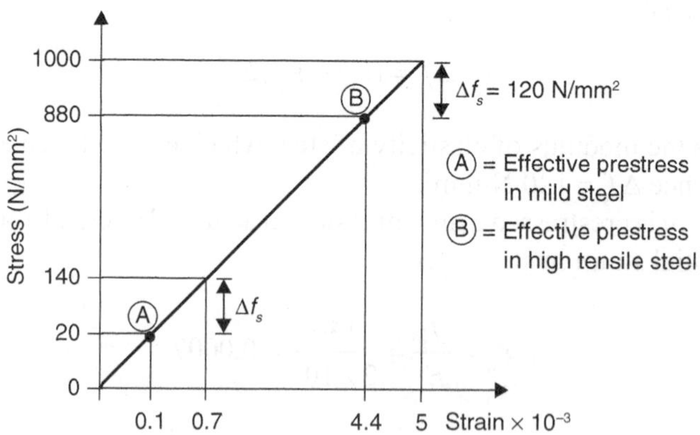

Fig. 2.3 Loss of stress in steel due to shrinkage and creep of concrete

2.2.3 Types of Prestressing Steel

Prestressing steel or tendons usually takes one of the following forms:

- Wires (confirming to IS: 1785 (Part I)[2], IS:1785 (Part II)[3] and IS:6003[4])
- Strands (confirming to IS: 6006[5] and IS:14268[6])
- Bars (confirming to IS: 2090[1])
- Cables

Basically there are only two types of prestressing steel, viz. wires and bars. In order to reduce the number of units to be handled during tensioning process, wires are grouped together to form strands or cables. The fabrication of strands is similar to the process of rope making. Strands consist of a single layer or multiple layers of wires laid in helices over a central core wire. Among the different types of strands, seven wire strands are commonly used for prestressed concrete. A prestressing cable consists of a group of strands or wires kept parallel to each other. Wires and strands are usually used for pre-tensioned concrete. For post-tensioned members, strands, bars and cables are used commonly.

High Tensile Steel Wires

High tensile steel wires are available in nominal diameters ranging from 2.5 mm to 8 mm. The wires are manufactured by drawing round rods through dies. In this drawing process, cold work is done on steel which increases the strength of steel. The wires thus obtained are known as the *as-drawn* wires, which are wound on to capstans of diameters 0.6–0.7 m.

---

[2] IS:1785 (Part I) – 1983 (Reaffirmed 2003), *Specification for Plain Hard-drawn Steel Wire for Prestressed Concrete*: Part I – Cold Drawn Stress-relieved Wire, Bureau of Indian Standards, Manak Bhavan, New Delhi.

[3] IS:1785 (Part II) – 1983 (Reaffirmed 2003), *Specification for Plain Hard-drawn Steel Wire for Prestressed Concrete*: Part II – As-drawn Wire, Bureau of Indian Standards, Manak Bhavan, New Delhi.

[4] IS 6003:1983 (Reaffirmed 2003), *Specification for Indented Wire for Prestressed Concrete*, Bureau of Indian Standards, Manak Bhavan, New Delhi.

[5] IS 6006:1983 (Reaffirmed 2003), *Specification for Uncoated Stress Relieved Strand for Prestressed Concrete*, Bureau of Indian Standards, Manak Bhavan, New Delhi.

[6] IS 14268:1995 (Reaffirmed 2003), *Uncoated Stress Relieved Low Relaxation Seven-ply Strand for Prestressed Concrete – Specification*, Bureau of Indian Standards, Manak Bhavan, New Delhi.

The minimum breaking load required for drawn wires as per IS:1785 (Part II) – 1983 are given in Table 2.1. The 0.2 percent proof load of these wires should not be less than 75 percent of the minimum breaking load.

Table 2.1 Minimum breaking load of as-drawn steel wires

Nominal Diameter (mm)	2.5	3.0	4.0	5.0
Minimum Breaking Load (kN)	8.84	12.48	21.55	30.83

The as-drawn wires will not pay out straight from the coils and hence they are only suitable for pre-tensioned concrete and for manufacturing of wire-wound pipes. The wires can be pre-straightened so that they can be threaded through ducts in post-tensioned concrete by heating them for a short time. Such wires are known as *stress-relieved* wires or *normal-relaxation* wires. The process of stress relieving removes the stresses induced by cold work and increases the ductility of wires. These pre-straightened wires are supplied in coils and they pay out straight. Thus they are easier to handle. For certain applications, it is necessary to reduce the losses in prestress due to relaxation of steel (see Section 4.33). In such cases, the as-drawn wires are heated while subjected to high tension. Such wires are known as *stabilized* wires or *low-relaxation* wires.

The minimum breaking load required for stress-relieved wires according to IS:1785 (Part I) – 1983 are given in Table 2.2. The 0.2 percent proof load should not be less than 85 percent of the minimum breaking load.

Table 2.2 Minimum breaking load of stress-relieved wires

Nominal Diameter (mm)	4.0	5.0	7.0	8.0
Minimum Breaking Load (kN)	21.55	30.83	56.57	69.12

In the case of pre-tensioned concrete members, the prestressing force is transferred to concrete by bond between steel and concrete. To improve the bond with concrete, the wires are sometimes indented in the form of small depressions at close regular spacing. The indentations are placed in two lines, diametrically opposite and the opposite indentations are staggered so that no two indentations are opposite to each other. The two common types of indentations are depicted in Fig. 2.4.

The minimum breaking load for indented wires (IS:6003–1983) are given in Table 2.3. The 0.2 percent proof load of indented wires should not be less than 85 percent of the minimum breaking load.

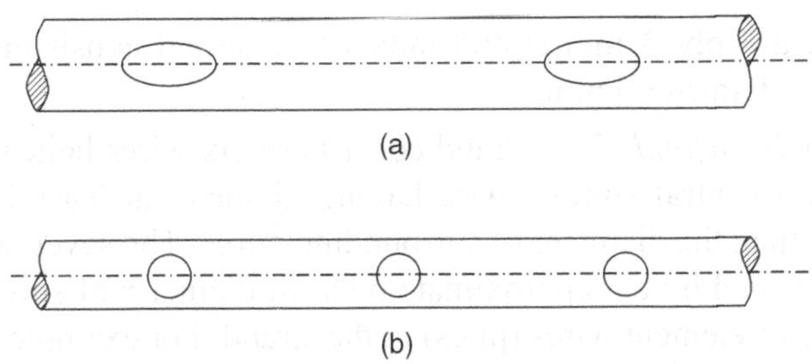

Fig. 2.4 Common types of indented wires
(a) elliptical indentation and (b) round indentation

Table 2.3 Minimum breaking load of indented wires

Nominal Diameter (mm)	3.0	4.0	5.0
Minimum Breaking Load (kN)	13.18	21.55	30.83

High Tensile Steel Strands

Strands are produced by spinning several wires in a helical form. The different types of strands are (Fig. 2.5):

- *Two wire strand:* In this type of strand, two wires are spun together to form the strand. They are designated by the number of element wires (*plies*) and the diameter of wire in the strand. For example, a 2-ply 2 mm strand means a strand containing two element wires of diameter 2 mm each.
- *Three wire strand:* In this case, three wires are spun together to form the strand and they are designated similar to the two wire strand.

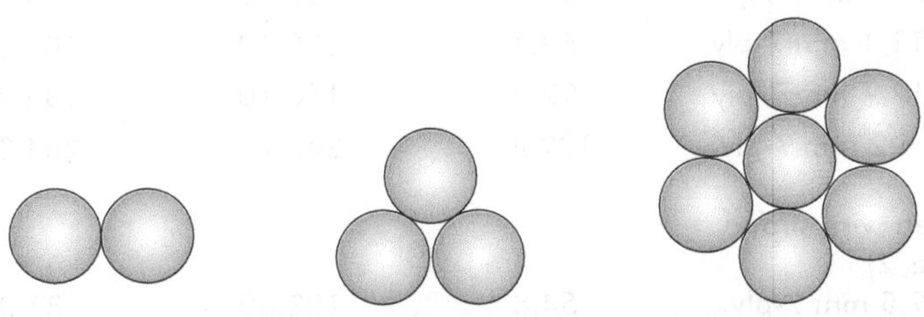

Fig. 2.5 Cross-section of different types of prestressing strands
(a) two wire strand, (b) three wire strand and (c) seven wire strand

Hence, a 3-ply 3 mm strand indicates a strand consisting of three wires of diameter 3 mm.

- *Seven wire strand:* This strand consists of six wires helically wound around a central straight wire having diameter at least 1.5 percent greater than the diameter of surrounding wires. The seven wire strand is designated by the approximate over all diameter of strand and the number of element wires (plies) in the strand. For example, a 6.3 mm 7-ply strand means a strand of approximate overall diameter 6.3 mm and consisting of 7 wires.

The minimum breaking load and 0.2 percent proof load for uncoated stress relieved strands (IS:6006–1983) and uncoated stress relieved low relaxation seven-ply strands (IS 14268:1995) are tabulated in Tables 2.4 and 2.5.

Table 2.4 Properties of stress relieved strands

No.	Type and Designation	Nominal Cross-sectional Area of Strand (mm²)	Minimum Breaking Load (kN)	Minimum 0.2 Percent Proof Load (kN)
1.	Two wire strand			
	(a) 2-ply 2 mm	6.28	12.75	10.84
	(b) 2-ply 3 mm	14.14	22.50	21.67
2.	Three wire strand			
	3-ply 3 mm	21.21	38.25	32.46
3.	Seven wire strand (*class* 1)			
	(a) 6.3 mm 7-ply	23.2	40.00	34.00
	(b) 7.9 mm 7-ply	37.4	64.50	54.70
	(c) 9.5 mm 7-ply	51.6	89.00	75.60
	(d) 11.1 mm 7-ply	69.7	120.10	102.30
	(e) 12.7 mm 7-ply	92.9	160.10	136.20
	(f) 15.2 mm 7-ply	139.4	240.20	204.20
4.	Seven wire strand (*class* 2)			
	(a) 9.5 mm 7-ply	54.8	102.30	87.00
	(b) 11.1 mm 7-ply	74.2	137.90	117.20
	(c) 12.7 mm 7-ply	98.7	183.70	156.10
	(d) 15.2 mm 7-ply	140.0	260.70	221.50

Table 2.5 Properties of stress relieved relaxation seven-plystrands

Class	Nominal Diameter of Strand (mm)	Nominal Area of Strand (mm²)	Minimum Breaking Load (kN)	Minimum 0.2 Percent Proof Load (kN)
I	9.5	51.6	89.0	80.1
	11.1	69.7	120.1	108.1
	12.7	92.9	160.1	144.1
	15.2	139.4	240.2	216.2
II	9.5	54.8	102.3	92.1
	11.1	74.2	137.9	124.1
	12.7	98.7	183.7	165.3
	15.2	140.0	260.7	234.6

High Tensile Steel Bars

High tensile steel bars are available in the following nominal sizes: 10, 12, 16, 20, 22, 25 and 32 mm. As per IS:2090–1983, the minimum tensile strength of bars is 980 N/mm^2, and the 0.2 percent proof stress should not be less than 80 percent of the minimum specified tensile strength.

2.2.4 Modulus of Elasticity of Prestressing Steel

In the absence of experimental data, the values of modulus of elasticity of prestressing steel E_p can be chosen from the values given in Table 2.6.

Table 2.6 Modulus of elasticity of prestressing steel (Cl. 4.5.3.1 of the Code)

Type of Steel	Modulus of Elasticity *Ep* (N/mm²)
Plain cold-drawn wires	2.10×10^5
High tensile steel bars	2.00×10^5
Strands	1.95×10^5

2.2.5 Maximum Initial Prestress in Tendon

As per Cl. 18.5.1 of the Code, the maximum tensile stress during prestressing (f_{pi}) should not exceed 80 percent of the ultimate tensile strength (f_p) of the prestressing steel. Thus,

$$f_{pi} \leq 0.8 f_p \tag{2.1}$$

2.3 Concrete

2.3.1 Need for High Strength Concrete

In a prestressed concrete structure, concrete of higher compressive strength is required due to the following reasons:

- The large prestressing force from the high tensile steel tendons is transferred to concrete by bearing in post-tensioned members and through bond in pre-tensioned members. The high bearing stress in post-tensioned members can be resisted by using high strength concrete thereby reducing the size of the required bearing area. In the case of pre-tensioned members, the bond length required to transfer prestress is reduced by using high strength concrete.
- Increasing the permissible compressive stress of concrete helps to reduce the size of cross-section of the member. This leads to reduced dead load, and hence structures having longer spans are practically possible by using prestressed concrete.
- High strength concrete has a high tensile strength which delays the formation of flexural and diagonal tension cracks.
- Concrete having high strength is less liable to shrinkage cracks which occur in low strength concrete before the application of prestress.
- High strength concrete has a higher modulus of elasticity which helps in reducing the prestress loss due to elastic shortening of concrete and creep in concrete.
- Higher modulus of elasticity of concrete reduces the deflection associated with slender members.

Due to all these reasons, the design codes of various countries prescribe a certain minimum compressive strength for concrete used in prestressed concrete construction.

2.3.2 Compressive Strength of Concrete

The term compressive strength refers to the uniaxial compressive strength of concrete obtained by testing 28-days old 150 mm concrete cubes as per IS:516–1959[7]. The loading is strain controlled applied at a rate of 0.001 mm/mm per minute. This strain rate represents static loads applied for short

[7] IS 516:1959 (Reaffirmed 2004), *Methods of Tests for Strength of Concrete*, Bureau of Indian Standards, Manak Bhavan, New Delhi.

duration. The maximum stress obtained from the test is reported as the cube strength of concrete.

Concrete is a heterogeneous material and hence the strength of concrete obtained from the same mix is subject to large variation. To account for this variation in strength, the compressive strength of concrete is specified in terms of *characteristic strength* (f_{ck}), which is defined as the strength of material below which not more than 5 percent of the test results are expected to fall. The concept of characteristic strength is based on the assumption of normal distribution of the strength of concrete cubes obtained from experiments. The characteristic strength (f_{ck}) is related to the *mean strength* (f_{cm}) and *standard deviation* (σ) by the relation:

$$f_{ck} = f_{cm} - 1.65\sigma \tag{2.2}$$

The characteristic strength is represented by the *grade of concrete* which is designated by a number which denotes f_{ck} in N/mm^2 preceded by the letter M which refers the mix. For example, M 30 grade concrete represents a concrete mix designed for a characteristic strength of 30 N/mm^2.

The Code (Cl. 5.1) recommends a minimum grade of M 30 for post-tensioned members and M 40 for pre-tensioned members. The maximum grade of concrete indicated in the Code is M 60. The Code (Cl. 22.8.2.1) also recommends that the strength of concrete at the time of transfer (f_{ci}) should be at least equal to half the characteristic compressive strength of concrete. Hence,

$$f_{ci} \geq 0.5 f_{ck} \tag{2.3}$$

2.3.3 Tensile Strength of Concrete

The tensile strength of concrete is only about 8–15 percent of its compressive strength. Due to the difficulty in conducting direct tension tests on concrete specimens, indirect tension tests like the cylinder splitting test and flexure test are usually used to find the tensile strength.

Cylinder Splitting Test

In the cylinder splitting test (conducted as per IS 5816:1999[8]), a standard plain concrete cylinder (diameter = 150 mm, length = 300 mm) is loaded across a diameter till failure occurs by splitting across a vertical plane

[8] IS 5816:1999 (Reaffirmed 2004), *Splitting Tensile Strength of Concrete-Method of Test*, Bureau of Indian Standards, Manak Bhavan, New Delhi.

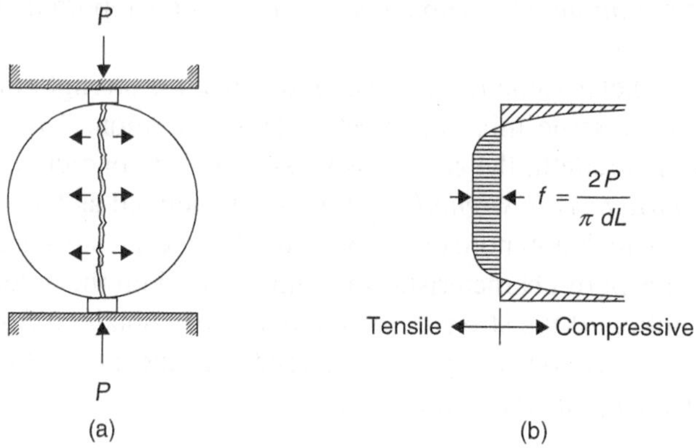

Fig. 2.6 Cylinder splitting test for tensile strength (a) splitting test and (b) distrubution of horizontal stress

(Fig. 2.6(a)). If the cylinder behaves as a linear elastic body, then the horizontal stress distribution along the vertical diameter would be as shown in Fig. 2.6(b). From the stress distribution, the splitting tensile strength f_{ct} is obtained as:

$$f_{ct} = \frac{2P}{\pi dL} \tag{2.4}$$

where P is the load at failure. d and L are the diameter and length of the concrete cylinder specimen.

The splitting tensile strength of concrete generally lies between 1/8 to 1/12 of the cube strength.

Flexure Test

In the flexure test (as per IS: 516-1959), a standard plain concrete beam is loaded up to failure as shown in Fig. 2.7. The maximum tensile stress reached at the extreme fibre is known as the *modulus of rupture* (f_{cr}) and is calculated as:

$$f_{cr} = \frac{M}{Z} \tag{2.5}$$

where M is the bending moment at the section where failure occurs and Z is the elastic section modulus.

It should be remembered that modulus of rupture is only a hypothetical stress calculated by assuming linear elastic behaviour till failure.

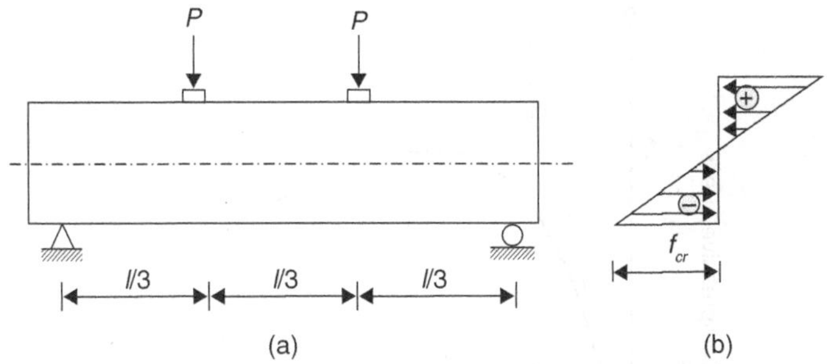

Fig. 2.7 Flexure test for tensile strength (a) loading arrangement and (b) longitudinal stress distribution at failure

The modulus of rupture overestimates the actual flexural strength of concrete and is only used as a comparative measure. Modulus of rupture is usually about 1.5 times the cylinder splitting strength of concrete.

The Code (Cl. 5.2.1) recommends the following equation to find f_{cr}:

$$f_{cr} = 0.7\sqrt{f_{ck}} \tag{2.6}$$

where f_{ck} is the characteristic compressive strength in N/mm^2.

2.3.4 Modulus of Elasticity of Concrete

The Young's modulus of elasticity is defined as the ratio of uniaxial stress to uniaxial strain within the linear elastic range. It is needed for the estimation of prestress losses and for calculating deflections of members.

A typical short-term stress-strain curve of concrete obtained from uniaxial compression test is shown in Fig. 2.8. In the figure, f_c is the maximum stress and ε_{cu} is the failure strain of concrete. An idealized form of this stress-strain curve is used for design purposes (see Chapter 3). From the stress-strain curve, it can be seen that up to a stress level of one-third of f_c, the stress-strain relation is approximately linear. The modulus of elasticity E_c can be taken as the slope of the line AB in Fig. 2.8. This slope is known as the *secant modulus*.

IS 456:2000[9] (Cl. 6.2.3.1) recommends the following empirical equation for E_c:

$$E_c = 5000\sqrt{f_{ck}} \text{ N/mm}^2 \tag{2.7}$$

where f_{ck} is in N/mm^2.

[9] IS 456:2000, *Plain and Reinforced Concrete – Code of Practice*, Bureau of Indian Standards, Manak Bhavan, New Delhi.

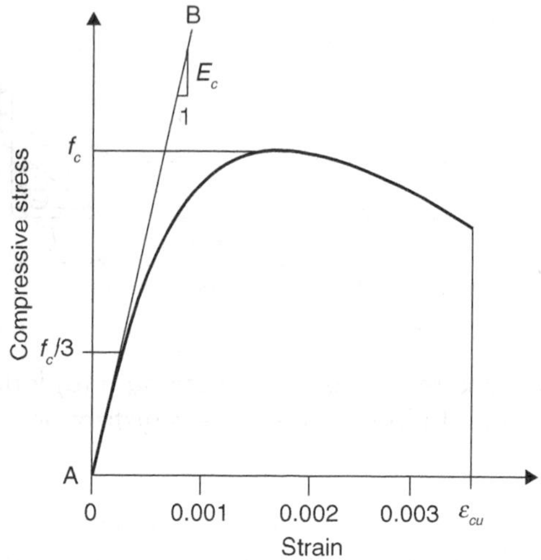

Fig. 2.8 Stress-strain curve for concrete

For the analysis and design calculations, this empirical equation is used to find E_c.

Comments

The Code (Cl. 5.2.3.1) and the earlier edition of IS: 456 recommend a higher value for modulus of elasticity ($E_c = 5700\sqrt{f_{ck}}$ N/mm^2). But the use of a lower value of E_c results in conservative or larger estimate of prestress losses and deflections. Hence in this book, Eq. (2.7) is used for finding E_c.

2.3.5 Time-dependent Deformation of Concrete

The time dependent deformation of concrete due to creep and shrinkage results in increased deflections of members and losses of prestress. In addition to this, shrinkage may cause cracking leading to serviceability or durability problems. The phenomenon of creep and shrinkage is briefly discussed in this section, and their effects on prestress losses and deflections are discussed in Chapters 4 and 8.

Creep of Concrete

When a concrete specimen is subjected to *sustained* compressive loading, the specimen shows a immediate strain followed by additional deformation,

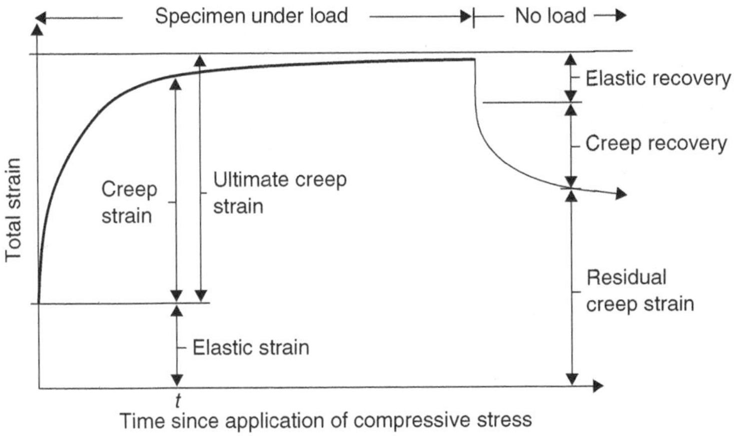

Fig. 2.9 Typical strain-time curve for concrete in uniaxial compression

which though increase at a diminished rate may become several times of the immediate strain (Fig. 2.9). The immediate strain is known as the *elastic strain* and the time-dependent strain is termed as *creep strain* or *creep*. The maximum value of creep is called the *ultimate creep strain.*

Upon unloading the specimen, the part of the strain immediately recovered is called the *elastic recovery*, and the delayed recovery is referred as *creep recovery*. The elastic recovery is less than elastic strain and the creep recovery is much less than creep.

Shrinkage of Concrete

A concrete mix usually contains more water than is required for hydration of cement. This excess water evaporates with time leading to volume contraction which is known as shrinkage. The amount of shrinkage depends on the environmental conditions, size and shape of concrete specimens and is independent of the external load acting on the member.

chapter 3

Limit State Design

3.1 Introduction

The basic design requirements of structures are that they should be *safe* from *collapse*, *serviceable* during their design life and should be *economic*.

In the beginning, design Codes for concrete structures were based on *Working Stress Method* (WSM) in which safety is ensured by restricting the stresses in materials caused due to *service* or *working loads* (i.e., the loads acting on the structure during its normal use). The *permissible* or *working stresses* in materials are obtained by dividing the material strength by the *factor of safety*. Since the permissible stresses of the materials are very small compared to their material strength, in WSM, the materials are assumed to be linearly elastic.

Though the concept of WSM is simple, it has some limitations, which are listed below:

- The important assumption of linear elastic behaviour of materials obtained by limiting the stresses within permissible stresses is not found to be true. The long-term effects like creep, shrinkage and other factors like stress concentration causes significant redistribution of stresses and increases the stresses in materials beyond the linear elastic range.
- From experimental studies, it is observed that the ratio between the ultimate load and service load is not same as the ratio of material

strength to permissible stress (i.e., the factor of safety). Hence by using WSM, it is not possible to predict the ultimate load.

- In WSM, it is not possible to differentiate between loads having different degrees of uncertainties acting simultaneously on a structure (for example a structure subjected to dead load and earthquake load)

In spite of these limitations, WSM has the advantage of its simplicity, both in concept and in application. The design using WSM usually results in relatively larger sections and hence the structural members have better serviceability behaviour.

The second design approach known as the *Ultimate Load Method* (ULM) was developed to overcome the limitations of WSM. In ULM, safety is ensured by magnifying the service load and checking the capacity of the structure at this magnified load known as the *ultimate load* or the *collapse load*. The magnification factor is called the *load factor*, defined as the ratio of the ultimate load to service load. The load factor is a measure of the reserve strength available in the structure, and the designer will have the confidence that the structure designed using ULM can safety withstand the service load.

In ULM, the design is done at the collapse stage. Hence the non-linear stress–strain curve of the material is used. By assigning separate load factors to different loads, in ULM, it is possible to distinguish different types of loads. Thus most of the shortcomings of WSM were overcome by using ULM.

But the designs using ULM resulted in relatively slender sections. Hence even though, the member is safe at ultimate load, it may not satisfy the serviceability criteria (i.e., the member will have large deflection and cracking under service load).

From the above discussions, it can be seen that the structures designed using WSM have satisfactory performance at service loads. But by using WSM, it is not possible to find the ultimate capacity of the structure. In ULM, though the design is safe at ultimate load, it may not meet the serviceability requirements. Thus a ideal design approach is the one which considers both safety at ultimate load and serviceability at service load. This conclusion lead to the development of the *Limit State Method* (LSM), which was first introduced in 1950's in the concrete design Codes of the former U.S.S.R.

In the LSM recommended by the Code, adequate safety at ultimate loads and adequate serviceability at service loads are ensured by using *partial safety factors* for loads and materials. The partial safety factors are determined by using the concepts of *structural reliability*. They take into account of the uncertainties in loads and material strengths, and the different types of failures (*limit states*).

3.2 Limit States

A structure is designed to satisfy a number of functions. A limit state is the situation when the structure fails to satisfy any one of the specific functions for which it is designed. For concrete structure, two limit states are considered:

1. *Ultimate Limit State (ULS) or Limit State of Collapse:* This limit state is reached when the structure or a part of the structure collapses. The ultimate limit states considered for design are strength, overturning, sliding, buckling, fatigue, etc.
2. *Serviceability Limit State (SLS):* This limit state is reached when the structure though safe is not suitable for everyday use due to excessive deflection, cracking, vibration, etc.

From the definition of ULS and SLS, it can be seen that ULS is more critical than SLS. Hence the partial safety factors applied for ULS are greater than that for SLS.

The aim of LSM is to ensure that the structure being designed will not reach any limit state. For this to happen, it is necessary to consider each limit state and design (or check) accordingly. Generally prestressed concrete members are designed for the serviceability limit state of cracking and checked for the ultimate strength limit state. This is in contrast with the design of reinforced concrete members in which the members are initially designed for ultimate strength limit state and later checked for serviceability limit states.

3.3 Characteristic and Design Load

To account for the variation in load, the Code (Cl. 20.2) introduces the concept of *characteristic load* (F) defined as the load that has a 95 per-cent probability of not being exceeded during the life of the structure. In the absence of statistical data, the dead, live and wind loads specified in

IS 875: (Parts 1-3)-1987[1,2,3] and seismic loads as per IS 1893 (Part 1): 2002[4] are taken as the characteristic load.

The *design load* (or *factored load*) F_d is taken as the product of characteristic load F and *partial safety factor for load* γ_f.

$$F_d = \gamma_f F \tag{3.1}$$

The partial safety factor γ_f is introduced to account for:

- Possible unusual increase in load not considered while arriving at characteristic load
- Inaccurate assessment of the effects of loading
- Variation in dimensions of members
- Importance of the limit state and probability of occurrence of the load combination

The recommended values of γ_f (Cl. 20.4.2 of the Code) for estimating ultimate load effect (UL) and service load effect (SL) are:

Ultimate Limit State:

$$\text{UL} = 1.5\ (\text{DL} + \text{LL}) \tag{3.2a}$$

$$\text{UL} = 1.5\ (\text{DL} + \text{QL}) \text{ or } (0.9\ \text{DL} + 1.5\ \text{QL}) \tag{3.2b}$$

$$\text{UL} = 1.2\ (\text{DL} + \text{LL} + \text{QL}) \tag{3.2c}$$

Serviceability Limit State:

$$\text{SL} = 1.0\ (\text{DL} + \text{LL}) \tag{3.3a}$$

$$\text{SL} = 1.0\ (\text{DL} + \text{QL}) \tag{3.3b}$$

$$\text{SL} = (1.0\text{DL} + 0.8\ \text{LL} + 0.8\text{QL}) \tag{3.3c}$$

---

[1] IS: 875 (Part 1)–1987 (Reaffirmed 2003), *Code of Practice for Design Loads (other than earthquake) for Buildings and Structures* : Part 1 Dead Loads – Unit Weights of Building Materials and Stored Materials, Bureau of Indian Standards, Manak Bhavan, New Delhi.

[2] IS: 875 (Part 2)–1987 (Reaffirmed 2003), *Code of Practice for Design Loads (other than earthquake) for Buildings and Structures*: Part 2 Imposed Loads, Bureau of Indian Standards, Manak Bhavan, New Delhi.

[3] IS: 875 (Part 3)–1987 (Reaffirmed 2003), *Code of Practice for Design Loads (other than earthquake) for Buildings and Structures*: Part 3 Wind Loads, Bureau of Indian Standards, Manak Bhavan, New Delhi.

[4] IS 1893 (Part 1): 2002, *Criteria for Earthquake Resistant Design of Structures*: Part 1, General Provisions and Buildings, Bureau of Indian Standards, Manak Bhavan, New Delhi.

In the above equations, DL is the characteristic dead load, LL is the characteristics live load and QL is the characteristic wind/earthquake load.

Since the consequence of collapse of a structure is very critical, γ_f for ULS is higher than that used for SLS. The probability of DL, LL and QL simultaneously reaching their characteristic values is very low. To account for this, the Code recommends a lower value of γ_f for this combination (Eqs. 3.2c and 3.3c).

3.4 Characteristic and Design Strength of Material

The characteristic strength of material f (Cl. 20.1 of the Code) is defined as the value below which not more than 5 percent of the test results are expected to fall. For a normal distribution of test results with standard deviation σ, the characteristic strength f is related to the mean strength f_m as (Fig. 3.1):

$$f = f_m - 1.65\sigma \tag{3.4}$$

The characteristic strength of concrete (f_{ck}) is taken as the 28-day cube strength, for prestressing steel the characteristic strength (f_p) is taken as the ultimate tensile stress and for reinforcing steel, characteristic strength (f_y) is taken as the yield stress/0.2 percent proof stress below which not more than 5 percent of test results fall.

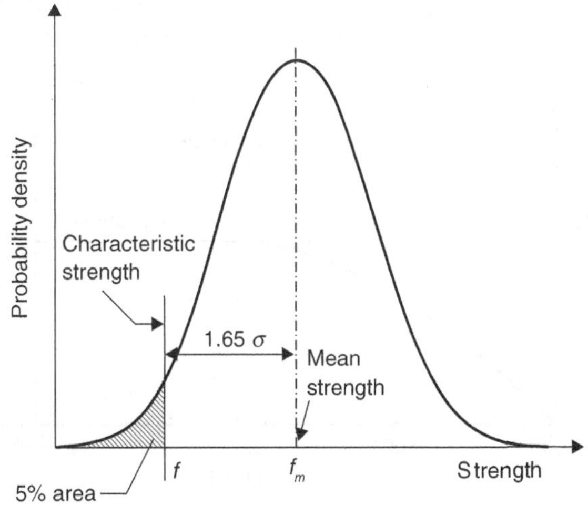

Fig. 3.1 Characteristic strength

The *design strength* (or *factored strength*) f_d is taken as (Cl. 20.3.1 of the Code):

$$f_d = \frac{f}{\gamma_m} \tag{3.5}$$

where γ_m is the *partial safety factor for the material*.

γ_m takes into account of difference between the laboratory and field strength of materials, methods of manufacture of materials, accuracy of methods used to test materials, etc. Hence γ_m for concrete made at the site will be higher than that for steel manufactured in the factory with good quality control. The recommended values of γ_m by the Code (Cl. 20.4.1) are given in Table 3.1.

Table 3.1 Partial safety factor for materials

	γ_m	
Material	**ULS**	**SLS**
Concrete	1.5	1.0
Steel	1.15	1.0

3.5 Characteristic and Design Stress–Strain Curves

3.5.1 Stress–Strain Curves for Concrete

The characteristic and design stress–strain curves for concrete in flexural compression (Cl. 22.1.1 of the Code) are shown in Fig. 3.2. The stress–

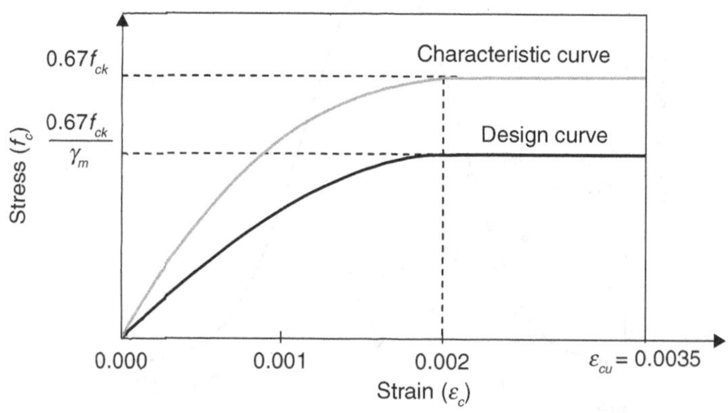

Fig. 3.2 Stress–strain curves for concrete in flexural compression

strain curves consist of a parabola up to a strain of 0.002 and a straight line thereafter up to the ultimate strain (ε_{cu}) of 0.0035.

It is important to note that the maximum stress in concrete in a structure is limited to $0.67 f_{ck}$. This is due to the fact that the compressive strength of concrete in a actual structure will be less than the cube strength of concrete. Many factors like duration of loading, size effect, multi-axial state of stress, etc. are responsible for this difference in strength.

The design stress–strain curve is obtained by dividing the ordinates of characteristic stress–strain curve by the partial safety factor for material γ_m. Thus for ULS (γ_m = 1.5), the maximum design stress works out to $0.45 f_{ck}$. The design stress–strain curve at ULS is given by:

$$f_c = \begin{cases} 0.45 f_{ck}\left[2\left(\dfrac{\varepsilon_c}{0.002}\right) - \left(\dfrac{\varepsilon_c}{0.002}\right)^2\right] & \text{for } \varepsilon_c \le 0.002 \\ 0.45 f_{ck} \quad \text{for } 0.002 < \varepsilon_c \le 0.0035 \end{cases} \tag{3.6}$$

3.5.2 Stress–Strain Curves for Prestressing Steel

The characteristic stress–strain curve for prestressing steel recommended by the Code is depicted in Fig. 3.3. The details of salient points (A to E in Fig. 3.3) in the characteristic stress–strain curve for different types of tendons are given in Table 3.2.

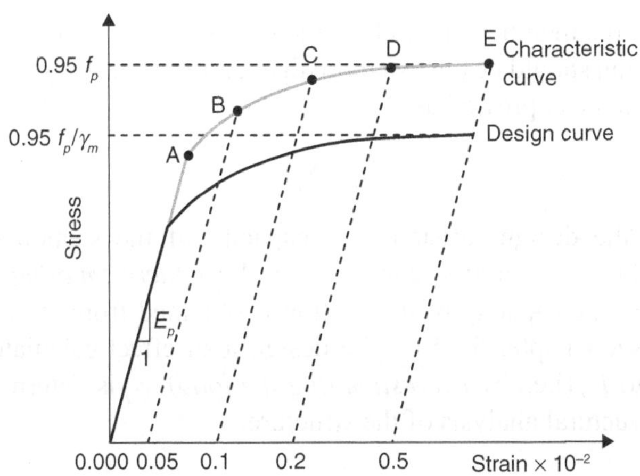

Fig. 3.3 Stress–strain curves for prestressing steel

Table 3.2 Salient points in characteristic stress–strain curves of prestressing steel

		Characteristic Stress	
Point	**Inelastic Strain**	**Wires (Stress Relieved), Strands, Bars**	**Wires (as-drawn)**
A	0.0000	0.80 f_p	0.70 f_p
B	0.0005	0.84 f_p	0.77 f_p
C	0.0010	0.87 f_p	0.81 f_p
D	0.0020	0.90 f_p	0.85 f_p
E	0.0050	0.95 f_p	0.95 f_p

The design stress–strain curve of prestressing steel is obtained from the characteristic stress–strain curve by dividing with the partial safety factor γ_m (γ_m =1.15 for ULS). From Fig. 3.3, it can be observed that unlike in concrete (Fig. 3.2), where γ_m was applied at all stress levels, γ_m for prestressing steel is applied only in the inelastic region. This is because the stress–strain curve of concrete is highly dependent on the compressive strength of concrete (including the initial regions due to the dependence of E_c on f_{ck}). But in the case of steel, E_p does not depend on f_p. Hence there is no need for using γ_m in the elastic region of prestressing steel.

3.6 Design Requirements as per LSM

The design of the member should be done such that the resistance or capacity of the section should be always more than or at the most equal to the load effect. This can be expressed as:

$$R_d \geq S_d \tag{3.7}$$

where R_d is the design resistance or capacity of the section computed using the reduced material strength f_d (i.e., by *underestimating the material strength*). The capacity of the section is obtained from the analysis of the section (see Chapter 5). S_d is the design load effect calculated for the enhanced load F_d (i.e., by *overestimating the load)*. S_d is determined from the overall structural analysis of the structure.

3.7 Limit State Design of Prestressed Concrete Members

The design of prestressed concrete members at SLS and ULS are covered in the remaining chapters of the book. Since the design at SLS is important; the design requirements at SLS are briefly discussed in this section.

Design Criteria at SLS

The design of a prestressed concrete member at SLS should consider three design criteria; namely

1. Limits on compressive stress
2. Crack control
3. Deflection control.

Limits on Compressive Stress: The compressive stress in concrete is limited to prevent the development of longitudinal cracks, micro cracks and to prevent excessive loss of prestress due to creep. The maximum allowable bending compressive stress in concrete recommended by the Code (Cl. 22.8) at transfer (f_{ct}) and at service condition (f_{cw}) is given in Tables 3.3 and 3.4. The Code identifies two zones, Zone I and Zone II for recommending allowable stresses at service condition. Zone I represents the location where the compressive stresses are not likely to increase, and Zone II represents the

Table 3.3 Allowable bending compressive stress in concrete at transfer (f_{ct})

Grade of Concrete		M 30	M 35	M 40	M 45	M 50	M 55	M 60
$\frac{f_{ct}}{f_{ci}}$	Post-tensioned work	0.540	0.512	0.483	0.455	0.427	0.398	0.370
	Pre-tensioned work	–	–	0.510	0.493	0.475	0.458	0.440

Table 3.4 Allowable bending compressive stress in concrete at service condition (f_{cw})

Grade of Concrete		M 30	M 35	M 40	M 45	M 50	M 55	M 60
$\frac{f_{cw}}{f_{ck}}$	Zone I	0.410	0.400	0.390	0.380	0.370	0.360	0.350
	Zone II	0.340	0.328	0.317	0.305	0.293	0.282	0.270

locations where the compressive stresses are likely to increase in service condition.

The allowable direct compressive stress in concrete (except in regions immediately behind the anchorage) is taken as 0.8 times of that in bending.

Crack Control: Cracking in concrete are not allowed in type 1 and type 2 members at SLS. This condition automatically sets a limit for allowable tensile stresses in prestressed concrete members. The Code (Cl. 22.7.1) does not permit the occurrence of tensile stresses in type 1 members. In the case of type 2 members, tensile stresses are permitted but there should be no visible cracking. The allowable tensile stress in type 2 members is limited to 3 MPa. The allowable stress can be increased to 4.5 MPa for temporary loads, provided the stresses caused by permanent loads are compressive in nature.

Cracking is permitted in type 3 members at SLS. But the cracks should not affect the appearance and durability of the members. In these members, the surface width of cracks should not exceed 0.2 mm in general case and 0.1 mm for members located in aggressive environment. For type 3 members, the Code recommends hypothetical allowable tensile stresses for a given crack width w_{cr} (Fig. 3.4). The hypothetical tensile stresses in *cracked beams* are calculated by assuming the sections to be *uncracked.*

This method of specifying allowable hypothetical tensile stresses is a indirect way of controlling the width of cracks. Therefore the calculation of crack width is not required for type 3 members. The allowable tensile stress in Fig. 3.4 should be multiplied by the depth factor given in Table 3.5.

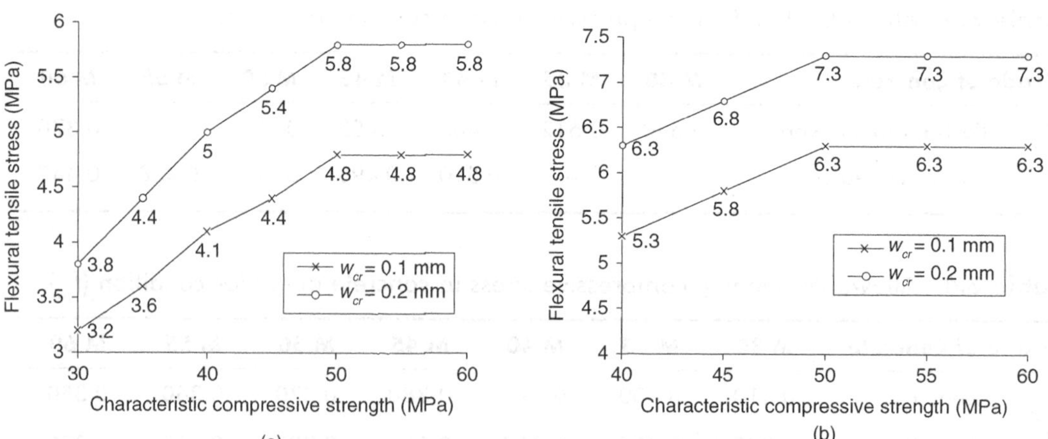

Fig. 3.4 Allowable hypothetical flexural tensile stress in concrete for type 3 members (a) for members with grouted post-tensioned tendon and pre-tensioned tendon, (b) for members with pre-tensioned tendon kept close to tension face

Table 3.5 Depth factor

Depth of Member (mm)	≤ 200	400	600	800	≥ 1000
Depth Factor	1.1	1.0	0.9	0.8	0.7

The allowable hypothetical tensile stress can be increased further if additional nontensioned reinforcement is distributed in the tension zone and placed close to the tension face. The stress can be increased by 4 MPa for members with pre-tensioned or grouted post-tensioned tendons and by 3 MPa for other members for 1 percent of additional reinforcement. Further increase in hypothetical stress is permitted for higher percentages of reinforcement subject to the condition that the maximum hypothetical tensile stress is limited to $0.25 f_{ck}$.

Control of Deflection: The deflection of members under service loads should be restricted. Otherwise excessive deflection leads to damage of finishes, partitions, claddings, etc. The maximum allowable deflections in prestressed concrete flexural members (Cl. 19.3.1 of the Code) are given in Table 3.6.

Table 3.6 Allowable deflection

Load Condition	**Allowable Deflection**
Final deflection including all effects (creep, shrinkage, temperature etc.) measured from the cast level	Span/250
Deflection including all effects after erection of partitions and application of finishes	Span/350 or 20 mm
Total upward deflection or camber	Span/300

chapter 4

Losses in Prestress

4.1 Introduction

It is found that the prestressing force in prestressed concrete members decreases with time. Hence the actual prestressing force will not be equal to the initial prestressing force (P_i). This decrease in prestressing force is known as losses in prestress.

In the analysis of prestressed concrete members (Chapter 5), it is necessary to check the stresses in concrete at different loading stages like transfer loading stage and at service condition. Hence it is necessary to determine the losses in prestressing force at these stages.

The losses in prestressing force can be classified based on the time at which they occur. The losses that occur during the prestressing process are known as *immediate losses* and the losses which occur after the prestress is transferred to concrete are called as *time dependent losses*. The prestressing force at different loading stages is indicated in Fig. 4.1.

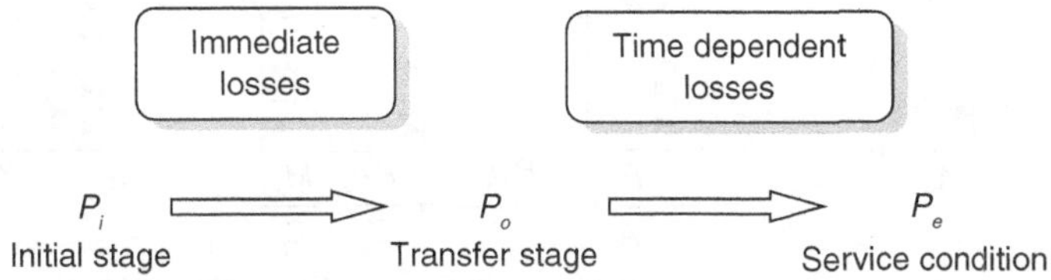

Fig. 4.1 Prestressing force at different loading stages

In the figure, P_i is the initial prestressing force, i.e., the jacking force in post-tensioned members, or the prestressing force applied on bulk heads or on self-straining frame in pre-tensioned members. P_0 is the prestressing force after immediate losses, and P_e is the final or effective value of prestressing force after all losses (both immediate and time dependent losses).

In pre-tensioned members, immediate losses occur due to elastic shortening of concrete. While in post-tensioned members, elastic shortening of concrete, slip in anchorage and the friction between the tendons and ducts contribute to immediate losses. Time dependent losses are caused by relaxation of steel and due to creep and shrinkage of concrete.

The recommendations to determine the various losses in prestressed concrete members are given in Cl. 18.5.2 of the Code.

4.2 Immediate Losses

4.2.1 Loss Due to Elastic Shortening of Concrete

At the time of transfer of prestress from tendon to concrete, the concrete undergoes a reduction in length. To maintain deformation compatibility, the prestressing steel also shortens, resulting in the loss in prestress.

Pre-tensioned Member

In a pre-tensioned member, the tendons are stressed before casting of concrete. At the time of release of tendon, the concrete is subjected to a compressive force P_i. The stress distribution at a section in concrete due to eccentric prestressing is illustrated in Fig. 4.2.

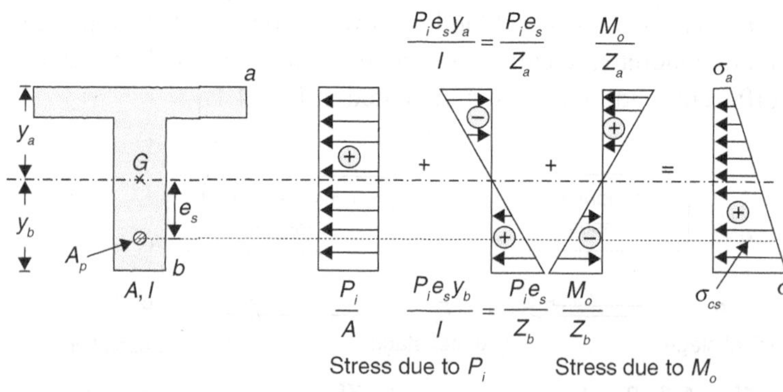

Fig. 4.2 Stresses in pre-tensioned beam during elastic shortening

The stress in concrete at the level of prestressing steel (σ_{cs}) is*:

$$\sigma_{cs} = \frac{P_i}{A} + \frac{(P_i e_s)e_s}{I} - \frac{M_o e_s}{I}$$

where e_s is the eccentricity of the prestressing steel and M_0 is the bending moment at the section during the release of tendon, i.e., at transfer. A and I are the gross cross-sectional area and moment of inertia of the section.

The strain in concrete (ε_c) due to σ_{cs} is:

$$\varepsilon_c = \frac{\sigma_{cs}}{E_{ci}}$$

E_{ci} is the modulus of elasticity of concrete at the time of transfer of prestress.

The prestressing steel undergoes a loss of strain ($\Delta\varepsilon_p$) equal to ε_c. The loss of prestress due to elastic shortening of concrete ($\Delta f_{p,\,es}$) is:

$$\Delta f_{p,es} = \Delta\varepsilon_p E_p = \varepsilon_c E_p = \sigma_{cs}\frac{E_p}{E_{ci}}$$

$$= \frac{E_p}{E_{ci}}\left[\frac{P_i}{A} + \frac{(P_i e_s)e_s}{I} - \frac{M_o e_s}{I}\right] \qquad (4.1)$$

In the above equation, E_p is the modulus of elasticity of prestressing steel.

Post-tensioned Member

For a post-tensioned member having a single tendon, or with several tendons tensioned simultaneously, elastic shortening of concrete occurs during the jacking process. But the decrease in prestress can be compensated by applying additional jacking force. Hence in these cases, there is no loss in prestress due to elastic shortening.

But loss in prestress occurs in the case of members having multiple tendons stretched sequentially. This is explained by considering a beam having two tendons (Fig. 4.3).

In this case, first tendon 1 is stressed to a force $P_{i,\,1}$ and anchored. This does not cause any loss in prestress. Next, when tendon 2 is stressed to a force $P_{i,\,2}$, it causes shortening of concrete at the level of tendon 1 leading to loss of prestress in tendon 1. The stress in concrete at the level of tendon 1, due to $P_{i,\,2}$ is:

* The reader may refer the procedure to find combined stresses due to prestress and applied moment (Eq. (5.4)) discussed in Section 5.2.1.

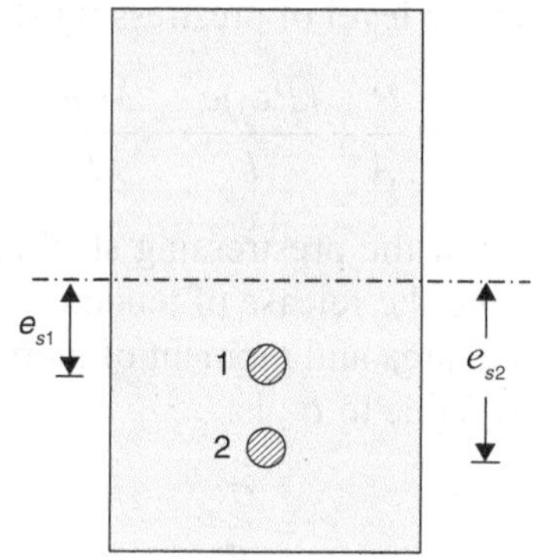

Fig. 4.3 Post-tensioned beam having two tendons

$$\sigma_{cs,1} = \frac{P_{i,2}}{A} + \frac{(P_{i,2}\, e_{s2})e_{s1}}{I}$$

The loss of prestress in tendon 1 is:

$$(\Delta f_{p,es})_1 = \frac{E_p}{E_{ci}}\sigma_{cs,1} = \frac{E_p}{E_{ci}}\left[\frac{P_{i,2}}{A} + \frac{(P_{i,2}\, e_{s2})e_{s1}}{I}\right] \tag{4.2}$$

But there is no loss in prestress in tendon 2.

The above approach can be extended to determine the loss in prestress in post-tensioned members having multiple tendons stretched sequentially.

Comments

- At transfer loading stage, no external load acts on the prestressed concrete member. But the eccentricity of prestressing force causes the beam to bend upwards (Fig. 4.4). Hence at transfer, the bending moment M_o (Eq. (4.1)) will be caused due to the self-weight of the member.

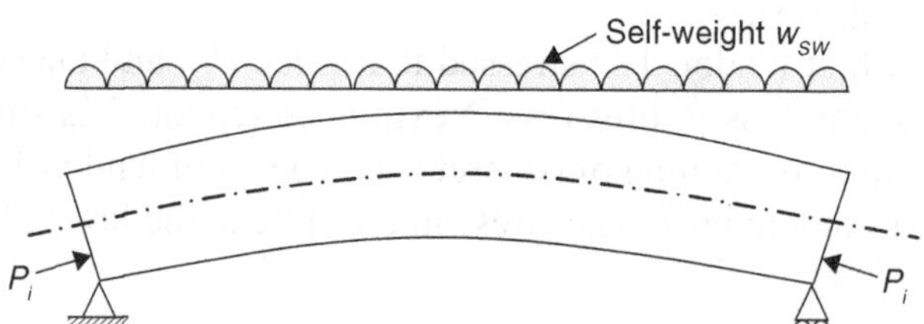

Fig. 4.4 Deformation of the beam at transfer

- Usually in a prestressed concrete member, e_s and M_0 vary from section to section. Hence the loss in prestress due to elastic shortening of concrete is calculated using the average value of σ_{cs}.
- In Eqs. (4.1) and (4.2), compressive stresses are taken as positive and tensile stresses as negative. Further sagging bending moment is positive. The tendon eccentricity e_s is taken as positive, if it is below the centroid and taken as negative, if it is measured above the centroid. More details on calculation of stresses and sign conventions are discussed in Chapter 5.

Example 4.1

Determine the loss in prestress due to elastic shortening of concrete for the pre-tensioned beam in Fig. 4.5.

Data: Cross-sectional area of prestressing steel A_p $= 2100$ mm²
Initial stress in prestressing steel f_{pi} $= 1360$ N/mm²
Compressive strength of concrete at transfer $f_{ci} = 30$ N/mm²
Modulus of elasticity of prestressing steel E_p $= 2.1 \times 10^5$ N/mm²

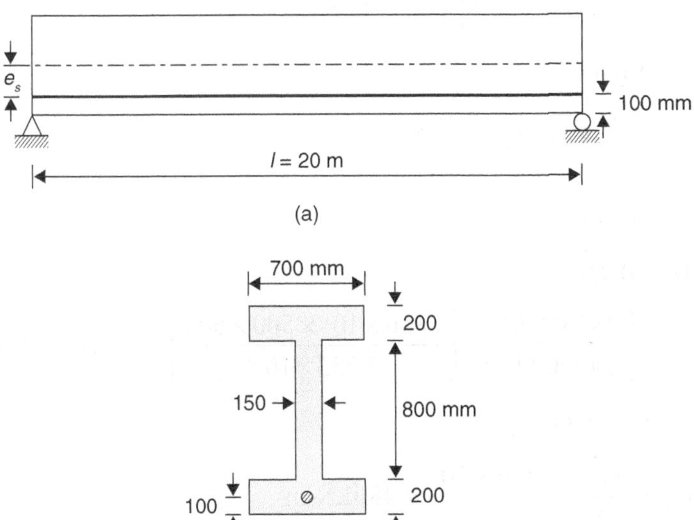

Fig. 4.5 Details of pre-tensioned beam (a) elevation and (b) cross-section

Solution:

Step 1: *Calculation of section properties*

- Area of cross-section of the beam A

  $A = (2 \times 700 \times 200) + (150 \times 800) = 400000\ \text{mm}^2$

- Moment of inertia of the section I

$$I = 2\left[\frac{700 \times 200^3}{12} + (700 \times 200 \times 500^2)\right] + \left[\frac{150 \times 800^3}{12}\right]$$

$$= 7.733 \times 10^{10}\ \text{mm}^4$$

- Tendon eccentricity e_s

  $e_s = (1200/2) - 100 = 500\ \text{mm}$

- Distributed load due to self-weight w_{SW}

  Density of concrete $\gamma_c = 24\ \text{kN/m}^3$

  $w_{SW} = \gamma_c A = 24 \times 400000 \times 10^{-6} = 9.6\ \text{kN/m}$

Step 2: *Calculation of stress in concrete at the level of tendon* σ_{cs}

- Initial prestressing force P_i

$$P_i = A_p \times f_{p_i} = \frac{2100 \times 1360}{1000} = 2856\,\text{kN}$$

- Calculation of σ_{cs}

$$\sigma_{cs} = \frac{P_i}{A} + \frac{(P_i e_s)e_s}{I} - \frac{M_o e_s}{I}$$

σ_{cs} *at support*

$M_o = 0$ kN-m

$$\sigma_{cs} = \left[\frac{2856 \times 10^3}{400000}\right] + \left[\frac{2856 \times 10^3 \times 500 \times 500}{7.733 \times 10^{10}}\right] - 0 = 16.37\,\text{N/mm}^2$$

σ_{cs} *at mid-span*

$$M_o = \frac{w_{SW} l^2}{8} = \frac{9.6 \times 20^2}{8} = 480\,\text{kN-m}$$

$$\sigma_{cs} = \left[\frac{2856 \times 10^3}{400000}\right] + \left[\frac{2856 \times 10^3 \times 500 \times 500}{7.733 \times 10^{10}}\right] - \left[\frac{480 \times 10^6 \times 500}{7.733 \times 10^{10}}\right]$$

$$= 13.27\ \text{N/mm}^2$$

Average value of σ_{cs}

$$\sigma_{cs} = \left[\frac{16.37+13.27}{2}\right] = 14.82\,\text{N/mm}^2$$

Step 3: *Calculation of elastic shortening loss*

$$E_p = 2.1 \times 10^5\ \text{N/mm}^2$$

$$E_{ci} = 5000\sqrt{f_{ci}} = 5000\sqrt{30} = 27386.13\,\text{N/mm}^2$$

- Loss of prestress due to elastic shortening

$$\Delta f_{p,es} = \frac{E_p}{E_{ci}}\sigma_{cs}$$

$$= \frac{2.1\times10^5\times14.82}{27386.13} = 113.64\,\text{N/mm}^2$$

Example 4.2

A simply supported post-tensioned beam 300 mm wide and 600 mm deep, and span of 20 m is prestressed by successive tensioning and anchoring of three cables A, B and C. Cable A is parabolic with an eccentricity of 100 mm above the centroidal axis at support and 100 mm below the centroidal axis at the mid-span. Cable B is also parabolic with zero eccentricity at the support and 100 mm below the centroidal axis at the centre of the span. Cable C is straight and is kept 100 mm below the centroidal axis. The cables are tensioned one-by-one in the order A, B and C. Determine the loss in prestress in the cables due to elastic shortening of concrete for the following data.

Data: Cross-sectional area of each cable $A_p = 400\ \text{mm}^2$

$f_{pi} = 1300\ \text{N/mm}^2$

$f_{ci} = 30\ \text{N/mm}^2$

$E_p = 2.1 \times 10^5\ \text{N/mm}^2$

Solution:

Step 1: *Calculation of section properties*

Cross-sectional area of beam $A = 300 \times 600 = 180000\ \text{mm}^2$

Moment of inertia of the section $I = \dfrac{300\times600^3}{12} = 54\times10^8\ \text{mm}^4$

- Tendon eccentricity e_s

  e_s for the three cables are tabulated below.

Table 4.1 Tendon eccentricity

	Tendon Eccentricity (mm)	
Cable	**At Support**	**At Mid-span**
A	$e_{sA} = -100$	$e_{sA} = 100$
B	$e_{sB} = 0$	$e_{sB} = 100$
C	$e_{sC} = 100$	$e_{sC} = 100$

Step 2: *Calculation of loss due to elastic shortening*

- Initial prestressing force P_i

$$P_i = A_p \times f_{p_i} = \frac{400 \times 1300}{1000} = 520\,\text{kN}$$

In this example P_i is same for all cables, i.e.,

$$P_{i,A} = P_{i,B} = P_{i,C} = 520 \text{ kN}$$

- Modular ratio m

$$m = \frac{E_p}{E_{ci}} = \frac{2.1 \times 10^5}{5000\sqrt{30}} = 7.67$$

- Prestress loss due to tensioning cable A

  When cable A is tensioned and anchored, there will be no loss due to elastic shortening of concrete.
- Prestress loss due to tensioning cable B

  Tensioning and anchoring cable B causes prestress loss in cable A. Stress in concrete at the level of cable A due to tensioning cable B ($\sigma_{cs,A}$):

$$\sigma_{cs,A} = \frac{P_{i,B}}{A} + \frac{(P_{i,B}\, e_{sB})e_{sA}}{I}$$

$\sigma_{cs,A}$ *at support*

$$\sigma_{cs,A} = \left[\frac{520 \times 10^3}{180000}\right] + \left[\frac{520 \times 10^3 \times 0 \times (-100)}{54 \times 10^8}\right] = 2.89 \text{ N/mm}^2$$

$\sigma_{cs,A}$ at mid-span

$$\sigma_{cs,A} = \left[\frac{520 \times 10^3}{180000}\right] + \left[\frac{520 \times 10^3 \times 100 \times 100}{54 \times 10^8}\right] = 3.85 \text{ N/mm}^2$$

Average value of $\sigma_{cs,A}$

$$\sigma_{cs,A} = \left[\frac{2.89 + 3.85}{2}\right] = 3.37 \text{ N/mm}^2$$

Prestress loss in cable A, $(\Delta f_{p,es})_A$:

$$(\Delta f_{p,es})_A = \frac{E_p}{E_{ci}} \sigma_{cs,A} = 7.67 \times 3.37 = 25.85 \text{ N/mm}^2$$

- Prestress loss due to tensioning cable C

  Tensioning and anchoring cable C causes loss of prestress in cable A and B.

  Stress in concrete at level of cable A due to tensioning cable C ($\sigma_{cs,A}$):

$$\sigma_{cs,A} = \frac{P_{i,C}}{A} + \frac{(P_{i,C}\, e_{sC})e_{sA}}{I}$$

$\sigma_{cs,A}$ at support

$$\sigma_{cs,A} = \left[\frac{520 \times 10^3}{180000}\right] + \left[\frac{520 \times 10^3 \times 100 \times (-100)}{54 \times 10^8}\right] = 1.93 \text{ N/mm}^2$$

$\sigma_{cs,A}$ at mid-span

$$\sigma_{cs,A} = \left[\frac{520 \times 10^3}{180000}\right] + \left[\frac{520 \times 10^3 \times 100 \times 100}{54 \times 10^8}\right] = 3.85 \text{ N/mm}^2$$

Average value of $\sigma_{cs,A}$

$$\sigma_{cs,A} = \left[\frac{1.93 + 3.85}{2}\right] = 2.89 \text{ N/mm}^2$$

Prestress loss in cable A, $(\Delta f_{p,es})_A$:

$$(\Delta f_{p,es})_A = \frac{E_p}{E_{ci}} \sigma_{cs,A} = 7.67 \times 2.89 = 22.22 \text{ N/mm}^2$$

Stress in concrete at level of cable B due to tensioning cable C ($\sigma_{cs,B}$):

$$\sigma_{cs,B} = \frac{P_{i,C}}{A} + \frac{(P_{i,C}\, e_{sC}) e_{sB}}{I}$$

$\sigma_{cs,B}$ at support

$$\sigma_{cs,B} = \left[\frac{520\times10^3}{180000}\right] + \left[\frac{520\times10^3\times100\times0}{54\times10^8}\right] = 2.89\,\text{N/mm}^2$$

$\sigma_{cs,B}$ at mid-span

$$\sigma_{cs,B} = \left[\frac{520\times10^3}{180000}\right] + \left[\frac{520\times10^3\times100\times100}{54\times10^8}\right] = 3.85\ \text{N/mm}^2$$

Average value of $\sigma_{cs,B}$

$$\sigma_{cs,B} = \left[\frac{2.89+3.85}{2}\right] = 3.37\,\text{N/mm}^2$$

Prestress loss in cable B, $(\Delta f_{p,es})_B$:

$$(\Delta f_{p,es})_B = \frac{E_p}{E_{ci}}\sigma_{cs,B} = 7.67\times3.37 = 25.85\ \text{N/mm}^2$$

- Total loss in prestress due to elastic shortening
  Prestress loss in cable A, $(\Delta f_{p,es})_A$ $= 25.85 + 22.22 = 48.07$ N/mm^2
  Prestress loss in cable B, $(\Delta f_{p,es})_B$ $= 25.85$ N/mm^2
  Prestress loss in cable C, $(\Delta f_{p,es})_C$ $= 0$ N/mm^2

4.2.2 Loss Due to Friction

This loss occurs only in post-tensioned members due to friction between the duct and tendon. Due to this friction, the prestressing force is not constant along the length of the member. The two important sources of friction in post-tensioned member are:

1. Curvature of tendon, and
2. Unintentional deviation of duct from the design profile known as *wobble*.

Loss due to friction is highly dependent on the curvature of the tendon, i.e., greater the curvature, more will be the loss. The prestressing force (P_x)

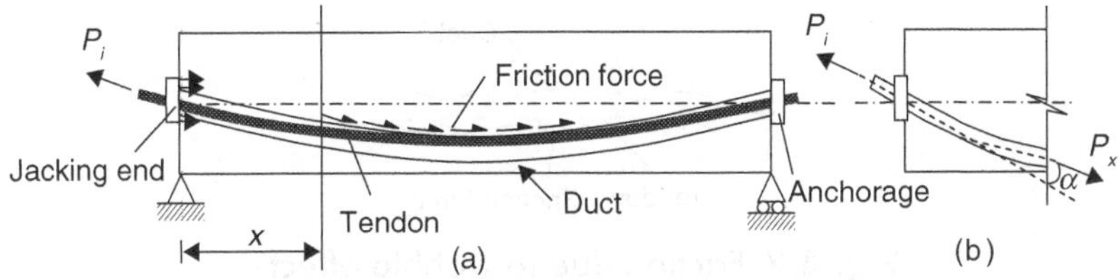

Fig. 4.6 Friction losses due to curvature

in a section at a distance x from the jacking end considering the curvature effect[1] (Fig. 4.6) is:

$$P_x = P_i\, e^{-\mu\alpha}$$

where μ is the friction coefficient between the tendon and duct, α is the cumulative change in slope in radians between the jacking end and the section. As shown in Fig. 4.6 (b), P_x and P_i will be at tangent to the cable.

In the absence of experimental data, the friction coefficient μ can be determined based on code recommendations (Table 4.2).

Table 4.2 Coefficient of friction μ (Cl. 18.5.2.6 of the code)

Type of Interface	μ
Steel moving on smooth concrete	0.55
Steel moving on steel fixed to duct	0.30
Steel moving on lead	0.25

Loss due to wobble increase with distance from the jacking end and is independent on the curvature of the tendon (Fig. 4.7). Due to wobble effect, the prestressing force at a distance x (x in metres) from the jacking end is:

$$P_x = P_i\, e^{-kx}$$

where k is the wobble or wave coefficient which depends on the quality of workmanship, distance between the supports to tendon, stiffness of the duct and the amount of vibration used when placing the concrete. As per the recommendations in the Code (Cl. 18.5.2.6), the value of k lies between 15×10^{-4} and 50×10^{-4} per metre.

---

[1] For derivation of this equation, the reader can refer the topic 'belt friction' in books on Engineering Mechanics.

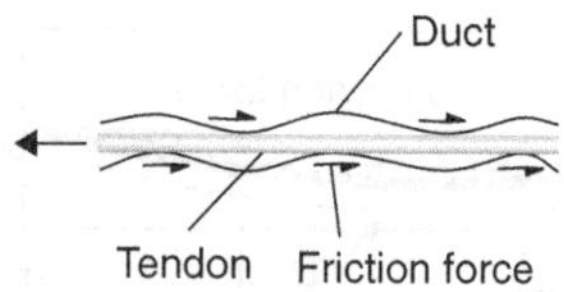

Fig. 4.7 Friction due to wobble effect

Combining both curvature and wobble effect, the prestressing force at a section is:

$$P_x = P_i\, e^{-(\mu\alpha + kx)} \tag{4.3}$$

Hence the total loss in prestressing force due to friction (ΔP_{fr}) is:

$$\Delta P_{fr} = P_i - P_x = P_i\,(1 - e^{-(\mu\alpha + kx)}) \tag{4.4}$$

Comments

- α is obtained by adding the absolute values (i.e., without considering the sign) of all angle changes from the jacking end. This is illustrated by finding the prestressing force at different sections in the beam shown in Fig. 4.8.

  The prestressing force (considering only curvature effect) at sections A, B and C are:

  $$P_A = P_i e^{-\mu\theta_1}$$

  $$P_B = P_A\, e^{-\mu\theta_2} = P_i e^{-\mu\theta_1}\, e^{-\mu\theta_2} = P_i e^{-\mu(\theta_1+\theta_2)}$$

  $$P_C = P_B\, e^{-\mu\theta_3} = P_i e^{-\mu(\theta_1+\theta_2)} e^{-\mu\theta_3} = P_i e^{-\mu(\theta_1+\theta_2+\theta_3)}$$

- If $(\mu\alpha + kx) < 0.2$, Eq. (4.3) can be simplified using Taylor series expansion as:

  $$P_x = P_i\,[1 - (\mu\alpha + kx)] \tag{4.5}$$

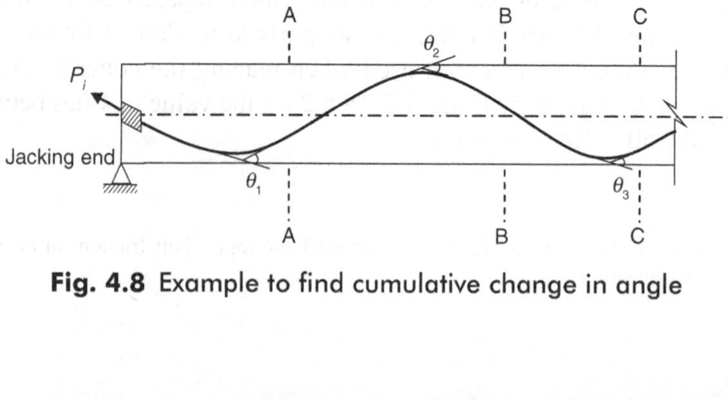

Fig. 4.8 Example to find cumulative change in angle

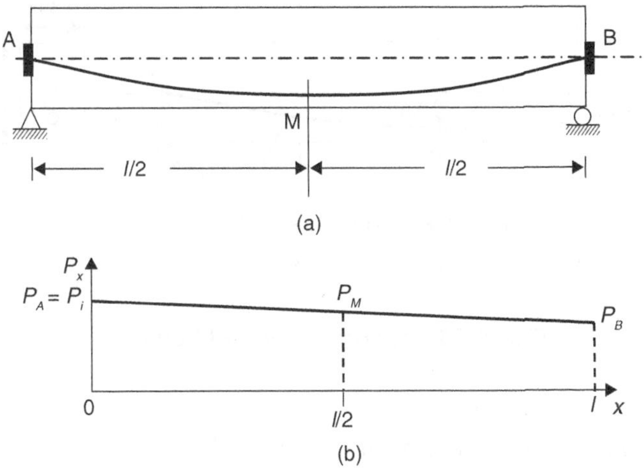

Fig. 4.9 (a) Member having tendon of single curvature and (b) force variation diagram (Tensioned from end A)

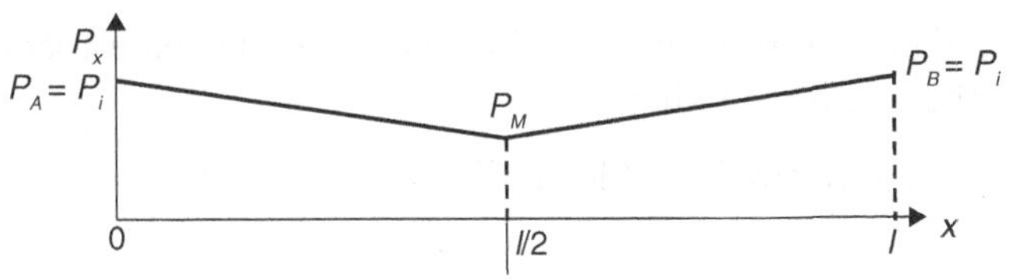

Fig. 4.10 Force variation diagram (tensioned from both ends A and B)

Hence for a tendon with single curvature (Fig. 4.9(a)), the prestressing force can be assumed to decrease linearly from the jacking end. This variation of force in the prestressed concrete member (*force variation diagram*) is shown in Fig. 4.9(b).

- Friction losses can be very large in members having tendons of large curvature. In such cases, the loss can be decreased by tensioning the tendons from both ends. The force variation diagram for the beam in Fig. 4.9(a) tensioned from both ends is shown in Fig. 4.10.

Example 4.3

Determine the prestressing loss due to friction for the post-tensioned beam (Fig. 4.11(a)) at sections 1, 2 and 3 if a) prestressing force is applied at left-hand end; b) prestressing force is applied at both ends.

Data:

$$A_p = 1200 \text{ mm}^2, f_{pi} = 1300 \text{ N/mm}^2, \mu = 0.35, k = 0.0015\text{/m},$$
$$E_p = 2.1 \times 10^5 \text{ N/mm}^2$$

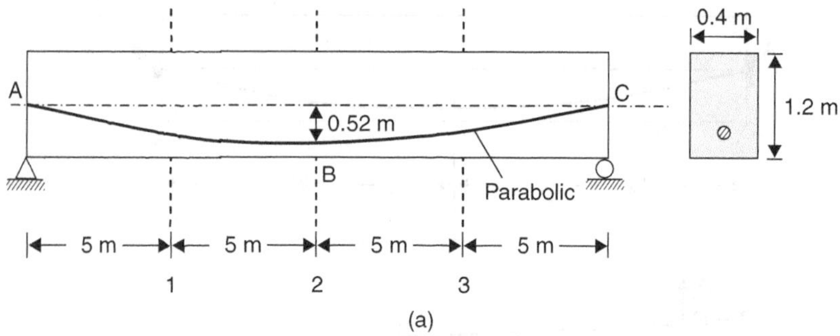

Fig. 4.11 (a) Post-tensioned beam

Solution:

Step 1: *Calculation of initial prestressing force P_i*

$$P_i = A_p \times f_{pi} = \frac{1200 \times 1300}{1000} = 1560 \, \text{kN}$$

Step 2: *Calculation of* α (change in slope between the nearest jacking end and the section in radians)

- Properties of parabolic cable profile (Fig. 4.11(b))

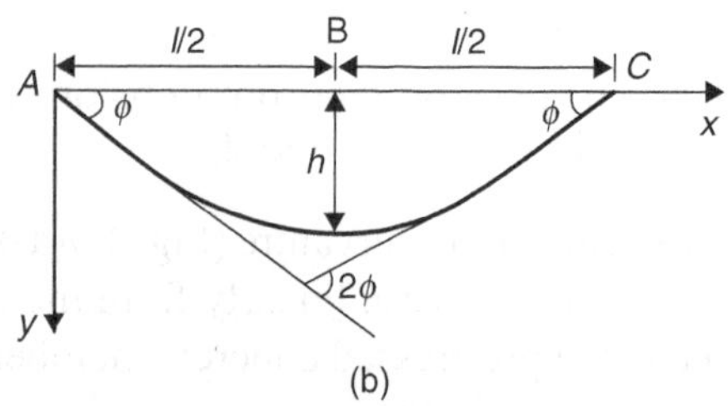

Fig. 4.11 (b) Parabolic cable profile

The equation of the cable profile is:

$$y = \frac{4h}{l^2}(lx - x^2)$$

where $h = 0.52$ m and $l = 20$ m

$$\text{slope } \phi = \left.\frac{dy}{dx}\right|_{x=0} = \frac{4h}{l} = \frac{4 \times 0.52}{20} = 0.104 \text{ radians}$$

Case A: Prestressing force applied at left-hand end.

In this case, the change in slope (α) varies linearly from 0 to 2ϕ from $A(x = 0)$ to $C(x = l)$.

$$\alpha \text{ at section 1 } (x = 0.25l) = 0.25l \times \frac{2\phi}{l} = 0.5\phi = 0.052 \text{ radians}$$

$$\alpha \text{ at section 2 } (x = 0.5l) = 0.5l \times \frac{2\phi}{l} = \phi = 0.104 \text{ radians}$$

$$\alpha \text{ at section 3 } (x = 0.75l) = 0.75l \times \frac{2\phi}{l} = 1.5\phi = 0.156 \text{ radians}$$

Case B: Prestressing force applied at both ends A and C

In this case, α varies from 0 to ϕ from A to B and also from C to B

α at section 1 ($x = 0.25l$) = 0.052 radians

α at section 2 ($x = 0.5l$) = 0.104 radians

α at section 3 ($x = 0.25l$) = 0.052 radians

Note: x is measured from the nearest jacking end. For section 3 in Case B, x is measured from right-hand end.

Step 3: *Calculation of friction losses, ΔP_{fr}*

$$\Delta P_{fr} = P_i(1 - e^{-(\mu\alpha + kx)})$$

Case A: Prestressing force applied at left-hand end.

- at section 1 ($x = 5$ m)

$$\Delta P_{fr} = 1560[1 - \exp(-[(0.35 \times 0.052) + (0.0015 \times 5)])] = 39.58 \text{ kN}$$

- at section 2 ($x = 10$ m)

$$\Delta P_{fr} = 1560[1 - \exp(-[(0.35 \times 0.104) + (0.0015 \times 10)])] = 78.16 \text{ kN}$$

- at section 3 ($x = 15$ m)

$$\Delta P_{fr} = 1560[1 - \exp(-[(0.35 \times 0.156) + (0.0015 \times 15)])] = 115.76 \text{ kN}$$

Case B: Prestressing force applied at both ends

- at section 1 ($x = 5$m): $\Delta P_{fr} = 39.58$ kN
- at section 2 ($x = 10$m): $\Delta P_{fr} = 78.16$ kN
- at section 3 ($x = 5$m): $\Delta P_{fr} = 39.58$ kN

4.2.3 Loss Due to Slip in Anchorage

During the transfer of prestressing force from the jack to the anchorage, the tendon gets shortened due to slip occurring in anchorage at the jacking end. This cause a loss in force and this decrease in prestressing force is known as loss due to slip in anchorage.

The amount of slip depends on the type of anchorage and is usually specified by the manufacturer. The slip (Δ_s) varies from zero in the case of systems having threaded bars and nuts to 8 mm in wedge-type strand anchorages. The slip occurring in a wedge-type anchorage is shown in Fig. 4.12.

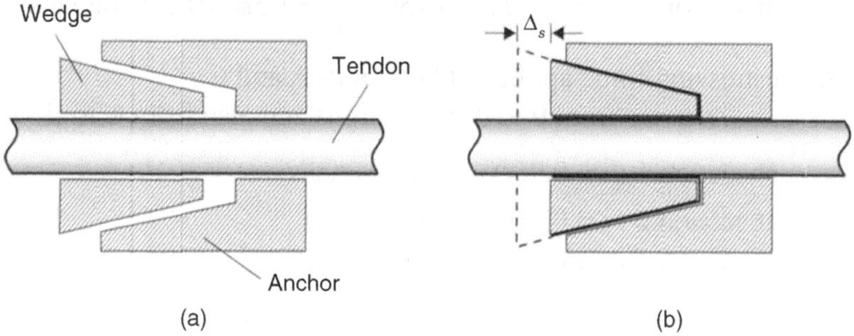

Fig. 4.12 Slip in a wedge-type anchorage system (a) before anchoring and (b) after anchoring

The effect of anchorage slip reduces due to friction between the tendon and duct. Due to this, beyond a certain length from the jacking end, the effect due to slip can be neglected. This length is known as the *setting length* (l_{set}) and it can be determined using the force variation diagram (Fig. 4.13).

In Fig. 4.13, the line AD represents the prestressing force variation using Eq. (4.5). AC represents the loss in prestressing force due to anchorage slip at the jacking end (ΔP_A). In the region AB, due to slip at the anchorage, the tendon moves opposite to the direction of tensioning. Assuming Eq. (4.5) can be used regardless of the direction of movement of tendon, the force variation in this region is represented by the line CB. Thus line CB is the

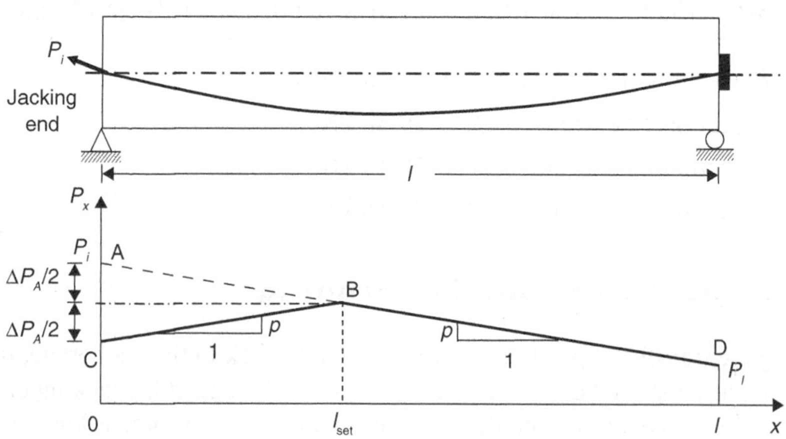

Fig. 4.13 Force variation diagram after slip in anchorage

mirror image of line AB. The slope of both lines is equal to p, which represents the loss in prestress for unit length due to friction.

Beyond B, the effect of anchorage slip can be neglected. If ε_j is the loss of strain in tendon due to slip at the jacking end, then the anchorage slip Δ_s is:

$$\Delta_s = \frac{1}{2}\varepsilon_j l_{set} \Rightarrow \varepsilon_j = 2\Delta_s / l_{set}$$

The loss in tendon force at anchorage ΔP_A is:

$$\Delta P_A = \varepsilon_j A_p E_p = \frac{2\Delta_s}{l_{set}} A_p E_p \tag{4.6}$$

From Fig. 4.13, the slope of line p can be expressed as:

$$p = \frac{\Delta P_A / 2}{l_{set}} \tag{4.7}$$

Substituting ΔP_A (from Eq. (4.6)) in the above equation gives:

$$l_{set} = \sqrt{\frac{\Delta_s A_p E_p}{p}} \tag{4.8}$$

From Eq. (4.7),

$$\Delta P_A = 2pl_{set} \tag{4.9}$$

After finding the setting length l_{set} (Eq. (4.8)), ΔP_A can be determined using Eq. (4.9).

Comments

The loss due to anchorage slip will be significant in a short prestressed member. In such cases, the loss can be compensated by initial overstressing the tendon.

Example 4.4

Upon anchoring, the tendon in the post-tensioned beam of Example 4.3 undergoes an anchorage slip (Δ_s) of 6 mm. Determine the loss due to anchorage slip at sections 1, 2 and 3, if the tendon is jacked from left-hand end.

Solution:

Step 1: *Calculation of prestressing force considering the effect of friction*

The prestressing force at any section P_x is calculated using Eq. (4.5).

$$P_x = P_i\,[1 - (\mu\alpha + kx)]$$

At $x = 20$ m, $\alpha = 2\phi = 0.208$ radians.

$$P_x(x = 20 \text{ m}) = 1560\,(1 - [(0.35 \times 0.208) + (0.0015 \times 20)])$$
$$= 1399.63 \text{ kN}$$

Step 2: *Calculation of loss in prestress per unit length due to friction p*

$$p = \frac{P_i - P_x(x = l)}{l} = \frac{1560 - 1399.63}{20} = 8 \text{ kN/m} = 8 \text{ N/mm}$$

Step 3: *Calculation of setting length l_{set}*

$$l_{set} = \sqrt{\frac{\Delta_s A_p E_p}{p}} = \sqrt{\frac{6 \times 1200 \times 2.1 \times 10^5}{8}} = 13747.73 \text{ mm} = 13.75 \text{ m}$$

Step 4: *Calculation of loss in prestressing force due to slipping at the anchorage end ΔP_A*

$$\Delta P_A = 2pl_{set} = 2 \times 8 \times 13.75 = 220 \text{ kN}$$

Step 5: *Plotting the force variation diagram (Fig. 4.14)*

In Fig. 4.14, the loss due to anchorage slip at any section x is indicated as $\Delta P_A(x)$.

The y co-ordinates of points 1, 2, 3 and 4 are:

$P_1 = 1340 + (p \times 5) = 1380$ kN; where $p = 8$ kN/m
$P_2 = 1340 + (8 \times 10) = 1420$ kN
$P_3 = 1560 - (8 \times 5) = 1520$ kN
$P_4 = 1560 - (8 \times 10) = 1480$ kN

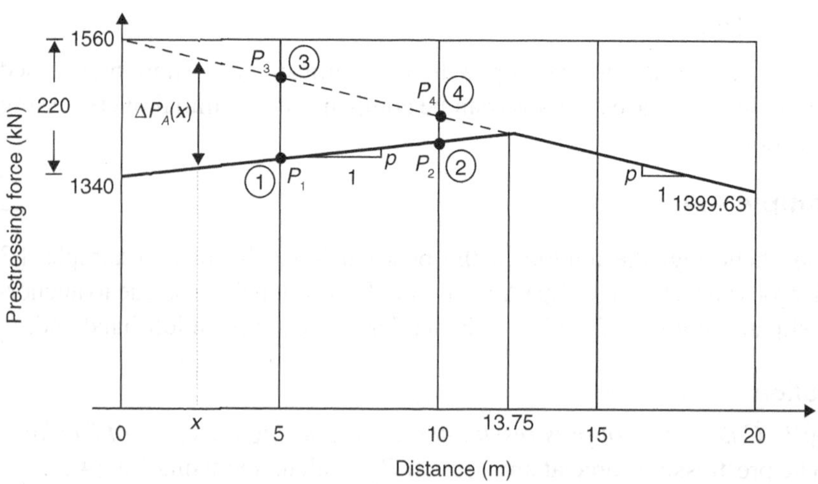

Fig. 4.14 Force variation diagram

Step 6: *Calculation of loss due to anchorage* $\Delta P_A(x)$

at section 1: $\Delta P_A\,(x = 5\text{ m}) = P_3 - P_1 = 1520 - 1380 = 140$ kN

at section 2: $\Delta P_A\,(x = 10\text{ m}) = P_4 - P_2 = 1480 - 1420 = 60$ kN

at section 3: $\Delta P_A\,(x = 15\text{ m}) = 0$ kN

(*Note:* There is no effect due to anchorage slip beyond $l_{set} = 13.75$ m)

Comments

- $\Delta P_A(x)$ varies linearly from 220 kN at $x = 0$m to 0 kN at $x = 13.75$m. As an alternate way, this information can be used to calculate $\Delta P_A(x)$ at sections 1 and 2.
- In the above example, no advantage is obtained if the tendon is tensioned from both ends. The force variation diagram for this case is shown in Fig. 4.15. Further in this case, if tensioned from both ends, the prestressing force at right half of the beam will be less than that obtained by tensioning only left-hand end.

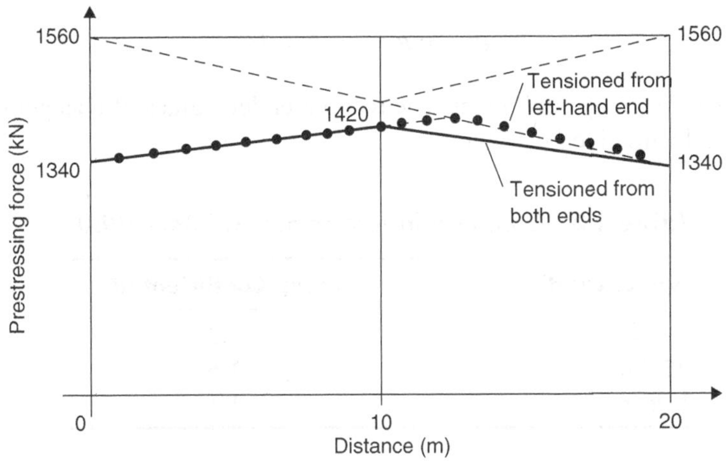

Fig. 4.15 Force variation diagram (tensioned from both ends)

4.3 Time Dependent Losses

4.3.1 Loss Due to Shrinkage of Concrete

Shrinkage in concrete denotes the contraction caused by loss of moisture. Due to shrinkage, the tendon in the member shortens, resulting in the loss of prestressing force. The loss in prestress due to shrinkage ($\Delta f_{p,\,sh}$) is:

$$\Delta f_{p,\,sh} = E_p \varepsilon_{sh} \qquad (4.10)$$

where ε_{sh} is the ultimate shrinkage strain. The recommended values of ε_{sh} as per the Code (Cl. 5.2.4.1) are:

- For pre-tensioned member, $\varepsilon_{sh} = 0.0003$
- For post-tensioned member, $\varepsilon_{sh} = 0.0002/\log_{10}(t + 2)$

where t is the age of concrete at transfer in days

4.3.2 Loss Due to Creep of Concrete

Creep refers to increase in strain occurring at constant stress. Similar to shrinkage, creep in concrete causes the tendons to slacken leading to loss of prestress.

The creep strain increases with time and reaches a ultimate value $\varepsilon_{cr,\,u}$. If the stress in concrete is less than one-third of the characteristic compressive strength, the ultimate creep strain $\varepsilon_{cr,\,u}$ is found to be directly proportional to elastic strain ε_c. Hence,

$$\varepsilon_{cr,u} \propto \varepsilon_c \Rightarrow \varepsilon_{cru} = \theta \varepsilon_c$$

where θ is the creep coefficient. The recommended values of θ as per Code (Cl. 5.2.5.1) are given in Table 4.3.

Table 4.3 Creep coefficient as per IS: 1343–1980

Age of Loading	Creep Coefficient (θ)
7 days	2.2
28 days	1.6
1 year	1.1

The elastic strain in concrete ε_c is:

$$\varepsilon_c = \frac{\sigma_{cs}}{E_c}$$

where σ_{cs} is the stress in concrete at the level of tendon caused due to prestress and sustained loads (usually the dead loads[2]). Thus the ultimate creep strain is:

[2] A small fraction of live loads can be also included as part of sustained load.

$$\varepsilon_{cr,u} = \theta\varepsilon_c = \theta\frac{\sigma_{cs}}{E_c} \tag{4.11}$$

where,

$$\sigma_{cs} = \frac{P_o}{A} + \frac{(P_o e_s)e_s}{I} - \frac{M_{su} e_s}{I} \tag{4.12}$$

In the above equation, P_o is the prestressing force after immediate losses and M_{su} is the bending moment due to sustained loads. The loss in prestress due to creep in concrete ($\Delta f_{p,\, cr}$) is:

$$\Delta_{fp,\, cr} = E_P\, \varepsilon_{cr,\, u} \tag{4.13}$$

4.3.3 Loss Due to Relaxation of Steel

Under constant strain, steel starts loosing stress with time. This is known as relaxation of steel, and is similar to creep, except that creep refers to increase in strain under constant stress while relaxation is the decrease in stress under constant strain.

The loss of prestress due to relaxation ($\Delta f_{p,\, re}$) depends on the stress at which steel is initially tensioned (Δf_{pi}), temperature and on the type of steel. In the absence of test data, $\Delta f_{p,\, re}$ can be calculated using Cl. 18.5.2.3 of the Code (Table 4.4).

Table 4.4 Loss in prestress due to relaxation of steel

Initial Stress in Pre-stressing Steel (f_{pi})	$\Delta f_{p,\, re}$ (N/mm²)
0.5 f_p	0
0.6 f_p	35
0.7 f_p	70
0.8 f_p	90

In the above table, f_p is the characteristic strength of prestressing steel.

Example 4.5

For the beam in Example 4.1, determine the prestress loss due to shrinkage, creep and steel relaxation for the following data:

Intensity of super imposed dead load (w_{SIDL}) = 5 kN/m
Characteristic strength of prestressing steel (f_p) = 1700 N/mm²
Compressive (cube) strength of concrete (f_{ck}) = 50 N/mm²
Age of loading = 28 days

Solution:

(a) Prestress loss due to shrinkage of concrete $\Delta f_{p, sh}$

Step 1: *Calculation of ultimate shrinkage strain ε_{sh}*

$$\varepsilon_{sh} = 0.0003 \text{ (for pre-tensioned member)}$$

Step 2: *Calculation of shrinkage loss $\Delta f_{p, sh}$*

$$\Delta f_{p, sh} = E_p \, \varepsilon_{sh} = 2.1 \times 10^5 \times 0.0003 = 63 \text{ N/mm}^2$$

(b) Prestress loss due to creep of concrete $\Delta f_{p, cr}$

Step 1: *Calculation of creep coefficient θ*

Age of loading = 28 days

$$\theta = 1.6 \text{ (from Cl. 5.2.5.1 of the Code)}$$

Step 2: *Calculation of σ_{cs} due to prestress and sustained loading*

Intensity of self-weight (w_{SW}) = 9.6 kN/m
Super imposed load (w_{SIDL}) = 5 kN/m
Intensity of dead load w_{DL} = $w_{SW} + w_{SIDL}$ = 9.6 + 5 = 14.6 kN/m
Intensity of Sustained load = Dead load = w_{DL} =14.6 kN/m

- Stress in prestressing steel immediately after transfer f_{po}

$$f_{po} = f_{p_i} - (\Delta f_p)_{il}$$

$(\Delta f_p)_{il}$ = Immediate losses = $\Delta f_{p, es}$ (loss due to friction and anchorage slip does not occur in pre-tensioned member)

$$f_{po} = 1360 - 113.64 = 1246.36 \text{ N/mm}^2$$

- Prestressing force after immediate losses P_o

$$P_o = A_p \times f_{po} = \frac{2100 \times 1246.36}{1000} = 2617.36 \text{kN}$$

- Stress in concrete at the level of tendon σ_{cs}

$$\sigma_{cs} = \frac{P_o}{A} + \frac{(P_o e_s) e_s}{I} - \frac{M_{su} e_s}{I}$$

σ_{cs} at support

$$M_{su} = 0$$

$$\sigma_{cs} = \left[\frac{2617.36 \times 10^3}{400000}\right] + \left[\frac{2617.36 \times 10^3 \times 500 \times 500}{7.733 \times 10^{10}}\right] - 0$$

$$= 15.01 \text{ N/mm}^2$$

σ_{cs} *at midspan*

$$M_{su} = \frac{w_{DL} l^2}{8} = \frac{14.6 \times 20^2}{8} = 730\,\text{kN-m}$$

$$\sigma_{cs} = \left[\frac{2617.36 \times 10^3}{400000}\right] + \left[\frac{2617.36 \times 10^3 \times 500 \times 500}{7.733 \times 10^{10}}\right] - \left[\frac{730 \times 10^6 \times 500}{7.733 \times 10^{10}}\right] = 10.29\ \text{N/mm}^2$$

Average value of σ_{cs}

$$\sigma_{cs} = \frac{15.01 + 10.29}{2} = 12.65\,\text{N/mm}^2$$

Step 3: *Calculation of ultimate creep strain* $\varepsilon_{cr,\,u}$

$$E_c = 5000\sqrt{f_{ck}} = 5000\sqrt{50} = 35355.34\,\text{N/mm}^2$$

$$\varepsilon_{cr,u} = \theta \frac{\sigma_{cs}}{E_c} = \frac{1.6 \times 10.29}{35355.34} = 4.657 \times 10^{-4}$$

Step 4: *Calculation of creep losses* $\Delta f_{p,\,cr}$

$$\Delta f_{p,\,cr} = E_p\, \varepsilon_{cr,\,u} = 2.1 \times 10^5 \times 4.657 \times 10^{-4} = 97.79\ \text{N/mm}^2$$

(c) Loss due to relaxation of steel $\Delta f_{p,re}$

$$\frac{f_{P_i}}{f_p} = \frac{1360}{1700} = 0.8$$

For $\frac{f_{P_i}}{f_p} = 0.8$, $\Delta f_{p,re} = 90$ N/mm^2 (from Cl. 18.5.2.3. of the Code)

4.4 Total Loss in Prestress

For initial design, it is necessary to have a rough estimate of the loss in prestress. If P_i is the initial prestressing force, the force after immediate losses P_o is βP_i and the final or effective prestressing force after all losses P_e is η P_i. Where β and η are loss ratios which takes into account of immediate and time dependent losses.

For both pre-tensioned and post-tensioned members, for initial design, β and η can be taken as 0.9 and 0.75 respectively.

PROBLEMS

4.1 Determine the loss in prestress due to elastic shortening of concrete for the pre-tensioned beam shown in Fig. 4.16. The beam is tensioned using 6 mm wires kept parallel to the centroidal axis.

Data:

$$f_{pi} = 1000 \text{ N/mm}^2, f_{ci} = 50 \text{ N/mm}^2, E_p = 2 \times 10^5 \text{ N/mm}^2$$

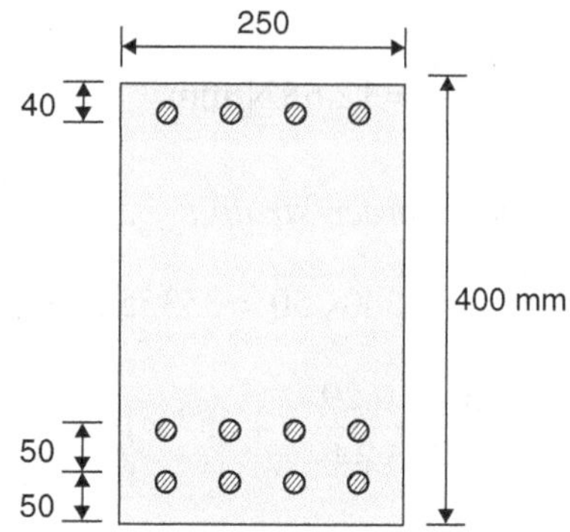

Fig. 4.16

4.2 Determine the loss in prestress due to friction at the anchorage end for the post-tensioned beam in Example 4.2, if the cables are tensioned only from one end. Assume $\mu = 0.35$ and $k = 0.0015$/m.

4.3 Find the prestress loss due to shrinkage of concrete if the beam is

(i) Pre-tensioned and

(ii) Post-tensioned.

The age of concrete at transfer is 7 days. The modulus of elasticity of prestressing steel is 2×10^5 N/mm^2.

chapter 5

Analysis of Sections

5.1 Introduction

The term 'analysis' in the context of structural design refers to determination of stresses in materials (or deflections, crack widths etc.) and finding the ultimate capacity of the section (in flexure, shear, torsion etc.). In order to find the stress or ultimate capacity, the cross-sectional details (including the area of steel) and the properties of materials should be known to begin with.

This chapter deals with the determination of flexural stress and ultimate moment capacity of the section. For this, three basic equations of mechanics are used.

1. Equilibrium equations
2. Constitutive relations
3. Compatibility conditions

Equilibrium Equations: At any section, the total compression force should be equal to the total tension force. The internal moment due to these forces should be equal to the moment caused by external loads.

Constitutive Relations: The material stress–strain relations (constitutive relations) used for analysis are already discussed in Chapters 2 and 3.

Compatibility Conditions: For analysis, it is assumed that strain varies linearly across the cross-section. The strain in steel in pre-tensioned and bonded post-tensioned members is taken to be same as the strain in the surrounding concrete at the level of steel.

The difference in the analysis of sections at serviceability and ultimate limit states lies only in the use of constitutive relations. In each case, different regions of the material stress–strain relations are used.

5.2 Analysis at Serviceability Limit State

The design of prestressed concrete member is usually governed by the permissible concrete stresses at serviceability limit state rather than by the strength at ultimate limit state. The member is designed to satisfy these stress limits and checked for the ultimate capacity. Hence determination of concrete stresses at serviceability limit state is a very important part in the design of prestressed concrete member.

The prestressed concrete member can be designed to remain as uncracked (type 1 and type 2) or as a cracked (type 3) member under service loads. Only the analysis of uncracked section is discussed in this chapter.

Loading Stages

Though the serviceability criteria should be checked at all loading stages, usually it will be sufficient to check at two critical loading conditions namely at transfer and at service condition.

The first loading condition known as the transfer loading stage occurs when the prestress from the tendon is transferred to concrete. At this stage, the concrete would not have reached its full strength (low concrete strength) and at the same time, the prestressing force has a high value as significant losses would not have occurred. Hence this loading stage is characterised by high prestressing force and low compressive strength of concrete.

Due to occurrence of immediate losses during the transfer stage, the concrete is never subjected to the initial prestressing force P_i. Hence at the transfer loading stage, the force in the tendon is P_o, which is the prestressing force after immediate losses.

The second loading condition is known as the service condition or final condition at which the member is subject to full service loads. At the time when the full live loads are applied, all losses in prestress would have taken place and concrete has reached its full strength. Thus this loading stage is characterized by minimum prestressing force (P_e) and maximum compressive strength of concrete.

The permissible stresses in concrete for these loading stages are given in Cl. 22.7 and 22.8 of the Code.

Assumptions for Analysis

The important assumptions used to determine stresses in concrete at serviceability limit state are:

1. Concrete is a homogeneous material.
2. Within the range of working stresses, both concrete and steel behaves as a linearly elastic material (i.e., stress–strain curve for both concrete and steel is linear).
3. Plane section before bending remains plane after bending (this is the compatibility condition which assumes linear variation of strain).
4. Up to the stage of visible cracking, the stress in the tendon does not change due to flexure caused by external loads.

The fourth assumption implies that, till the stage of visible cracking (tensile stress in concrete is less than modulus of rupture), the function of tendon is just to provide and maintain prestress in concrete. Any change in loading on the member causes a change of stress in concrete alone. After cracking, the tendon behaves similar to tensile reinforcement provided in reinforced concrete and assists in resisting the external load.

Sign Convention

The prestressing force is considered as positive when it causes compression in concrete. Sagging bending is treated as positive and tendon eccentricity is positive if it is below the centroidal axis. The sign conventions followed in this book are tabulated in Table 5.1.

Table 5.1 Sign convention

No.	Parameter	Positive	Negative
1.	Axial forces, stresses, strains	Compression	Tension
2.	Distances (e_s, e_{cc}, y)	Below centroidal axis	Above centroidal axis
3.	Transverse loads	Downward	Upward
4.	Shear force	Downward to right	Upward to right
5.	Bending moment	Sagging	Hogging
6.	Deflections	Downward	Upward
7.	Rotation or slope	Clockwise	Anti-clockwise

Methods to Calculate Elastic Stresses in Concrete

The elastic stress distribution in an uncracked concrete cross-section due to prestress, self-weight and applied loads can be determined using three different methods namely:

1. Combined load approach
2. Internal couple approach
3. Equivalent load approach.

These methods are discussed in detail in the sections to follow.

5.2.1 Combined Load Approach

In this method, the stress distribution at a section due to prestress and applied moment are determined separately and added to get the combined stress distribution. The calculation of stress distribution in concrete at any section 1–1 in a simply supported beam (Fig. 5.1) is explained below.

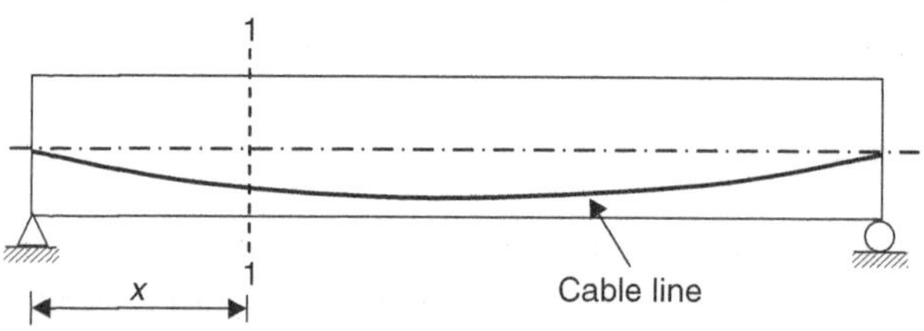

Fig. 5.1 Uncracked simply supported prestressed concrete beam

The section can be taken anywhere along the beam except near the supports. At section 1–1, the force in the tendon is P, which is inclined at an angle θ with the horizontal. The eccentricity of the tendon is e_s and the bending moment (sagging) due to the applied loads is M.

Stress Due to Prestress

The free body diagram of the left-hand side of section 1–1 (neglecting the external loads including self-weight) of the beam is shown in Fig. 5.2.

To maintain equilibrium at a section, the resultant of stress acting in concrete R (Fig. 5.2(a)) is equal in magnitude ($P = R$) and opposite in direction to the tendon force P. The horizontal component of R is H ($H = R \cos\theta$), which represents the resultant of longitudinal stresses across the section. Similarly, S the vertical component of R ($S = R \sin\theta$) represents the resultant of shear stresses acting on the section.

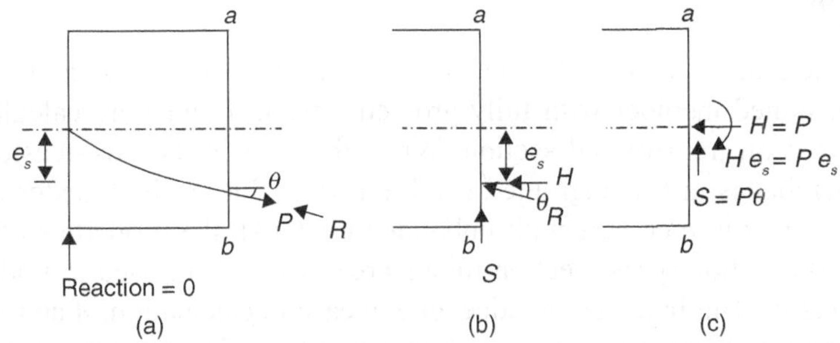

Fig. 5.2 Stress resultants in concrete due to prestress (a) force in tendon and concrete, (b) force components in concrete and (c) equivalent force system at centroid

Since the depth of the beam is small when compared to its length, the inclination of the tendon θ will be a small value. Hence,

$$H = R\cos\theta \approx R = P$$
$$S = R\sin\theta \approx R\theta = P\theta \tag{5.1}$$

To determine the longitudinal stress distribution, it is convenient to use the equivalent force system acting at the centroid of the section (Fig. 5.2(c)). Since S does not contribute to longitudinal stresses, its effects will not be further discussed in this chapter. The longitudinal stresses due to compressive force P and the hogging moment Pe_s are shown in Fig. 5.3.

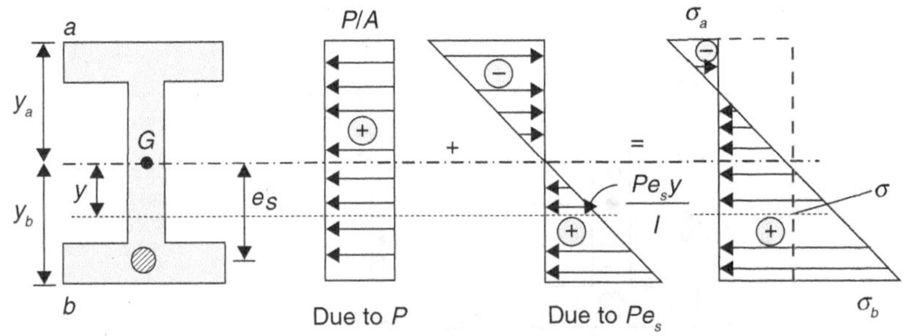

Fig. 5.3 Longitudinal stresses due to prestress

The stress (σ) at a distance y below the centroid G due to prestress is:

$$\sigma = \frac{P}{A} + \frac{Pe_s y}{I} \tag{5.2}$$

where A and I are the cross-sectional area and moment of inertia of the section respectively.

Comments

- When the tendon is bonded to concrete (pre-tensioned and post-tensioned member with fully grouted ducts), A and I are calculated based on transformed section. When the duct in the post-tensioned member is in the ungrouted condition (i.e., during the transfer condition or in members with unbonded tendons), the properties of net cross-section (gross section minus area of duct) are used to find the stresses. But in actual practice, due to ease in calculation, A and I are calculated based on gross section properties. The error involved in using gross section properties is usually very small.
- In the case of statically indeterminate structures, in addition to cross-sectional properties (A and I) and values of P and e_s at the section under consideration, the longitudinal stress depends on A, I, P and e_s at other sections in the member. Hence this method for finding stresses cannot be directly used for indeterminate structures.
- A prestressed concrete member may contain several cables. For convenience in analyzing the member, the cables are replaced by a single equivalent cable known as the *cable line* (Fig. 5.1), which is obtained by joining the line of action of the resultant prestressing force at all sections in the member (see Example 5.1).

Stresses Due to Applied Moment

Fig. 5.4 shows the longitudinal stress distribution due to the sagging bending moment M.

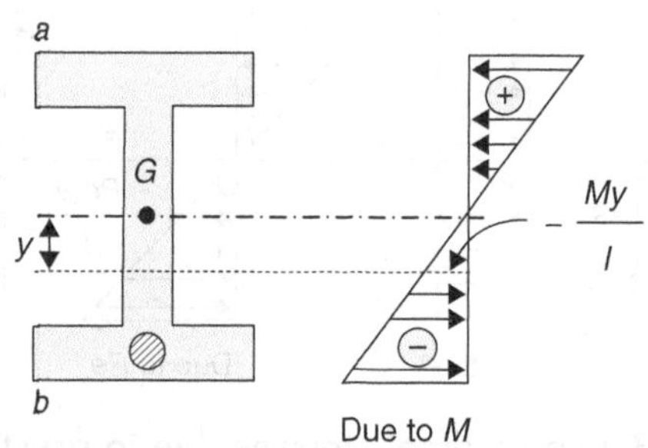

Fig. 5.4 **Longitudinal stresses due to applied moment**

Using bending theory, the stress at a level y below the centroid (G) due to M is:

$$\sigma = -\frac{M}{I}y \tag{5.3}$$

where the negative sign indicates the tensile nature of stress.

Combined Stresses Due to Prestress and Applied Moment

The total stress σ due to prestress and applied moment (Fig. 5.5) at y below the centroid is obtained by adding Eqs. (5.2) and (5.3).

$$\sigma = \frac{P}{A} + \frac{(Pe_s)y}{I} - \frac{My}{I} \tag{5.4a}$$

The total stress at the top fibre of the beam σ_a $(y = -y_a)$ is:

$$\sigma_a = \frac{P}{A} + \frac{(Pe_s)(-y_a)}{I} - \frac{M(-y_a)}{I}$$

$$\sigma_a = \frac{P}{A} - \frac{(Pe_s)}{Z_a} + \frac{M}{Z_a} \tag{5.4b}$$

Similarly the stress at the bottom fibre of the beam σ_b $(y = y_b)$ is:

$$\sigma_b = \frac{P}{A} + \frac{Pe_s y_b}{I} - \frac{M\,y_b}{I}$$

$$\sigma_b = \frac{P}{A} + \frac{Pe_s}{Z_b} - \frac{M}{Z_b} \tag{5.4c}$$

In these equations Z_a $(Z_a = I/y_a)$ and Z_b $(Z_b = I/y_b)$ are the section moduli for the top and bottom fibres of the beam.

From Fig. 5.5(a), it can be seen that the effect of sagging moment M is to rotate the stress line due to prestress (line ABD) in a clockwise direction about the point B (i.e., about the centroid). This results in an increase in compressive stress at the top fibre and a decrease in compressive stress at the bottom, while the compressive stress at B remains the same and is equal to P/A.

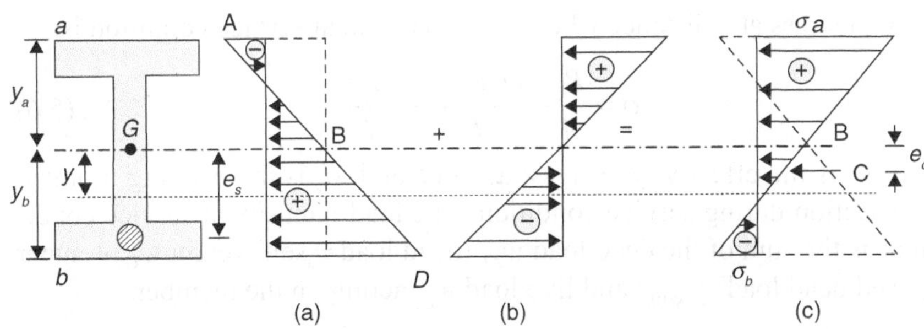

Fig. 5.5 Combined stresses due to prestress and applied moment (a) due to prestress, (b) due to M and (c) combined stresses

Combined Stresses at Transfer

The stress at a distance y below the centroid at transfer loading stage is:

$$\sigma = \frac{P_o}{A} + \frac{(P_o e_s)y}{I} - \frac{M_o y}{I} \tag{5.5}$$

where P_o is the prestressing force after immediate losses and M_o is the bending moment due to applied loads at transfer.

Due to the eccentricity of the prestressing force, the beam will deflect upward and the external moment M_o is due to the self-weight w_{SW} of the beam (Fig. 5.6). At this stage, the bottom fibre is in compression and top fibre in tension.

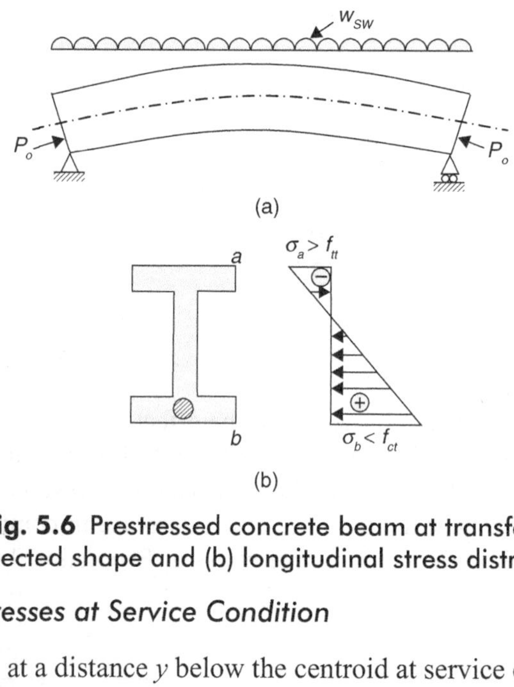

Fig. 5.6 Prestressed concrete beam at transfer (a) deflected shape and (b) longitudinal stress distribution

Combined Stresses at Service Condition

The total stress at a distance y below the centroid at service condition is:

$$\sigma = \frac{P_e}{A} + \frac{(P_e e_s)y}{I} - \frac{M_s y}{I} \tag{5.6}$$

where P_e is the effective prestressing force and M_s is the bending moment at the section during service condition. The load intensity at service condition w_s is the sum of the dead load w_{DL} (dead load = self weight w_{SW} + super imposed dead load* w_{SIDL}) and live load w_{LL} acting on the member.

---

* Superimposed dead load is the weight of non structural elements like partitions, ceilings, finishes etc.

At this loading stage, the bottom fibre of the beam is in tension and the top fibre in compression (Fig. 5.7).

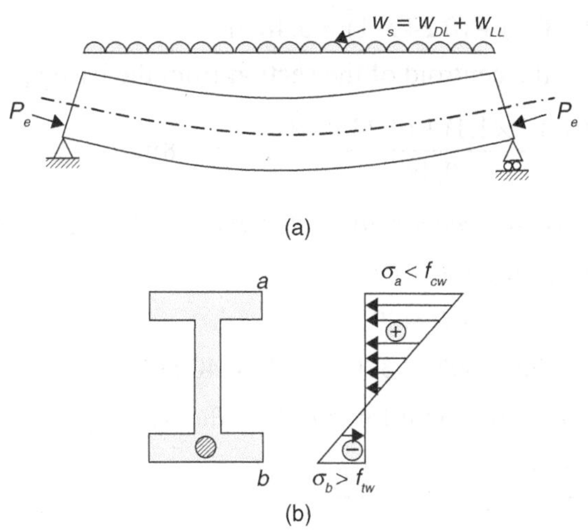

Fig. 5.7 Prestressed concrete member at service condition (a) deflected shape and (b) longitudinal stress distribution

Example 5.1

Determine the eccentricity of the cable line in the prestressed concrete beam section (Fig. 5.8(a)). The effective prestressing force in cables 1, 2, 3, 4 and 5 are 200 kN, 240 kN, 280 kN, 320 kN and 360 kN respectively.

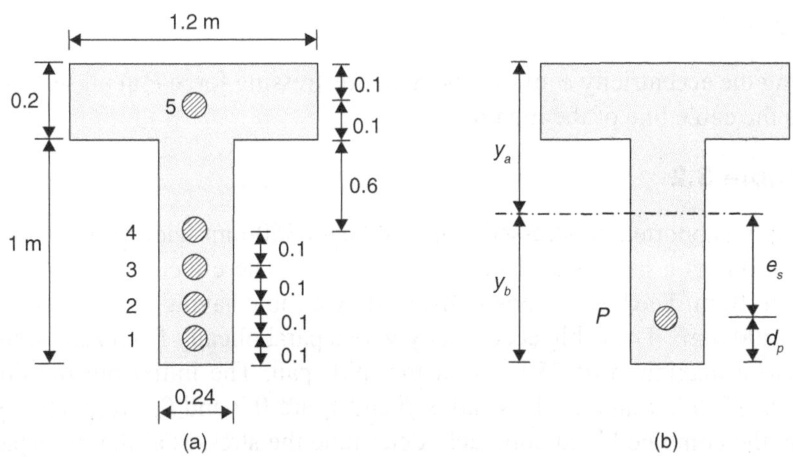

Fig. 5.8 Prestressed concrete beam section

Solution:

Step 1: *Calculation of section properties*

- Cross-sectional area A

$$A = (1.2 \times 0.2) + (0.24 \times 1) = 0.48\ \text{m}^2$$

- Distance of the centroid of the section from the base y_b

$$y_b = \frac{(1.2 \times 0.2 \times 1.1) + (0.24 \times 1 \times 0.5)}{0.48} = 0.8\,\text{m}$$

Step 2: *Calculation of eccentricity of the cable line e_s (Fig. 5.8(b))*

- Total prestressing force P

$$P = P_1 + P_2 + P_3 + P_4 + P_5$$
$$= 200 + 240 + 280 + 320 + 360 = 1400\ \text{kN}$$

- Distance of the resultant force P from the base (d_p)

$$d_p = \frac{P_1 d_1 + P_2 d_2 + P_3 d_3 + P_4 d_4 + P_5 d_5}{P}$$

where d_i is the distance of P_i from the base

$$d_p = \frac{(200 \times 0.1) + (240 \times 0.2) + (280 \times 0.3) + (320 \times 0.4) + (360 \times 1.1)}{1400}$$
$$= 0.483\ \text{m}$$

- Eccentricity of the cable line e_s

$$e_s = y_b - d_p$$
$$= 0.8 - 0.483 = 0.317\ \text{m}$$

Comment

Joining the eccentricity e_s of the resultant prestressing force P at all sections gives the cable line of the member.

Example 5.2

A simply supported prestressed concrete beam 350 mm wide and 800 mm deep supports a live load of intensity 25 kN/m. The effective span of the beam is 10 m. The beam is post-tensioned by a cable having cross-sectional area 1200 mm^2. The cable eccentricity varies parabolically from zero at the ends to a maximum of 250 mm at the mid-span. The initial prestressing force is 1500 kN and the loss ratios β and η are 0.9 and 0.7 respectively. Using the combined load approach, determine the stresses at the mid-span (i) at transfer and (ii) at service condition.

Solution:

Step 1: *Calculation of section properties*

- Cross-sectional area A

$$A = b\,D = 350 \times 800 = 280000 \text{ mm}^2$$

- Section modulus

$$Z_a = Z_b = \frac{1}{6} bD^2 = \frac{1}{6} \times 350 \times 800^2 = 3733.33 \times 10^4 \text{ mm}^3$$

- Self-weight of the member w_{SW}

  Unit weight of concrete $\gamma_c = 24$ kN/m^3

$$w_{SW} = \gamma_c A = 24 \times 280000 \times 10^{-6} = 6.72 \text{ kN/m}$$

Step 2: *Calculation of stresses at transfer*

Initial prestressing force $P_i = 1500$ kN

Prestressing force after immediate losses $P_o = \beta P_i = 0.9 \times 1500 = 1350$ kN

Span of beam $l = 10$ m

Tendon eccentricity at mid-span $e_s = 250$ mm

- Bending moment due to self-weight (at mid-span) M_o

$$M_o = \frac{w_{SW} l^2}{8} = \frac{6.72 \times 10^2}{8} = 84 \text{kN-m}$$

- Stress at bottom fibre σ_b

$$\sigma_b = \frac{P_o}{A} + \frac{P_o e_s}{Z_b} - \frac{M_o}{Z_b}$$

$$= \left[\frac{1350 \times 10^3}{280000}\right] + \left[\frac{1350 \times 10^3 \times 250}{3733.33 \times 10^4}\right] - \left[\frac{84 \times 10^6}{3733.33 \times 10^4}\right]$$

$$= 4.82 + 9.04 - 2.25 = 11.61 \text{ N/mm}^2$$

- Stress at top fibre σ_a

$$\sigma_a = \frac{P_o}{A} - \frac{P_o e_s}{Z_a} + \frac{M_o}{Z_a}$$

$$= 4.82 - 9.04 + 2.25 = -1.97 \text{ N/mm}^2$$

Step 3: *Calculation of stresses at service condition*

Effective prestressing force $P_e = \eta P_i = 0.7 \times 1500 = 1050$ kN

- Bending moment (at mid-span) at service condition M_s

  Live load intensity $w_{LL} = 25$ kN/m

Dead load intensity $w_{DL} = w_{SW} = 6.72$ kN/m
Load intensity at service condition $w_s = w_{DL} + w_{LL} = 6.72 + 25 = 31.72$ kN/m

$$M_s = \frac{w_s l^2}{8} = \frac{31.72 \times 10^2}{8} = 396.5 \text{ kN-m}$$

- Stress at bottom fibre σ_b

$$\sigma_b = \frac{P_e}{A} + \frac{P_e e_s}{Z_b} - \frac{M_s}{Z_b}$$

$$= \left[\frac{1050 \times 10^3}{280000}\right] + \left[\frac{1050 \times 10^3 \times 250}{3733.33 \times 10^4}\right] - \left[\frac{396.5 \times 10^6}{3733.33 \times 10^4}\right]$$

$$= 3.75 + 7.03 - 10.62 = 0.16 \text{ N/mm}^2$$

- Stress at top fibre σ_a

$$\sigma_a = \frac{P_e}{A} - \frac{P_e e_s}{Z_a} + \frac{M_s}{Z_a}$$

$$= 3.75 - 7.03 + 10.62 = 7.34 \text{ N/mm}^2$$

Example 5.3

For the type 1 post-tensioned beam in Fig. 5.9, determine the required prestressing force and its eccentricity at the mid-span during service condition. The effective span of the beam is 20 m and it has to carry a live load intensity of 20 kN/m. The compressive strength of M 30 grade concrete at transfer (f_{ci}) is 20 N/mm². Also check the stresses at transfer. The loss ratios β and η can be taken as 0.9 and 0.8 respectively.

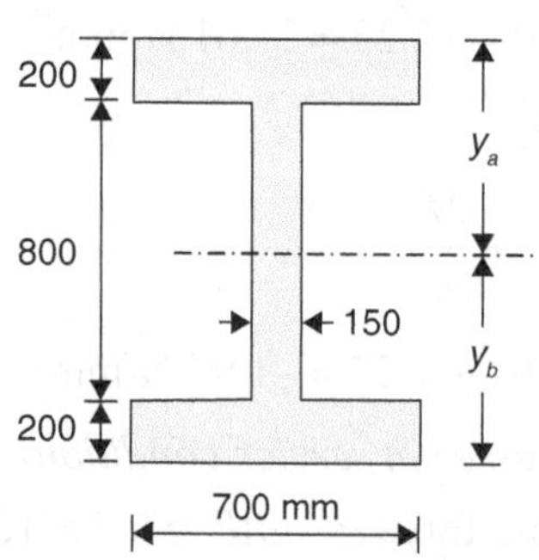

Fig. 5.9 Cross-section of the beam

Solution:

Step 1: *Calculation of section properties*

- Area of cross-section A

$$A = 2 \times (700 \times 200) + (800 \times 150) = 400000 \text{ mm}^2$$

- Moment of inertia I

$$I = 2\left[\left(\frac{700 \times 200^3}{12}\right) + (700 \times 200 \times 500^2)\right] + \left[\frac{150 \times 800^3}{12}\right]$$

$$= 7.733 \times 10^{10} \text{ mm}^4$$

- Section modulus Z

$$Z_a = I/y_a \,;\, Z_b = I/y_b$$

$$Z_a = Z_b = 7.733 \times 10^{10}/600 = 1.289 \times 10^8 \text{mm}^3$$

- Self-weight of the member w_{SW}

$$\gamma_c = 24 \text{ kN/m}^2$$

$$w_{SW} = \gamma_c A = 24 \times 400000 \times 10^{-6} = 9.6 \text{ kN/m}$$

Step 2: *Calculation of permissible stresses in concrete (Cl. 22.7 and 22.8 of the Code)*

- Permissible stresses at transfer

  Maximum permissible compressive stress at transfer f_{ct}

$$f_{ct} = 0.54 f_{ci} = 0.54 \times 20 = 10.8 \text{ N/mm}^2$$

  Maximum permissible tensile stress at transfer f_{tt}

$$f_{tt} = 0 \text{ N/mm}^2$$

- Permissible stresses at service condition

  Maximum permissible compressive stress at service f_{cw} (assuming zone I)

$$f_{cw} = 0.41 f_{ck} = 0.41 \times 30 = 12.3 \text{ N/mm}^2$$

  Maximum permissible tensile stress at service f_{tw}

$$f_{tw} = 0 \text{ N/mm}^2$$

Step 3: *Calculation of bending moment at mid-span*

- Bending moment at transfer M_o

$$M_o = \frac{w_{SW} l^2}{8} = \frac{9.6 \times 20^2}{8} = 480 \text{kN-m}$$

- Bending moment at service condition M_s

  Live load intensity $w_{LL} = 20$ kN/m

  Dead load intensity $w_{DL} = w_{SW} = 9.6$ kN/m

  Load intensity at service condition $w_s = w_{DL} + w_{LL} = 9.6 + 20 = 29.6$ kN/m

$$M_s = \frac{w_s l^2}{8} = \frac{29.6 \times 10^2}{8} = 1480 \text{ kN-m}$$

Step 4: *Calculation of P_e and e_s at service condition*

At service condition, the bottom fibre is in tension and the top fibre in compression (Fig. 5.7). Hence equating the extreme fibre stresses to its maximum permissible limits gives:

$$\sigma_b = \frac{P_e}{A} + \frac{P_e e_s}{Z_b} - \frac{M_s}{Z_b} = f_{tw} \quad \text{(A)}$$

$$\sigma_a = \frac{P_e}{A} - \frac{P_e e_s}{Z_a} + \frac{M_s}{Z_a} = f_{cw} \quad \text{(B)}$$

By adding Eqs. (A) and (B), and using the condition $Z_a = Z_b$, P_e is obtained as:

$$P_e = \frac{A}{2}(f_{cw} + f_{tw})$$

$$= \frac{40000(12.3 + 0)}{2 \times 1000} = 2460 \text{ kN}$$

The tendon eccentricity can be calculated using either Eq. (A) or (B). From Eq. (A),

$$\left[\frac{2460 \times 10^3}{4 \times 10^5}\right] + \left[\frac{2460 \times 10^3 \times e_s}{1.289 \times 10^8}\right] - \left[\frac{1480 \times 10^6}{1.289 \times 10^8}\right] = 0$$

By solving the above equation, e_s is obtained as 279.38 mm

Step 5: *Check for stresses at transfer*

Initial prestressing force $P_i = \frac{P_e}{\eta} = \frac{2460}{0.8} = 3075 \text{ kN}$

Prestressing force after immediate losses $P_o = \beta P_i = 0.9 \times 3075$
$= 2767.5$ kN

- Stress at bottom fibre σ_b

$$\sigma_b = \frac{P_o}{A} + \frac{P_o e_s}{Z_b} - \frac{M_o}{Z_b}$$

$$= \left[\frac{2767.5 \times 10^3}{4 \times 10^5}\right] + \left[\frac{2767.5 \times 10^3 \times 279.38}{1.289 \times 10^8}\right] - \left[\frac{480 \times 10^6}{1.289 \times 10^8}\right]$$

$= 6.92 + 6 - 3.72 = 9.2$ N/mm^2 (< 10.8 N/mm^2, OK)

- Stress at top fibre σ_a

$$\sigma_a = \frac{P_o}{A} - \frac{P_o e_s}{Z_a} + \frac{M_o}{Z_a}$$

$= 6.92 - 6 + 3.72 = 4.64$ N/mm^2 (> 0 N/mm^2, OK)

5.2.2 Internal Couple Approach

The longitudinal stress distribution due to the combined effect of prestress and applied moment can be replaced by the resultant compressive force C located at an eccentricity e_{cc} from the centroid (Fig. 5.5). The point at which C acts is known as the *centre of compression*. The force C acting at the centre of compression can be replaced by a statically equivalent force system acting at the centroid (Fig. 5.10).

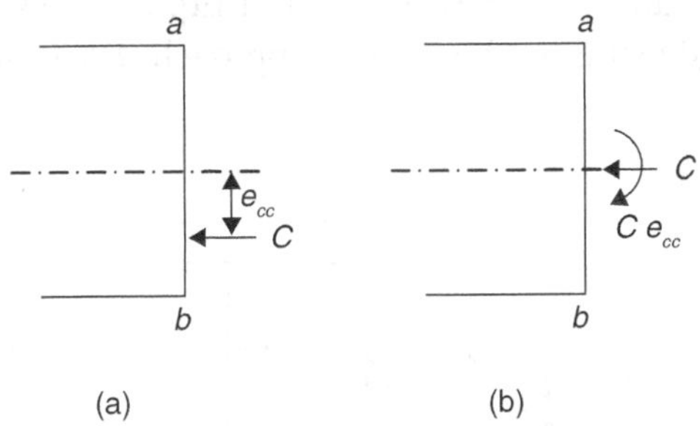

Fig. 5.10 Stress resultants in concrete (a) resultant compressive force in concrete and (b) equivalent force system at centroid

The longitudinal stress distribution caused by this equivalent force system is shown in Fig. 5.11. The stress (σ) at a distance y below the centroid (G) is:

$$\sigma = \frac{C}{A} + \frac{(C e_{cc})y}{I} \tag{5.7}$$

The compressive force C is located at a distance z from the prestressing force P (Fig. 5.12). Equilibrium of the free body diagram in Fig. 5.12(a) requires:

$$C = P \tag{5.8a}$$

$$M = C z = P z \tag{5.8b}$$

where M is the moment due to applied load.

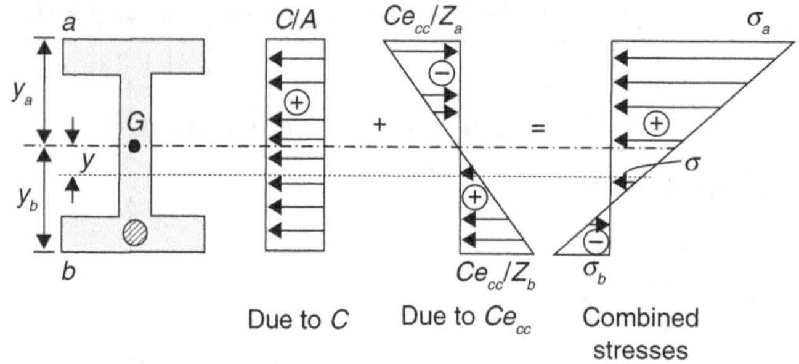

Fig. 5.11 Longitudinal stresses due to resultant compressive force

From Fig. 5.12(b), the eccentricity of the centre of compression e_{cc} is:

$$e_{cc} = e_s - z \tag{5.9}$$

Hence,

$$C\,e_{cc} = C\,e_s - C\,z$$

$$C\,e_{cc} = P\,e_s - M \qquad \text{(using Eq. (5.8))}$$

Substituting $C = P$ and $C\,e_{cc} = (P\,e_s - M)$ in Eq. (5.7) results in the same equation developed using combined load approach (Eq. (5.4a)).

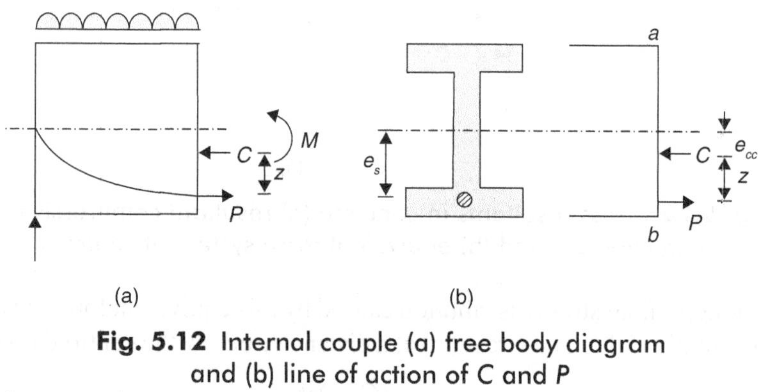

Fig. 5.12 Internal couple (a) free body diagram and (b) line of action of C and P

Comments

In an uncracked prestressed concrete member, the magnitude of P and hence C do not change with change in applied loading (fourth assumption in section 5.2). Hence as the moment M increases, the lever arm z increases ($z = M/P$) to resist M. But in the case of reinforced concrete member, the lever arm remains a constant and forces in concrete and steel reinforcement will increase with increase in *M*. However, after cracking, the behaviour of prestressed concrete member will be similar to that of a reinforced concrete member.

5.2.2.1 Kern Points

If the resultant compressive force *C* lies within a certain zone of the cross-section, no tensile stress is developed in the section. This zone is known as the *kern* of the section and the extreme limits of kern are called the *kern points*.

The distance of the upper and lower kern points from the centroid of the section are denoted as k_a and k_b. When *C* is located at the upper kern point ($e_{cc} = -k_a$), the stress at the bottom fibre of the section (σ_b) will be equal to zero (Fig. 5.13(a)). Similarly C at lower kern point ($e_{cc} = k_b$) results in zero stress at the top fibre (σ_a) of the section (Fig. 5.13(b)). This information is used to find k_a and k_b.

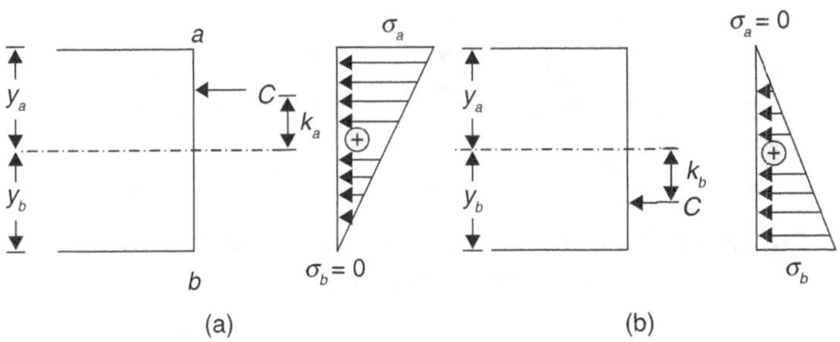

Fig. 5.13 Longitudinal stress distribution when C is at kern point (a) C at upper kern point and (b) C at lower kern point

Upper Kern Distance k_a: When C is at the upper kern point, $\sigma_b = 0$. Hence,

$$\sigma_b = \frac{C}{A} + \frac{C(-k_a)y_b}{I} = 0 \qquad (e_{cc} = -k_a;\ y = y_b)$$

$$\Rightarrow \qquad k_a = \frac{I}{Ay_b} = \frac{Ar^2}{Ay_b} \qquad (I = Ar^2,\ r = \text{radius of gyration})$$

$$k_a = \frac{r^2}{y_b} \tag{5.10}$$

Lower Kern Distance k_b: When C is at the lower kern point, $\sigma_a = 0$. Hence,

$$\sigma_a = \frac{C}{A} + \frac{Ck_b(-y_a)}{I} = 0 \qquad (e_{cc} = k_b;\ y = -y_a)$$

$$\Rightarrow \qquad k_b = \frac{I}{Ay_a} = \frac{Ar^2}{Ay_a}$$

$$k_b = \frac{r^2}{y_a} \tag{5.11}$$

5.2.2.2 Pressure Line

Pressure line in a prestressed concrete member is the locus of the centre of compression. It is also known as the *thrust* line or *C-line.*

Fig. 5.14 shows the pressure line for a simply supported prestressed concrete beam subjected to central concentrated load. The distance of the centre of compression from the tendon (z) at any section 1–1 can be found out using Eq. (5.8b).

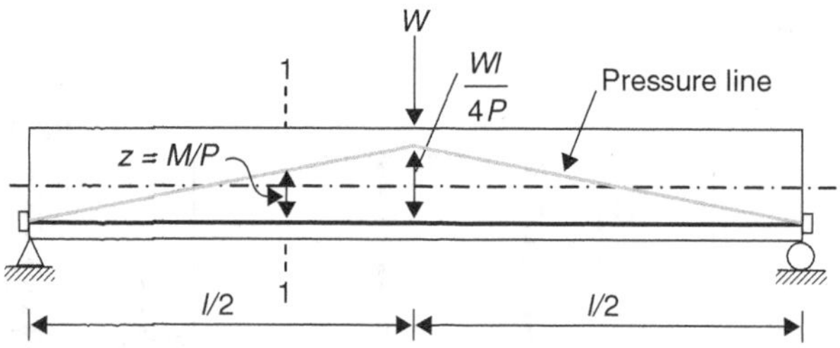

Fig. 5.14 Pressure line for a simply supported beam with concentrated load

At the supports ($M = 0$), the pressure line coincides with the centroid of the tendon. For the beam in Fig. 5.14, the bending moment M increases linearly from the supports to the mid-span. Hence z also has a linear variation.

The pressure line is very useful to check whether the resultant compressive force C lies within the kern at all sections so that no tensile stresses are developed in the member (for type 1 member).

Example 5.4

Determine the stresses in concrete at the mid-span of the prestressed concrete beam in Example 5.2 using the internal couple approach.

Solution:

Step 1: *Calculation of stresses at transfer*

Resultant compressive force $C = P_o = 1350$ kN

Tendon eccentricity $e_s = 250$ mm

Bending moment $M = M_o = 84$ kN-m

Section modulus $z_a = z_b = 3.733 \times 10^7$ mm^3
Cross-sectional area $A = 280000$ mm^2

- Lever arm z

$$z = \frac{M}{C} = \frac{84 \times 10^6}{1350 \times 10^3} = 62.22\,\text{mm}$$

- Eccentricity of the centre of compression e_{cc} (Fig. 5.15(a)).

$$e_{cc} = e_s - z = 250 - 62.22 = 187.78 \text{ mm}$$

- Stress at bottom fibre of the section σ_b

$$\sigma_b = \frac{C}{A} + \frac{Ce_{cc}}{Z_b}$$

$$= \left[\frac{1350 \times 10^3}{280000}\right] + \left[\frac{1350 \times 10^3 \times 187.78}{3.733 \times 10^7}\right]$$

$$= 4.82 + 6.79 = 11.61 \text{ N/mm}^2$$

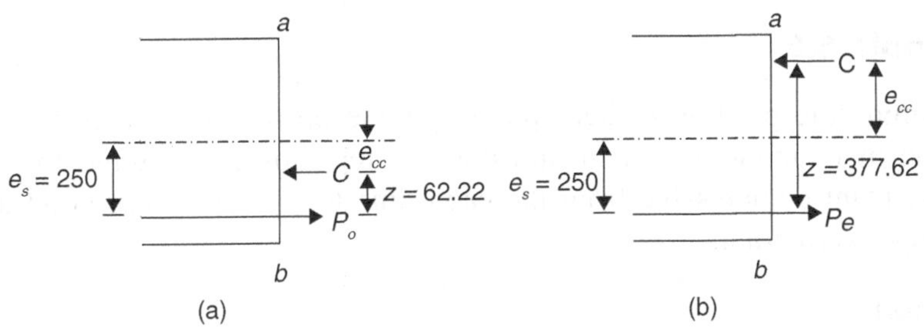

Fig. 5.15 Location of centre of compression (a) at transfer and (b) at service condition

- Stress at top fibre of the section σ_a

$$\sigma_a = \frac{C}{A} - \frac{Ce_{cc}}{Z_a}$$

$$= 4.82 - 6.79 = -1.97 \text{ N/mm}^2$$

Step 2: *Calculation of stresses at service stage*

Resultant compressive force $C = P_e = 1050$ kN
Bending moment $M = M_s = 396.5$ kN-m

- Lever arm z

$$z = \frac{M}{C} = \frac{396.5 \times 10^6}{1050 \times 10^3} = 377.62 \text{ mm}$$

- Eccentricity of the centre of compression e_{cc}

$$e_{cc} = e_s - z = 250 - 377.62 = -127.62 \text{ mm}$$

The negative value of e_{cc} indicates that C is located above the centroid (Fig. 5.15(b)).

- Stress at bottom fibre of the section σ_b

$$\sigma_b = \frac{C}{A} + \frac{Ce_{cc}}{Z_b}$$

$$= \left[\frac{1050 \times 10^3}{280000}\right] + \left[\frac{1050 \times 10^3 \times (-127.62)}{3.733 \times 10^7}\right]$$

$$= 3.75 - 3.59 = 0.16 \text{ N/mm}^2$$

- Stress at top fibre of the section σ_a

$$\sigma_a = \frac{C}{A} - \frac{Ce_{cc}}{Z_a}$$

$$= 3.75 + 3.59 = 7.34 \text{ N/mm}^2$$

Example 5.5

Determine the position of kern points for the prestressed concrete beam section in Example 5.1. Also find the bending moment required to shift the resultant compressive from the cable line to (i) lower kern point and (ii) upper kern point.

Solution:

Step 1: *Calculation of section properties*

Area of cross-section $A = 0.48$ m^2

Tendon eccentricity $e_s = 0.317$ m

- Distance of extreme fibre from the centroid

$y_b = 0.8$ m

$y_a = 1.2 - 0.8 = 0.4$ m

- Moment of inertia I

$$I = \left[\frac{1.2 \times 0.2^3}{12}\right] + [1.2 \times 0.2 \times 0.3^2] + \left[\frac{0.24 \times 1^3}{12}\right] + [0.24 \times 1 \times 0.3^2]$$

$$= 0.064 \text{ m}^4$$

- Radius of gyration r

$$r = \sqrt{\frac{I}{A}} = \sqrt{\frac{0.64}{0.48}} = 0.365 \text{ m}$$

$$\Rightarrow \qquad r^2 = 0.133 \text{ m}^2$$

Step 2: *Calculation of kern distance*

- Upper kern distance k_a

$$k_a = \frac{r^2}{y_b} = \frac{0.133}{0.8} = 0.166 \text{ m}$$

The upper kern point is at a distance of 0.166 m above the centroid of the section.

- Lower kern distance k_b

$$k_b = \frac{r^2}{y_a} = \frac{0.133}{0.4} = 0.333 \text{ m}$$

The lower kern point is at a distance of 0.333 m below the centroid.

Step 3: *Calculation of moment required to shift C to upper kern point*

$$e_{cc} = -k_a = -0.166 \text{ m}$$
$$C = P = 1400 \text{ kN}$$

- Lever arm z

$$z = e_s - e_c \qquad \text{(Eq. (5.9))}$$
$$= 0.317 - (-0.166) = 0.483 \text{ m}$$

- Bending moment M

$$M = C\,z = 1400 \times 0.483 = 676.2 \text{ kN-m (sagging moment)}$$

Thus a sagging moment of 676.2 kN-m is required to shift C to upper kern point.

Step 4: *Calculation of moment required to shift C to lower kern point*

$$e_{cc} = k_b = 0.333 \text{ m}$$

- Lever arm z

$$z = e_s - e_c$$
$$= 0.317 - 0.333 = -0.016 \text{ m}$$

- Bending moment M

$$M = C\,z = 1400 \times (-0.016) = -22.4 \text{ kN-m (hogging moment)}$$

Hence a hogging moment of 22.4 kN-m shifts C to lower kern point.

Example 5.6

A simply supported prestressed concrete beam 250 mm wide by 500 mm deep of effective span 9 m carries a distributed load of intensity 6 kN/m (including self-weight). The beam is prestressed by a straight tendon carrying a force of 200 kN located 100 mm below the centroidal axis of the beam. Plot the pressure line of the beam.

Solution:

Step 1: *Calculation of bending moment M*

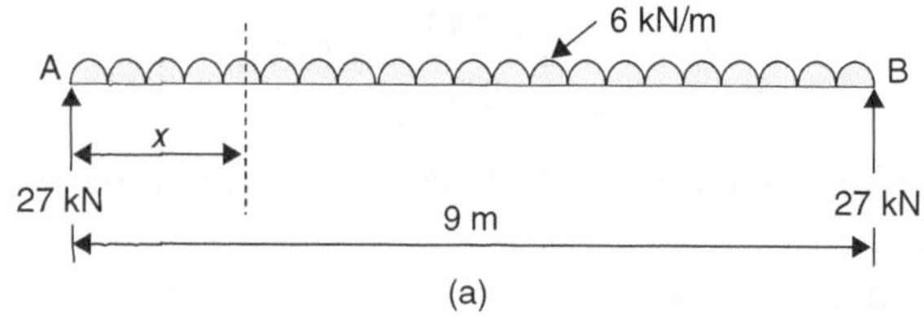

Fig. 5.16 (a) Bending moment calculation

Bending moment at a distance x metres from A (Fig. 5.16(a)) is:

$$M = 27x - 3x^2 \text{ kN-m}$$

Step 2: *Calculation of lever arm z*

Force in tendon $P = 200$ kN

$$z = \frac{M}{P} = \frac{27x - 3x^2}{200} = 0.135x - 0.015x^2 \text{ m}$$

The centre of compression lies at a distance z from the centroid of the tendon.

Step 3: *Plotting the pressure line of the beam*

By joining the centre of compression at all sections, the pressure line for the beam can be drawn and is shown in Fig. 5.16(b).

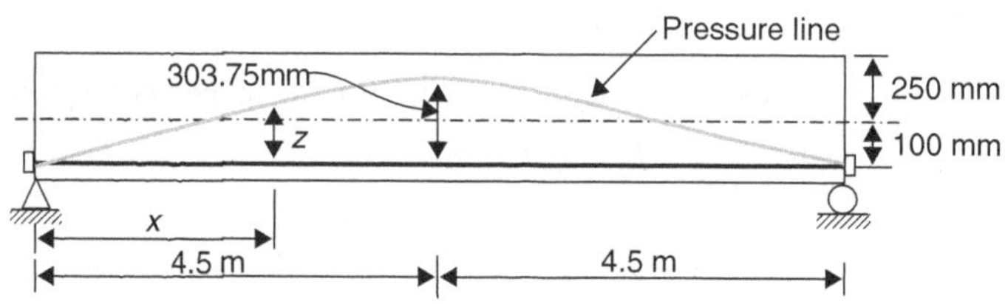

Fig. 5.16 (b) Pressure line for the beam

5.2.3 Concept of Equivalent Loads

For the analysis of prestressed concrete members, it is very convenient to replace the tendon by the forces exerted by tendon on concrete. These forces are known as *equivalent loads* and they occur at the anchorages and wherever there is a change in the direction of tendon.

Unlike Eq. (5.2) which can be used to find stresses only for determinate structures, this approach can be used to find the effect of prestress in both determinate and indeterminate structures.

The equivalent loads can be determined by first considering the free body diagram of the tendon. Using this, the force transferred from concrete to tendon and hence from tendon to concrete can be found out. This is the required equivalent load and they are applied in the member for analysis. The procedure for finding equivalent loads for some typical cases is discussed below.

Equivalent Loads at Anchorage

At the anchorage, the tendon exerts a force equal to the prestressing force on concrete and the direction of the force coincides with the tendon.

The beam in Fig. 5.17(a) is prestressed by a straight tendon. The free body diagram of the tendon and concrete for the left-hand side of section 1–1 are indicated in Figs. 5.17(b) and (c).

From Fig. 5.17(c), it can be seen that at anchorage A, the tendon exerts a force P at a eccentricity e_s on concrete. This equivalent load can be represented either by Fig. 5.17(d) or by Fig. 5.17(e).

The equivalent loads at the anchorage due to inclined tendon (Fig. 5.18(a)) can be represented either by Fig. 5.18(b) or (c). Since the inclination θ is small, using small angle approximation (i.e., $\cos\theta = 1$ and $\sin\theta = \theta$), the equivalent loads can be shown either as Fig. 5.18(d) or (e).

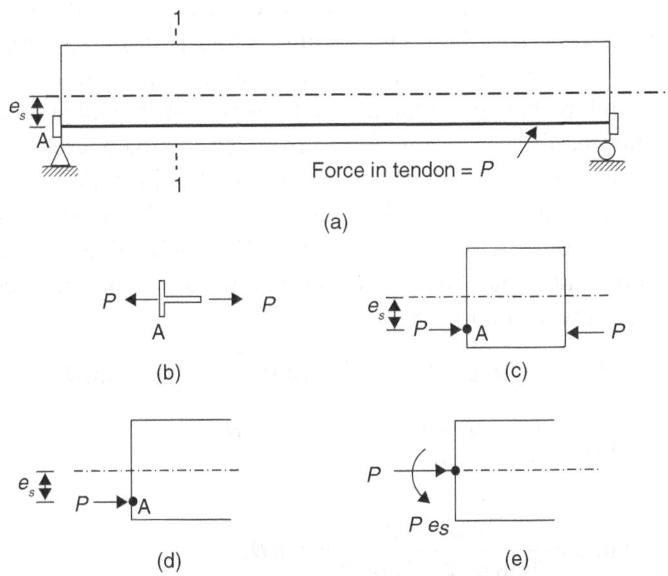

Fig. 5.17 Equivalent loads at anchorage due to straight tendon (a) prestressed concrete beam, (b) free body diagram of tendon, (c) free body diagram of concrete, (d) equivalent load at anchorage and (e) equivalent force system at centroid

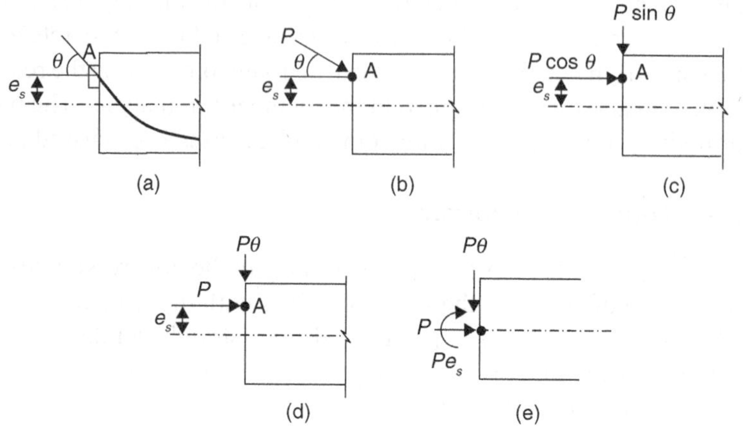

Fig. 5.18 Equivalent loads at anchorage due to inclined tendon (a) beam with inclined tendon, (b) equivalent load, (c) force components, (d) small angle approximation and (e) equivalent force system at centroid

Equivalent Loads at Kinks

In a pre-tensioned member, to change the eccentricity of tendons, they are pulled down at some specific points (points A and B in Fig. 5.19(a)). This process is known as *draping* or *harping* the tendons. These points where the slope of the tendon changes suddenly are known as *kinks* and at these locations, concentrated forces are transferred from tendon to concrete.

From the free body diagram of tendon at kink A (Fig. 5.19(b)), it can be seen that a change in direction of prestressing force P requires a force F transferred from concrete to steel. The force F is inclined at an angle ϕ to vertical, and the horizontal and vertical components of F are denoted as F_H and F_V respectively. From Fig. 5.19(b),

$$F_H = P - P\cos\theta = P(1-\cos\theta) \text{ and } F_V = P\sin\theta$$

$$\tan\phi = \frac{F_H}{F_V} = \frac{P - P\cos\theta}{P\sin\theta} = \frac{1-\cos\theta}{\sin\theta}$$

$$\tan\phi = \frac{2\sin^2\theta/2}{2\sin\theta/2\ \cos\theta/2} = \tan\theta/2$$

Hence $\phi = \theta/2$. Since θ is very small, the angle ϕ can be neglected. Thus the force F can be assumed to act in the vertical direction with a magnitude $P\theta$ (i.e., $F \approx F_V = P\sin\theta = P\theta$).

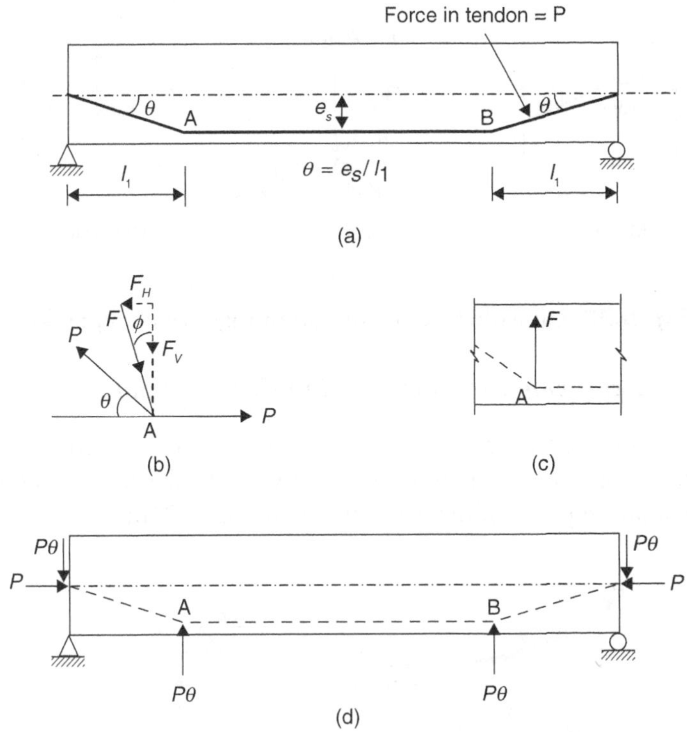

Fig. 5.19 Equivalent loads in member having draped tendon (a) prestressed concrete member, (b) free body diagram of tendon at A, (c) free body diagram of concrete at A and (d) equivalent loads

At A, the tendon exerts the reaction force F (equivalent load) on concrete in the upward direction (Fig. 5.19(c)). The equivalent loads for the given prestressed concrete beam is shown in Fig. 5.19(d).

The equivalent loads at some commonly occurring kinks are shown in Figs. 5.20(a) and (b).

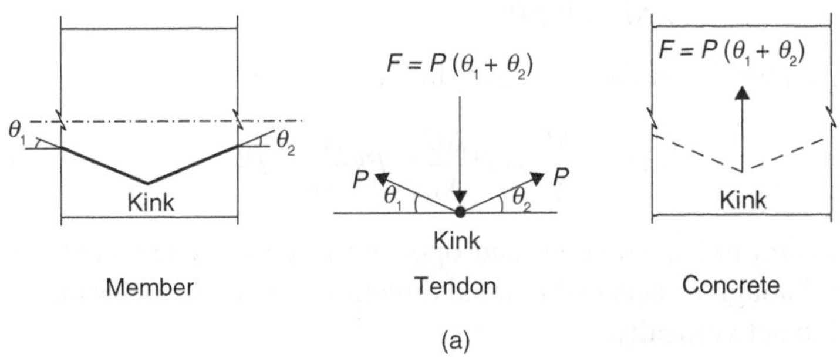

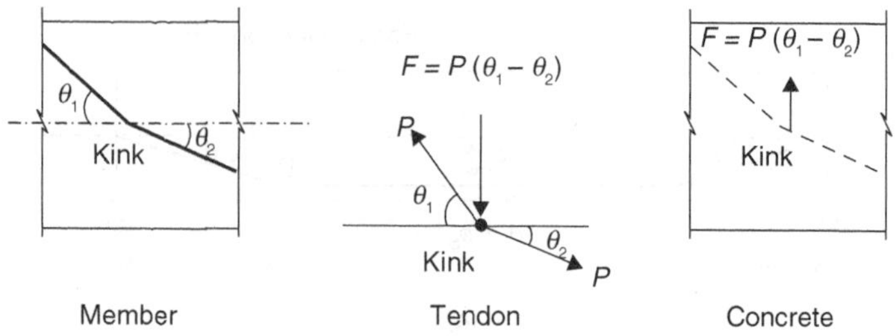

Fig. 5.20 Equivalent loads at commonly occurring kinks

Equivalent Loads Due to Cables Having Curved Profile

Most of the post-tensioned members have tendons with curved profiles. To find the equivalent loads due to curved cables, consider the free body diagram of a small segment of the cable shown in Fig. 5.21(a).

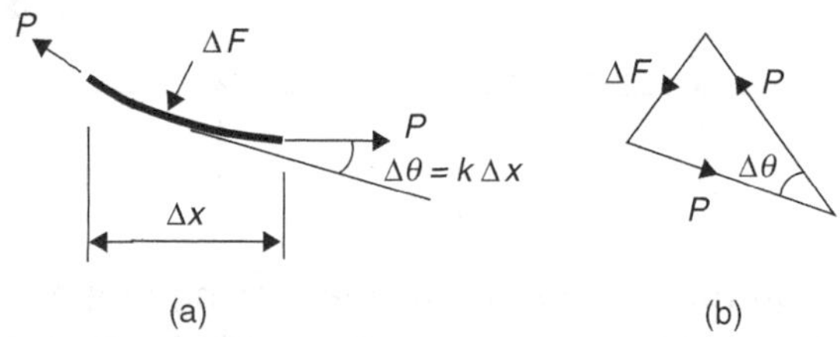

Fig. 5.21 Equivalent loads due to curved cables
(a) free body diagram of cable and (b) force triangle

The change of slope of the cable within the length Δx is $\Delta\theta$ which can be written as $k\,\Delta x$, where k is the curvature (change in slope per unit length) of the cable at the point considered.

To maintain force equilibrium, the cable is subjected to a radial force ΔF from concrete. The magnitude of ΔF can be determined from the force triangle (Fig. 5.21(b)).

$$\Delta F = P\,\Delta\theta$$

Hence the intensity of the force over the length Δx (w_p) is:

$$w_p = \frac{\Delta F}{\Delta x} = P\frac{\Delta\theta}{\Delta x} = Pk\frac{\Delta x}{\Delta x} = Pk \tag{5.12}$$

The tendon exerts an equal and opposite force w_p (equivalent load) on concrete. Though w_p acts in the radial direction, for small angles they can be assumed to act vertically.

Special Case: Equivalent loads due to parabolic cables

Due to constant curvature, the parabolic cable exerts a uniformly distributed equivalent load w_p.

The parabolic cable for the member in Fig. 5.22(a) has length l and sag h. The sag is the vertical offset at the mid-length of the cable measured relative to the line joining the ends.

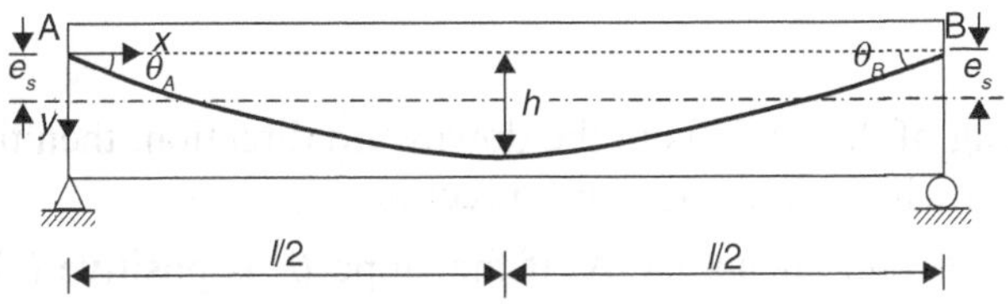

(a)

Fig. 5.22 (a) Prestressed concrete member having parabolic cable

The equation of the parabola is:

$$y = ax^2 + bx + c$$

where a, b and c are constants which can be determined from the three boundary conditions (i) $x = 0, y = 0$; (ii) $x = l/2, y = h$; (iii) $x = l, y = 0$.

Thus the equation of the cable is:

$$y = \frac{4h}{l^2}(lx - x^2) \tag{5.13}$$

The slope of the cable θ is:

$$\theta = \frac{dy}{dx} = \frac{4h}{l^2}(l - 2x) \tag{5.14}$$

The slope of the cable at A, θ_A is $\frac{4h}{l}$ (clockwise), and at B, θ_B is $-\frac{4h}{l}$ (anti-clockwise).

The curvature of the cable k is:

$$k = \frac{d\theta}{dx} = \left|\frac{d^2y}{dx^2}\right| = \frac{8h}{l^2} \tag{5.15}$$

The equivalent load w_p determined using Eq. (5.12) is:

$$w_p = Pk = \frac{8Ph}{l^2} \tag{5.16}$$

This force acts downwards and it is the force acting on the tendon. The tendon exerts a upward force on the concrete having same intensity. The equivalent loads for the beam is shown in Fig. 5.22(b).

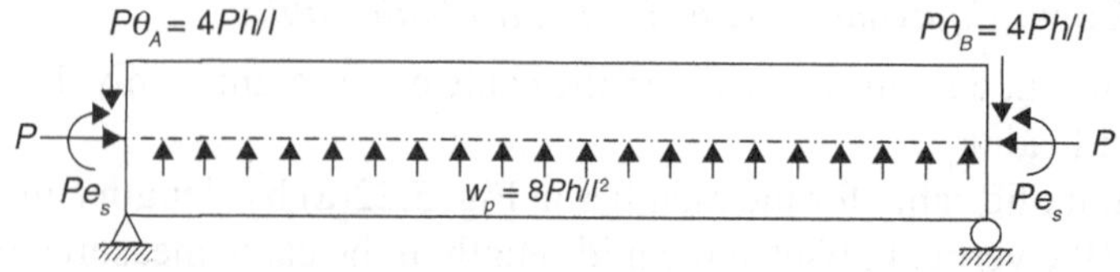

(b)

Fig. 5.22 (b) Equivalent loads due to parabolic cable

Comments

- If the sag of the cable is in the downward direction, then the equivalent load will be in the upward direction.
- At the left end anchorage A, if the slope θ_A is positive (clockwise), then the vertical equivalent load $P\theta_A$ will be acting in the downward direction. Similarly at the right end anchorage B, if the slope θ_B is negative (anti-clockwise), then the vertical equivalent load $P\theta_B$ will be acting in the downward direction.

Example 5.7

Determine the equivalent loads due to prestress for the prestressed concrete beam in Fig. 5.23(a). The force in the tendon is 1200 kN.

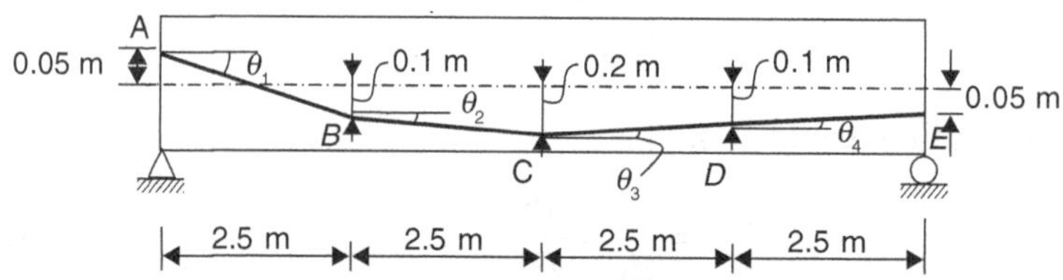

(a)

Fig. 5.23 (a) Prestressed concrete beam

Solution:

Step 1: *Calculation of inclination of tendon θ*

The magnitude of slope of tendon is given below:

$$\theta_1 = \frac{(0.1 + 0.05)}{2.5} = 0.06 \text{ radians}$$

$$\theta_2 = \frac{(0.2 - 0.1)}{2.5} = 0.04 \text{ radians}$$

$$\theta_3 = \frac{(0.2 - 0.1)}{2.5} = 0.04 \text{ radians}$$

$$\theta_4 = \frac{(0.1-0.05)}{2.5} = 0.02 \text{ radians}$$

Step 2: *Calculation of equivalent loads at the anchorage*

Anchorage A

$$P = 1200 \text{ kN};\ \theta_1 = 0.06 \text{ radians};\ e_s = 0.05 \text{ m (magnitude)}$$

The equivalent loads at anchorage A are given in Figs. 5.23(b), (c) and (d)

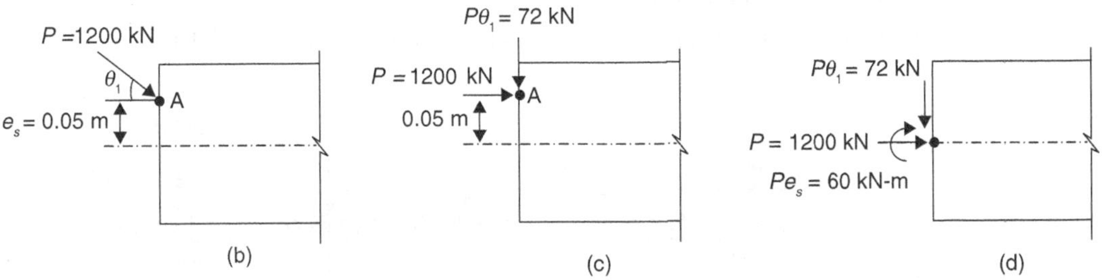

Fig. 5.23 (b) Equivalent load at A, (c) components and (d) equivalent load at the centroid

Anchorage E

$$P = 1200 \text{ kN};\ \theta_4 = 0.02 \text{ radians (magnitude)},\ e_s = 0.05 \text{ m}$$

The equivalent loads can be represented by either Fig. 5.23(e), (f) or (g)

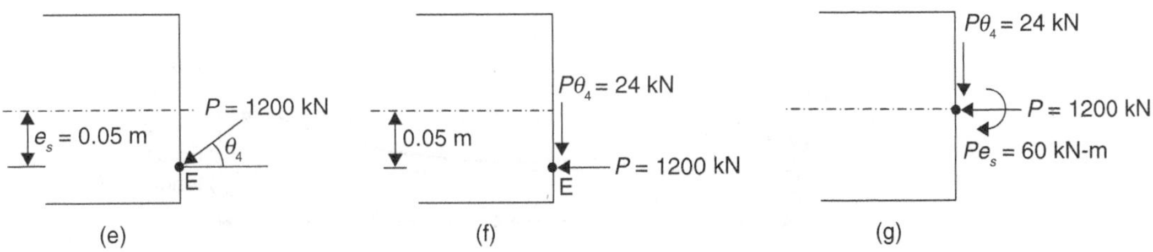

Fig. 5.23 (e) Equivalent load at E, (f) component and (g) equivalent load at centroid

Step 3: *Calculation of equivalent loads at the kinks*

The equivalent loads at kinks B, C and D are denoted as F_B, F_C, and F_D. They are obtained from the force body diagram of the kinks (Figs. 5.23(h), (i) and (j)). The forces F_B, F_C, and F_D in these figures are the forces exerted by concrete on tendon. The tendon applies an equal and opposite force on concrete.

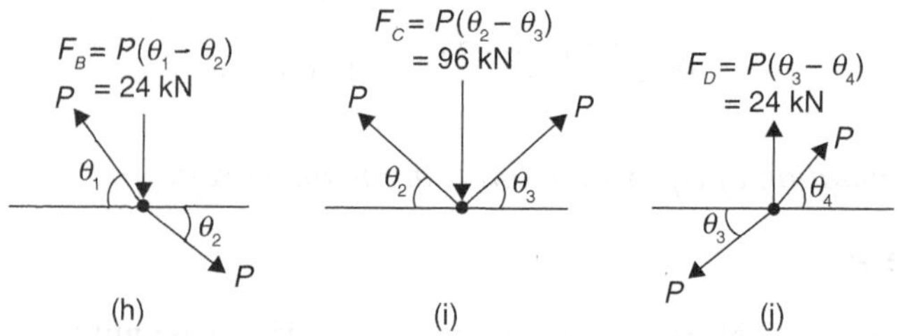

Fig. 5.23 (h) Kink B, (i) kink C, (j) kink D

Step 4: *Equivalent loads on the member*

The equivalent loads on the member are shown in Fig. 5.23(k)

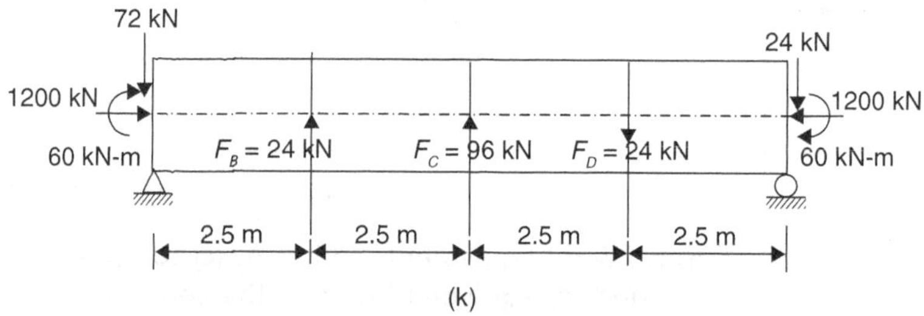

Fig. 5.23 (k) Equivalent loads

Example 5.8

Find the equivalent load due to prestress for the simply supported beam having parabolic profile (Fig. 5.24(a)).

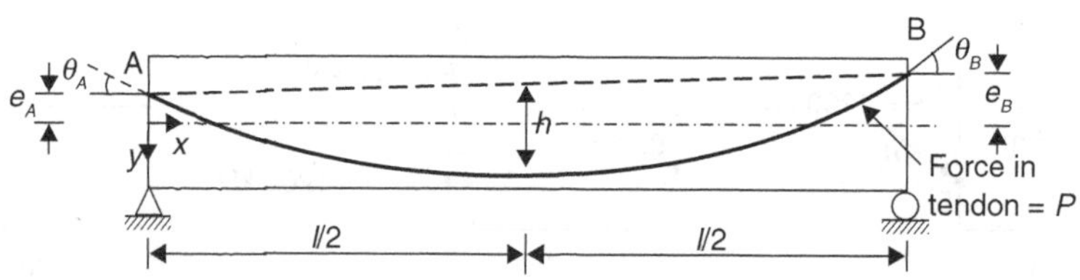

Fig. 5.24 (a) Beam with parabolic cable

Solution:

Step 1: *Calculation of geometric properties of the cable*

- Equation of the cable

  The general equation of the cable is:

$$y = ax^2 + bx + c$$

The constants a, b and c can be determined using the boundary conditions

$$(x = 0, y = -e_A), \left(x = l/2, y = h - \left[\frac{e_A + e_B}{2}\right]\right), (x = l, y = -e_B)$$

Thus the equation of the cable is:

$$y = \frac{4h}{l^2}(lx - x^2) - (e_B - e_A)\frac{x}{l} - e_A$$

- Slope of the cable θ

$$\theta = \frac{dy}{dx} = \frac{4h}{l^2}(l - 2x) - \frac{(e_B - e_A)}{l}$$

Slope at A,

$$\theta_A = \left.\frac{dy}{dx}\right|_{x=0} = \frac{4h}{l} - \frac{(e_B - e_A)}{l} = \frac{(4h + e_A - e_B)}{l}, \text{ clockwise}$$

Slope at B,

$$\theta_B = \left.\frac{dy}{dx}\right|_{x=l} = -\frac{4h}{l} - \frac{(e_B - e_A)}{l} = -\frac{(4h + e_B - e_A)}{l}, \text{ anticlockwise}$$

- Curvature of the cable k

$$\frac{d^2y}{dx^2} = \frac{-8h}{l^2}$$

$$k = \left|\frac{d^2y}{dx^2}\right| = \frac{8h}{l^2}$$

Step 2: *Calculation of equivalent loads*

Equivalent loads at the anchorage A and B are shown in Figs. 5.24(b) and (c)

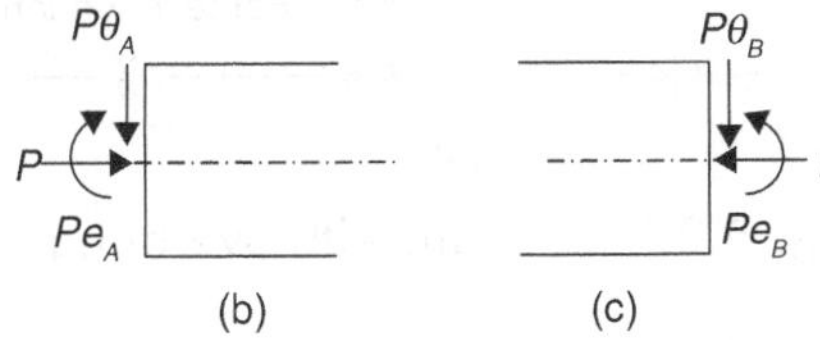

Fig. 5.24 (b) Equivalent loads at A and (c) equivalent loads at B

- Equivalent loads due to parabolic cable w_p

$$w_p = Pk = \frac{8Ph}{l^2}$$

The intensity of the equivalent loads is same as that given by Eq. (5.16). The equivalent loads for the complete beam is shown in Fig. 5.24(d).

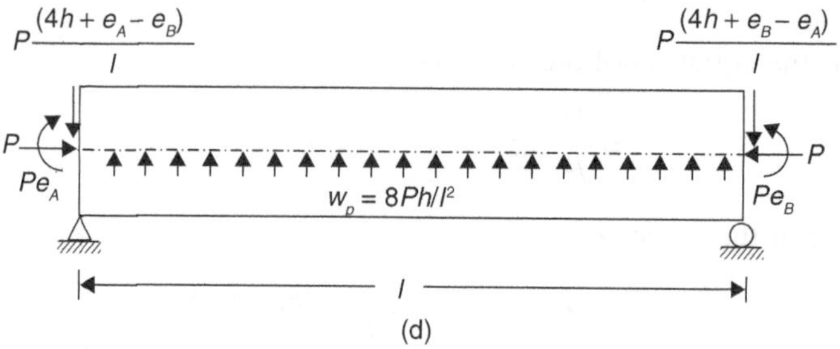

Fig. 5.24 (d) Equivalent loads due to prestress

Comments

From this example it can be seen that as long as the sag h and hence curvature ($k = 8h/l^2$) remains constant, the intensity of equivalent loading w_p is independent on the magnitudes of the end eccentricities e_A and e_B. Hence the cable can be adjusted by changing e_A and e_B without changing h. This adjustment known as *linear transformation* does not cause any change in w_p. This aspect is very useful to understand the effect of prestress in continuous members having parabolic cable profile.

Example 5.9

Determine the equivalent loads due to prestress for the beam with overhang shown in Fig. 5.25(a).

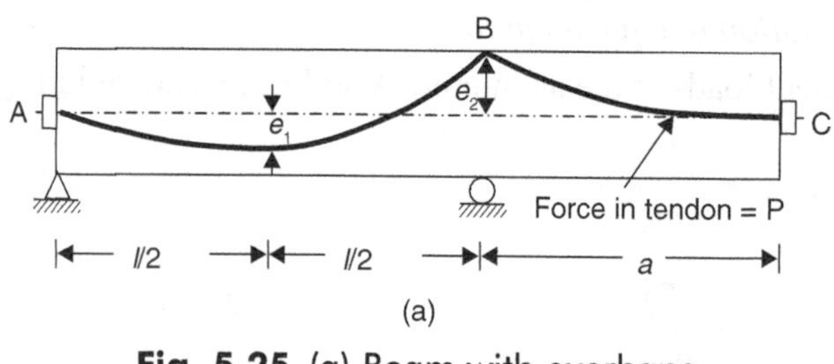

Fig. 5.25 (a) Beam with overhang

Solution:

Step 1: *Calculation of equivalent loads for span AB*

From Fig. 5.25(b), the sag of the cable in span AB, h_{AB} is:

$$h_{AB} = e_1 + \frac{e_2}{2}$$

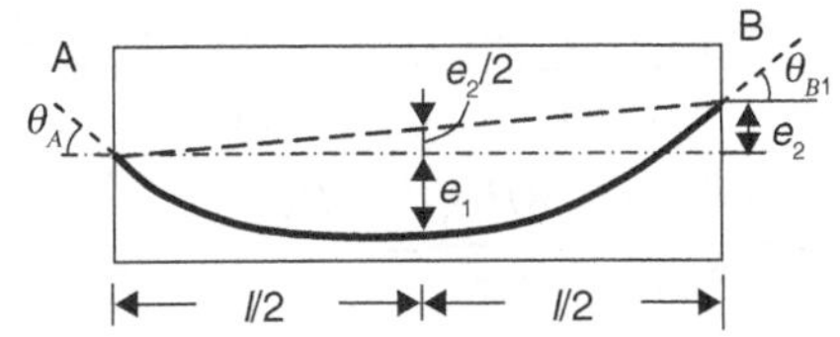

(b)

Fig. 5.25 (b) Geometric properties of cable in span AB

The slope of the cable at ends A and B (θ_A and θ_{B1}) can be determined using the equations derived in Example 5.8.

$$\theta_A = \frac{4h_{AB} + e_A - e_B}{l} = \frac{4h_{AB} - e_2}{l}, \ (e_A = 0;\ e_B = e_2)$$

$$\theta_{B1} = \frac{-(4h_{AB} + e_B - e_A)}{l} = \frac{-(4h_{AB} + e_2)}{l}$$

The intensity of the distributed load due to parabolic cable in the span AB, w_{p1} is:

$$w_{p1} = \frac{8P\,h_{AB}}{l^2}$$

Step 2: *Calculation of equivalent loads for span BC*
The sag of the cable BC is h_{BC} (Fig. 5.25(c)).

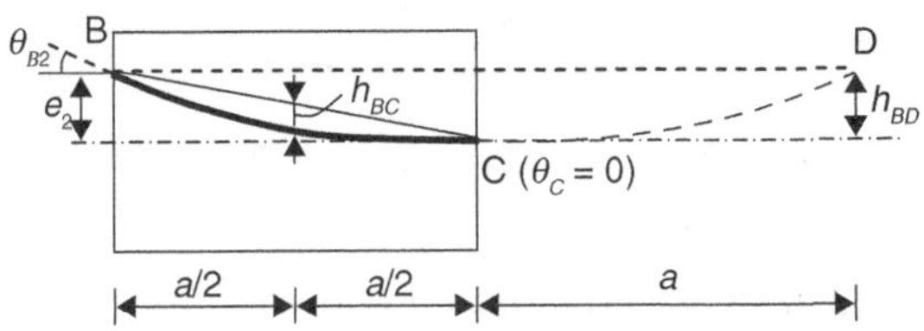

(c)

Fig. 5.25 (c) Geometric properties of cable in span BC

$$\text{Slope of tendon at B, } \theta_{B2} = \frac{4h_{BC} + e_B - e_C}{a} = \frac{(4h_{BC} + e_2)}{a}, \ (e_c = 0)$$

The intensity of the distributed load w_{p2} is:

$$w_{p2} = \frac{8P\,h_{BC}}{a^2}$$

The difficulty in using the above approach is that the sag h_{BC} should be determined using the cable profile. Alternatively, the cable BC can be considered to be the left half of the fictitious cable BD of length $2a$ and sag h_{BD} ($h_{BD} = e_2$). Hence the slope θ_{B2} and intensity of the load w_{p2} are:

$$\theta_{B2} = \frac{4h_{BD} + e_B - e_D}{(2a)} = \frac{4e_2 + e_2 - e_2}{2a} = \frac{2e_2}{a}$$

$$w_{p2} = \frac{8Ph_{BD}}{(2a)^2} = \frac{8Pe_2}{4a^2} = \frac{2Pe_2}{a^2}$$

The equivalent loads due to prestress for the complete beam is shown in Fig. 5.25(d).

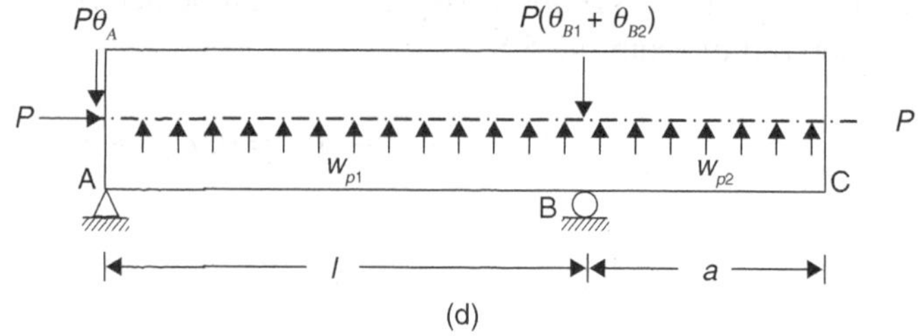

Fig. 5.25 (d) Equivalent loads

Example 5.10

Determine the stresses in concrete at the mid-span of the prestressed concrete beam (Fig. 5.26(a)) in Example 5.2 using equivalent load approach.

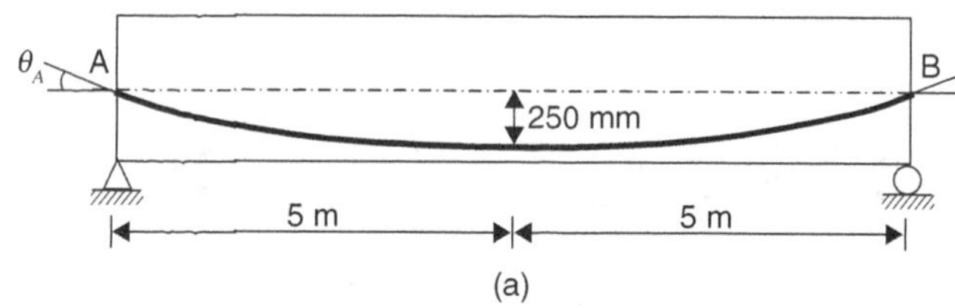

Fig. 5.26 (a) Prestressed concrete beam

Solution:

Step 1: *Calculation of equivalent loads due to prestressing force P*

Sag of the cable $h = 250$ mm

Slope of cable at A, $\theta_A = \frac{4h}{l} = \frac{4 \times 0.25}{10} = 0.1$ radians (clockwise)

Slope of cable at B, $\theta_B = -\frac{4h}{l} = -0.1$ radians (anti clockwise)

- Intensity of equivalent load due to parabolic cable (w_p)

  For a prestressing force P kN,

$$w_p = \frac{8Ph}{l^2} = \frac{8P \times 0.25}{10^2} = 0.02P \text{ kN/m}$$

The equivalent loads due to prestress are shown in Fig. 5.26(b).

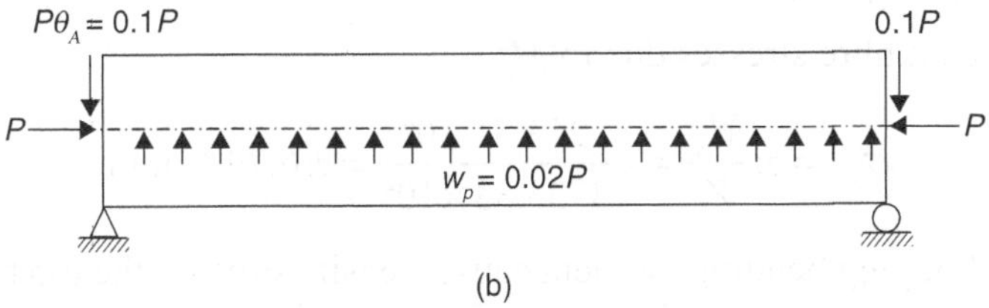

(b)

Fig. 5.26 (b) Equivalent loads due to prestressing force

Step 2: *Calculation of stresses at mid-span at transfer*

Area of cross-section $A = 280000$ mm^2

Section modulus $Z_a = Z_b = 3733.33 \times 10^4$ mm^3

Prestressing force $P = P_o = 1350$ kN

Intensity of applied load $w = w_{SW} = 6.72$ kN/m

The separate loads and the net loads acting on the beam at transfer are shown in Fig. 5.26(c).

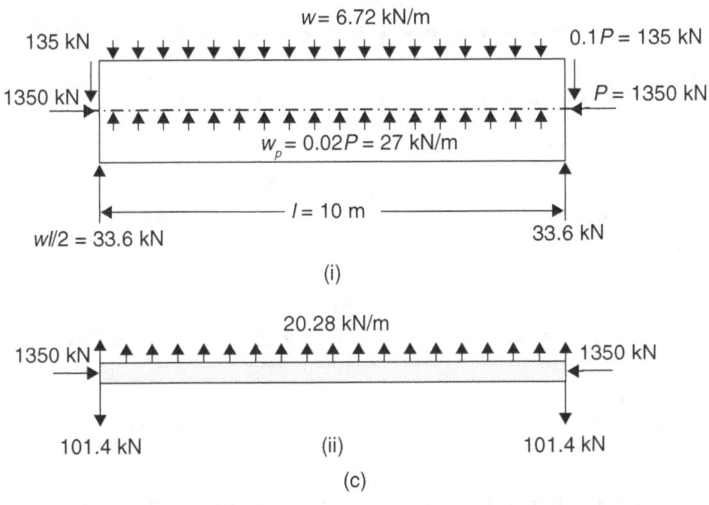

Fig. 5.26 (c) Total loads at transfer (i) prestress plus applied loads and (ii) net loads

- Bending moment at mid-span due to total (net) loads M_n

  $M_n = (-101.4 \times 5) + (20.28 \times 5^2/2) = -253.5$ kN-m (hogging moment)

- Axial stress due to P

$$\sigma_P = \frac{P}{A} = \frac{1350 \times 10^3}{280000} = 4.82 \text{ N/mm}^2$$

The axial force P causes compressive stress at both top and bottom fibre of the section

- Extreme fibre stresses due to M_n

$$\sigma_{Mn} = \pm\frac{M_n}{Z} = \pm\frac{253.5\times10^6}{3733.33\times10^4} = \pm 6.79 \text{ N/mm}^2$$

The hogging bending moment causes tensile stress at the top fibre and compressive stress at the bottom fire.

- Total extreme fibre stresses

Stress at the top fibre $\sigma_a = 4.82 - 6.79 = -1.97$ N/mm^2

Stress at the bottom fibre $\sigma_b = 4.82 + 6.79 = 11.61$ N/mm^2

Step 3: *Calculation of stresses at mid-span during service condition*

Prestressing force $P = P_e = 1050$ kN

Intensity of applied load $w = w_s = 31.72$ kN/m

The total load acting on the beam at service condition are shown in Fig. 5.26(d)

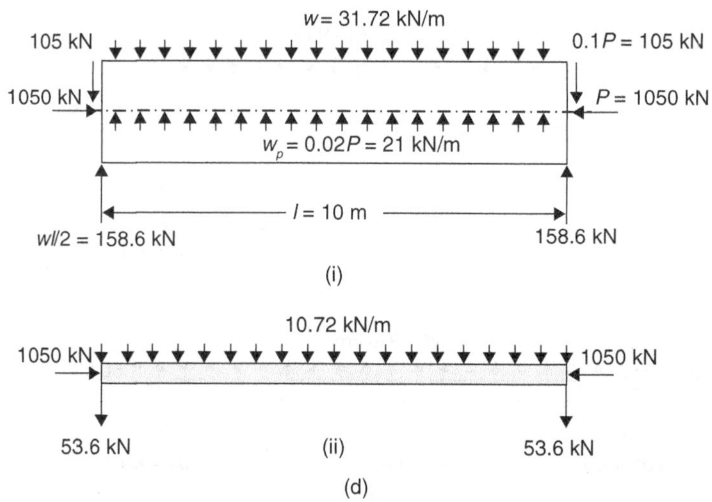

Fig. 5.26 (d) Total loads at service condition (i) prestress plus applied loads and (ii) net loads

- Bending moment at mid-span due to total loads M_n

$M_n = (53.6 \times 5) - (10.72 \times 5^2/2) = 134$ kN-m (sagging moment)

- Axial stress due to P

$$\sigma_P = \frac{P}{A} = \frac{1050\times10^3}{280000} = 3.75 \text{ N/mm}^2$$

The extreme fibre stress due to P will be compressive in nature

- Extreme fibre stresses due to M_n

$$\sigma_{Mn} = \pm\frac{M_n}{Z} = \pm\frac{134\times10^6}{3733.33\times10^4} = \pm\ 3.59\ \text{N/mm}^2$$

The sagging moment causes compressive stress at the top fibre and tensile stress at the bottom fibre.

- Total extreme fibre stresses

Stress at the top fibre $\sigma_a = 3.75 + 3.59 = 7.34$ N/mm^2

Stress at the bottom fibre $\sigma_b = 3.75 - 3.59 = 0.16$ N/mm^2.

5.3 Load Balancing

The concept of load balancing developed from the equivalent load approach refers to the application of prestressing force such that the equivalent loads due to prestress are equal and opposite to external loads. Then the external loads can be considered to be *balanced* by prestressing force.

In Fig. 5.27(a), w and w_p represents the load intensities of external loads and equivalent loads due to prestress. If $w = w_p$, then at each section, the bending moment and shear force due to external loads are balanced by those developed due to prestressing force. Hence the member is subjected to compressive force (P) acting at the anchorages. If the force P is located at the centroidal axis, then the member will not be subjected to bending. It will be only subjected to axial shortening.

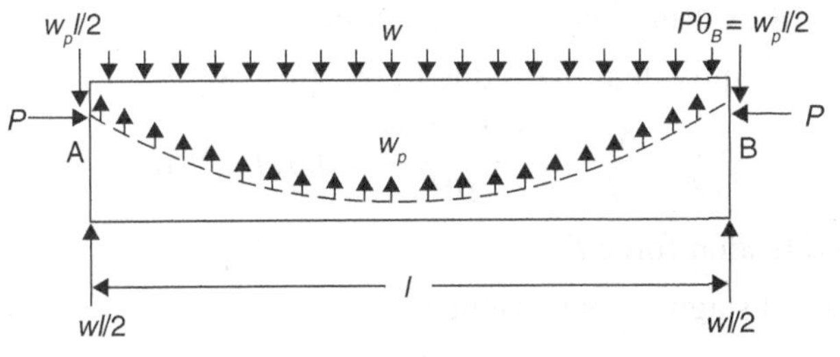

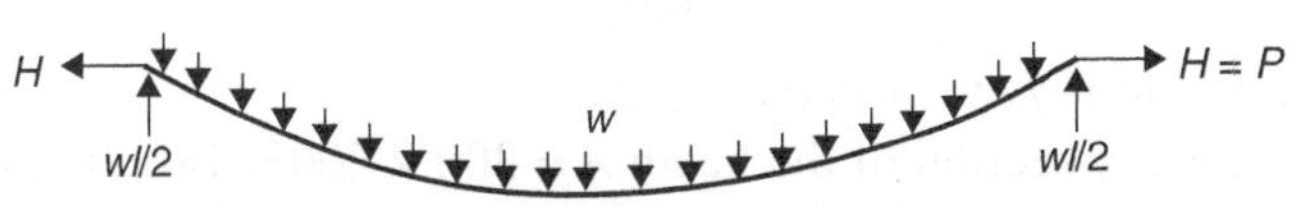

Fig. 5.27 Concept of load balancing (a) equivalent loads and external loads acting on the member and (b) cable assumed to carry w at load balanced condition

In the load balanced state, the prestressing tendon can be considered to carry the external load and the concrete can be imagined to provide the support reactions for the tendon at the anchorages (Fig. 5.27(b)).

The members designed at the load balanced condition will have negligible deflections and cracking. Hence such members will have good performance under serviceability conditions.

Example 5.11

For the beam in Fig. 5.28(a), determine the prestressing force required to balance a total applied load of 15 kN/m. Also determine the stress distribution at the mid-span.

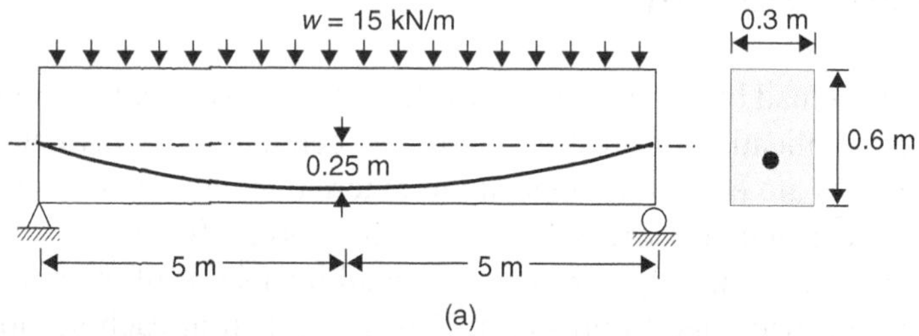

Fig. 5.28 (a) Prestressed concrete beam

Solution:

Step 1: *Calculation of required tendon force P for load balancing*

Sag of the cable $h = 0.25\ m$

Length of the cable $l = 10$ m

Intensity of external loads $w = 15$ kN/m

- Intensity of equivalent loads due to prestress w_p

$$w_p = \frac{8Ph}{l^2} = \frac{8P \times 0.25}{10^2} = 0.02P \text{ kN/m}$$

- Required tendon force P

At balanced stage, $w_p = w$. Hence,

$$0.02P = 15 \Rightarrow P = \frac{15}{0.02} = 750 \text{ kN}$$

Step 2: *Calculation of stress in concrete*

Area of cross-section of the beam $A = 300 \times 600 = 180000$ mm^2

The total load acting on the beam is shown in Fig. 5.28(b)

From the figure, it can be seen that the beam is subjected to a uniform compressive stress of magnitude P/A.

Hence stress at mid-span $\sigma_a = \sigma_b = \frac{P}{A} = \frac{750 \times 10^3}{180000} = 4.17$ N/mm^2

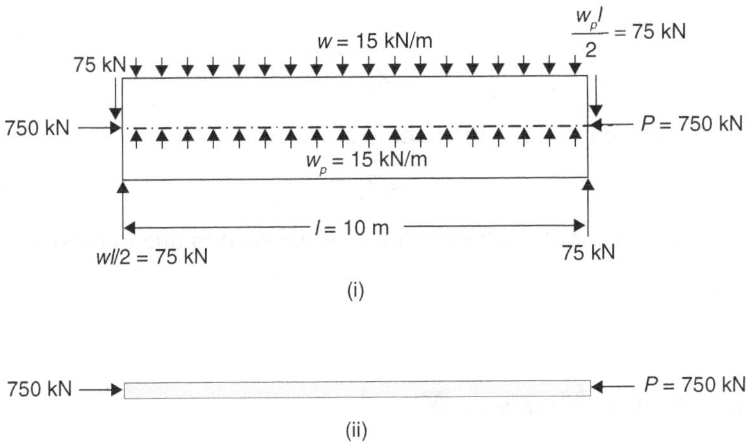

(b)

Fig. 5.28 (b) Total load acting on beam (i) prestress and applied load and (ii) net load on the beam

5.4 Decompression Moment (M_{dec})

Decompression moment M_{dec} is the applied moment that causes the stress (total stress including that caused due to prestress) at the extreme fibre of the section (σ_b) to be zero. σ_b is the stress in the extreme fibre nearest to the centroid of the prestressing force.

Fig. 5.29 shows the longitudinal stress distribution across the section 1–1 of the beam in Fig. 5.1 at $M_{dec.}$

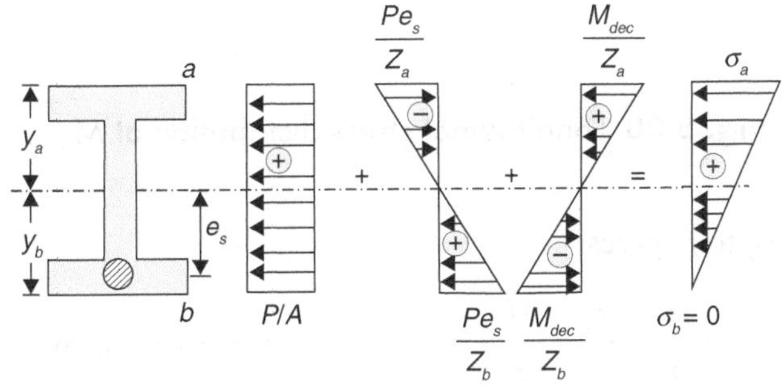

Fig. 5.29 Longitudinal stress distribution at M_{dec}

Equating the stress at the bottom fibre to zero gives:

$$\sigma_b = \frac{P}{A} + \frac{Pe_s}{Z_b} - \frac{M_{dec}}{Z_b} = 0$$

$$\Rightarrow \qquad \sigma_{bp} - \frac{M_{dec}}{Z_b} = 0$$

Hence the decompression moment is given as:

$$M_{dec} = Z_b \sigma_{bp} \tag{5.17}$$

where σ_{bp} is the stress at the bottom fibre of the section due to prestress and is equal to $\frac{P}{A} + \frac{Pe_s}{Z_b}$.

5.5 Cracking Moment M_{cr}

The first cracking in concrete occurs when the tensile stress at the extreme fibre σ_b reaches the tensile strength of concrete (i.e., the modulus of rupture f_{cr}). The applied moment at this stage is known as the *cracking moment* M_{cr}.

The longitudinal stress distribution at section 1–1 in Fig. 5.1 at M_{cr} is indicated in Fig. 5.30.

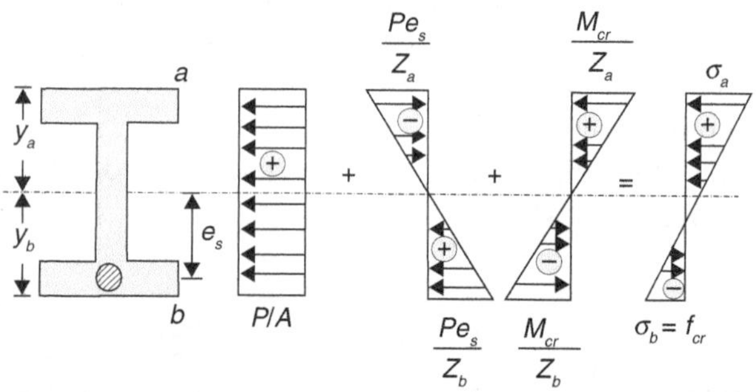

Fig. 5.30 Longitudinal stress distribution at M_{cr}

Equating σ_b to f_{cr} gives:

$$\sigma_b = \frac{P}{A} + \frac{Pe_s}{Z_b} - \frac{M_{cr}}{Z_b} = -f_{cr} \text{ (tensile stress is taken as negative)}$$

$$\Rightarrow \qquad \sigma_{bp} - \frac{M_{cr}}{Z_b} = -f_{cr}$$

Hence the cracking moment M_{cr} is:

$$M_{cr} = Z_b (\sigma_{bp} + f_{cr}) \tag{5.18}$$

As per Cl. 5.2.2 of the Code, the modulus of rupture can be determined using the equation:

$$f_{cr} = 0.7\sqrt{f_{ck}} \text{ N/mm}^2$$

where f_{ck} is the characteristic compressive strength of concrete in N/mm^2.

Example 5.12

Determine the decompression moment and cracking moment at the mid-span of the beam in Example 5.2. M50 grade concrete is used for construction.

Solution:

Step 1: *Calculation of stress at the bottom fibre due to prestress* σ_{bp}

Force in tendon $P = P_e = 1050$ kN

Cross-sectional area $A = 280000$ mm^2

Section modulus $Z_b = 3733.33 \times 10^4$ mm^3

Tendon eccentricity $e_s = 250$ mm

- Stress due to prestress σ_{bp}

$$\sigma_{bp} = \frac{P}{A} + \frac{Pe_s}{Z_b}$$

$$= \left[\frac{1050 \times 10^3}{280000}\right] + \left[\frac{1050 \times 10^3 \times 250}{3733.33 \times 10^4}\right]$$

$$= 10.78 \text{ N/mm}^2$$

Step 2: *Calculation of decompression moment* M_{dec}

$$M_{dec} = Z_b\, \sigma_{bp}$$

$$= \left[\frac{3733.33 \times 10^4 \times 10.78}{10^6}\right] = 402.45 \text{ kN-m}$$

Step 3: *Calculation of cracking moment* M_{cr}

Characteristic strength of concrete $f_{ck} = 50$ N/mm^2

- Modulus of rupture f_{cr}

$$f_{cr} = 0.7\sqrt{f_{ck}} = 0.7\sqrt{50} = 4.95 \text{ N/mm}^2$$

- Cracking moment M_{cr}

$$M_{cr} = Z_b\,[\sigma_{bp} + f_{cr}]$$

$$= \frac{3733.33 \times 10^4 [10.78 + 4.95]}{10^6}$$

$$= 587.25 \text{ N/mm}^2$$

5.6 Additional Stress in Tendon Due to Bending

The stress in the tendon of a unloaded member at any instant is obtained by deducting the losses in prestress from the initial prestress (f_{pi}). But in actual situation, the member is subjected to external loads at all stages (see Figs. 5.6 and 5.7). The effect of the external loads is to increase the stress in the tendon. Hence the stress in the tendon at any instant (f_{p1}) is:

$$f_{p1} = f_{pe} + f_{pm} \tag{5.19}$$

where f_{pe} is the effective prestress and f_{pm} is the increase in stress due to the applied bending moment M.

Example 5.13

Determine the additional stress in tendon due to external loads at the mid-span of the beam in Example 5.2. Assume modular ratio m ($m = E_p/E_c$) as 6.

Solution:

Step 1: *Calculation of effective prestress f_{pe}*

Cross-sectional area of the tendon $A_p = 1200$ mm^2

Effective prestressing force $P_e = 1050$ kN

$$\text{Effective prestress } f_{pe} = \frac{P_e}{A_p} = \frac{1050 \times 10^3}{1200} = 875 \text{ N/mm}^2$$

Step 2: *Calculation of f_{pm}*

- Stress in concrete at the level of tendon (σ_{cs}) due to M

  Intensity of external loads = live load intensity $w_{LL} = 25$ kN/m

  $$\text{Bending moment } M = \frac{w_{LL} l^2}{8} = \frac{25 \times 10^2}{8} = 312.5 \text{ kN-m}$$

  Tendon eccentricity $e_s = 250$ mm

  $$\text{Moment of inertia } I = \frac{350 \times 800^3}{12} = 1.4933 \times 10^{10} \text{ mm}^4$$

  $$\sigma_{cs} = \frac{M}{I} e_s = \frac{312.5 \times 10^6 \times 250}{1.4933 \times 10^{10}} = 5.23 \text{ N/mm}^2$$

- Increase in stress in tendon due to M, f_{pm}

  $$\text{Strain in concrete due to } \sigma_{cs},\ \varepsilon_c = \frac{\sigma_{cs}}{E_c}$$

  Increase in strain in tendon due to M, $\Delta\varepsilon_p = \varepsilon_c$ (due to compatibility of strains in bonded members)

$$f_{pm} = \Delta\varepsilon_p E_p = \varepsilon_c E_p = \sigma_{cs} \frac{E_p}{E_c} = \sigma_{cs}\ \text{m}$$

$$f_{pm} = 5.23 \times 6 = 31.38\ \text{N/mm}^2$$

Step 3: *Calculation of stress in tendon* f_{pl}

$$f_{pl} = f_{pe} + f_{pm}$$

$$= 875 + 31.28 = 906.38\ \text{N/mm}^2$$

Percentage increase in stress due to $M = \dfrac{31.38}{875} \times 100 = 3.59\%$

Comments

In this example, there is only 3.59% increase in the stress in tendon due to bending caused by external loads. Generally the increase in stress in tendon due to bending is small and hence usually ignored (see fourth assumption in analysis given in section 5.2).

5.7 Flexural Behaviour of Prestressed Concrete Member

5.7.1 Load Deflection Behaviour

A typical load-deflection curve of a simply supported prestressed concrete beam is depicted in Fig. 5.31. The longitudinal stress distribution in concrete at different stages of loading is also indicated in the figure.

When the beam is prestressed, due to the eccentricity of prestressing force, the beam will bend up. The upward deflection is treated as negative

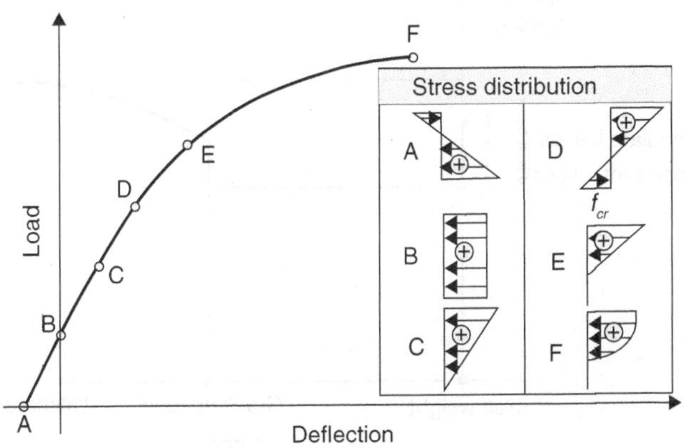

Fig. 5.31 Load-deflection curve

(Point A). At this stage, the stress at the bottom fibre is compressive and the top fibre is tensile in nature.

As the loading is increased, the compressive stress at the bottom fibre decreases. At the top fibre, the tensile stress reverses and becomes compressive. A particular loading stage is encountered, where at every section the bending moment due to external load is equal and opposite to the moment due to prestress. This stage is known as the load balanced stage (Point B) where there is no deflection and the section is subjected to a uniform compressive stress.

With increase in loading, the stress at bottom fibre becomes zero. This stage is known as the decompression stage (Point C). Beyond this stage, the stress at the bottom fibre becomes tensile in nature and reaches the modulus of rupture of concrete (Point D). On increasing the loading, cracks appear at the bottom of the beam and stress for some distance from the soffit becomes zero (Point E).

On further increase in load, the compressive stress distribution becomes nonlinear and when the compressive strain at the top fibre reaches the maximum value ε_{cu}, the concrete begins to crush (Point F). At this stage, the moment capacity of the section represents the ultimate capacity in bending.

5.7.2 Stress in Tendon

The variation of stress in tendon with applied load for a prestressed concrete beam with bonded tendon is shown in Fig. 5.32. In the initial stage, when the tendon is stretched using jacks, the stress in steel is f_{pi}. At the stage of transfer, the stress reduces to f_{po} due to the occurrence of immediate losses.

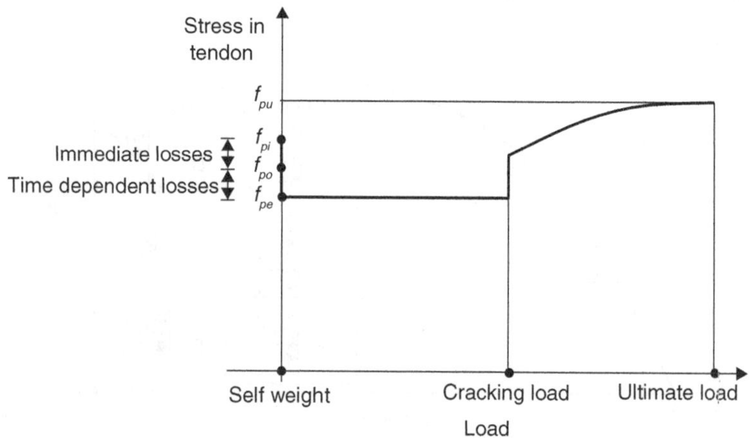

Fig. 5.32 Variation of tendon stress with load

It can be assumed that the time dependent losses also take place before the application of superimposed loads. Hence the stress is further decreased to the effective prestress f_{pe}.

With the application of superimposed dead load and live loads, there will be a increase in the stress in the tendon. The increase in the stress will be usually small and can be neglected (see Example 5.13). This stage continues till the cracking load is reached ($M = M_{cr}$). At this stage, there will be a sharp increase in the stress in the tendon as the tension resisted by concrete till cracking is suddenly transferred to tendon.

After cracking, there will be a significant increase in steel stress with increase in load. Finally at the ultimate load, the stress in the tendon reaches f_{pu}, the ultimate tensile stress in prestressing steel.

5.7.3 Modes of Failure in Flexure

A prestressed concrete flexural member can fail by any one of the following ways:

1. Failure caused due to fracture of steel. This type of failure occurs due to very small amount of reinforcement. The failure is brittle in nature and the Code (Cl. 18.6.3.3) avoids this mode of failure by specifying a minimum percentage of longitudinal reinforcement as 0.2 percent of the concrete area. For high yield strength deformed steel, the reinforcement can be reduced to 0.15 percent.
2. Failure initiated by yielding of steel followed by the rise of neutral axis and final failure due to crushing of concrete. The flexural members failing by this mode are said to be *under-reinforced.*
3. Failure by crushing of concrete and not preceded by yielding of steel. The members having this mode of failure are termed as *over-reinforced.*

5.7.4 Types of Section

- *Under-reinforced section:* In this type of section, the area of prestressing steel is such that at the ultimate stage, the steel yields before the compressive strain in concrete at the extreme fibre reaches the maximum value ε_{cu}.

  Since the steel has yielded, any increase in load will not appreciably increase the stress in tendon. Thus the tensile force in tendon T_p can be assumed to remain constant. Hence the increase in load can be resisted only by increasing the load arm z ($M = T_p z$), which causes the neutral axis to rise up. This results in the reduction of the area of concrete compression zone. To maintain the same compressive force in concrete C ($C = T_p$), the stress in concrete increases, resulting an increase in compressive strain. This process continues till the compressive strain in concrete at the extreme fibre reaches ε_{cu} leading to crushing of concrete.

The movement of neutral axis towards the compression zone is accompanied by development of wider and deeper tensile cracks and increased deflections. The failure of the member is ductile with sufficient warning (in terms of cracks and deflections) before failure. For this reason, this is the preferred mode of failure.

- *Over-reinforced section:* In a over-reinforced section, the area of steel is such that at the ultimate stage, the compressive strain in concrete at the extreme fibre reaches the maximum value ε_{cu} before yielding of steel. The failure of the member will be sudden and without any warning.
- *Balanced section:* In this type of section, the area of steel is such that at the ultimate stage, both the compressive strain in concrete at the extreme fibre reaches ε_{cu} and the yielding of steel occurs simultaneously.

5.8 Analysis at Ultimate Limit State

Satisfying stress limits at serviceability limit state does not guarantee adequate strength and safety of the member. For checking the safety of the member, it is necessary to analyse at the ultimate stage. Only by knowing the ultimate capacity of the member, adequate safety margin between the service load and ultimate load can be ensured.

Further it is necessary to check the ductility (i.e., under-reinforced or over-reinforced) of the section at ultimate limit state. Only ductile sections give sufficient warnings before failure. Also ductility is necessary for indeterminate structures to develop the predicted strength (by redistribution of moments).

Assumptions in Analysis

The basic assumptions used in the analysis at the ultimate limit state of flexure are listed below.

1. Plane sections normal to the axis remain plane after bending, i.e., the variation of strain across the cross-section is linear.
2. The maximum strain in concrete at the extreme compression fibre (ε_{cu}) is taken as 0.0035 in bending. This is because irrespective of whether

the beam is under-reinforced or over-reinforced, failure occurs by crushing of concrete. Hence the strain in the outermost compression fibre reaches ε_{cu}.

3. Concrete carries no tensile stress, i.e., the tensile strength of concrete is neglected.
4. The stress in concrete under compression and the tendon is obtained from the design stress–strain curves of the respective materials (see Chapter 3).

5.8.1 Analysis of Rectangular Sections with Bonded Tendons

The analysis of a prestressed concrete section at the ultimate limit state in flexure implies the determination of ultimate moment of resistance M_{uR} of the section. The ultimate moment of resistance of prestressed concrete sections with fully bonded tendons (i.e., pre-tensioned and bonded post-tensioned beam) can be determined using *strain compatibility method.*

Strain Compatibility Method

This is a iterative method to find M_{uR} and the final solution should satisfy

- The equilibrium equations
- Strain compatibility
- Material stress–strain relations.

Fig. 5.33 shows the distribution of longitudinal stresses and strains across the depth of a eccentrically prestressed rectangular concrete beam section at the ultimate limit state of collapse in flexure.

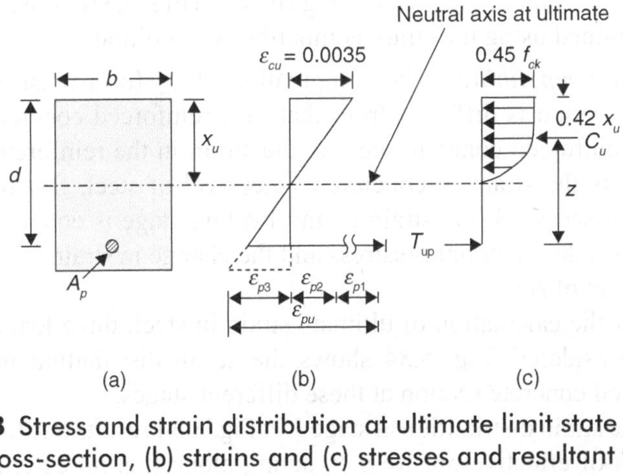

Fig. 5.33 Stress and strain distribution at ultimate limit state in flexure (a) cross-section, (b) strains and (c) stresses and resultant forces

- *Equilibrium equations:* The total compression (C_u) and tension (T_{up}) forces at the ultimate limit state are:

$$C_u = 0.36 f_{ck}\, b\, x_u$$

$$T_{up} = f_{pu} A_p \tag{5.20}$$

where b is the width of the beam, x_u is the depth of neutral axis, A_p is cross-sectional area of the tendon and f_{pu} is the ultimate tensile stress in tendon.

The force equilibrium in the horizontal direction gives:

$$C_u = T_{up}$$

$$0.36 f_{ck}\, b\, x_u = f_{pu} A_p$$

Using the above equation, the depth of neutral axis is obtained as:

$$x_u = \frac{f_{pu} A_p}{0.36 f_{ck}\, b} \tag{5.21}$$

The ultimate moment of resistance M_{uR} is obtained using the moment equilibrium equation:

$$M_{uR} = T_{up}\, z$$

where z is the lever arm and is equal to $(d - 0.42\, x_u)$, d being the effective depth (Fig. 5.33).

Hence,

$$M_{uR} = f_{pu} A_p (d - 0.42\, x_u) \tag{5.22}$$

The stress in the prestressing steel at collapse f_{pu} depends on the ultimate strain in the prestressing steel ε_{pu} (Fig. 5.33(b)) which can be determined using the strain compatibility condition.

- *Strain compatibility:* The computation of ε_{pu} for a prestressed concrete section is different from that of a reinforced concrete section. In a reinforced concrete section, the strain in the reinforcing steel is same as the strain in concrete at the level of steel. But for bonded prestressed steel, the strain at any loading stage is equal to the sum of strain due to initial prestress and the change in strain in concrete at the level of steel.

For the calculation of ultimate strain in steel, three loading stages are considered. Fig. 5.34 shows the strain distribution in the prestressed concrete section at these different stages.

The strain distribution at stage 1 in Fig. 5.34 results due to the application of effective prestress P_e acting alone (i.e. no external loads).

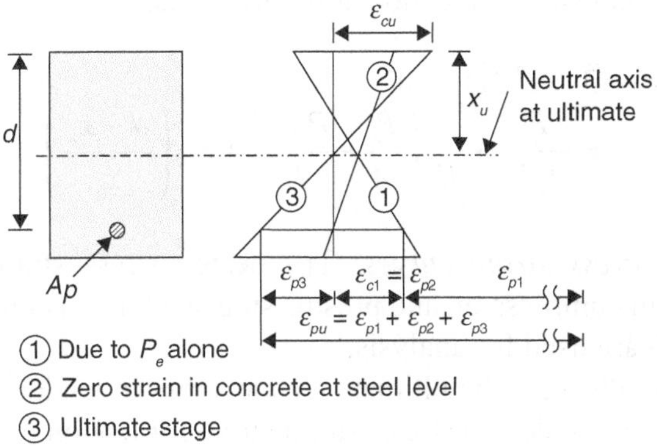

Fig. 5.34 Strain distribution at three different stages

At this stage, the strains in steel (ε_{p1}) and in concrete at the level of steel (ε_{c1}) are:

$$\varepsilon_{p1} = \frac{P_e}{A_p E_s}$$

$$\varepsilon_{c1} = \frac{1}{E_c}\left(\frac{P_e}{A} + \frac{P_e e_s^2}{I}\right)$$

Stage 2 represents the situation when the strain in concrete at the level of steel is zero ($\varepsilon_{c2} = 0$). Due to bond between steel and concrete, the increase in steel strain is equal to change in strain of concrete at the level of steel from Stage 1 to Stage 2. Hence,

$$\varepsilon_{p2} = |\varepsilon_{c2} - \varepsilon_{c1}| = \frac{1}{E_c}\left(\frac{P_e}{A} + \frac{P_e e_s^2}{I}\right)$$

The strain distribution at Stage 3 represents the ultimate loading condition. The strain at extreme compression fibre of the section is ε_{cu}. The strain in concrete at the level of steel (ε_{c3}) is:

$$\varepsilon_{c3} = \varepsilon_{cu}\left(\frac{d - x_u}{x_u}\right)$$

Hence the increase in strain in steel (ε_{p3}) is:

$$\varepsilon_{p3} = |\varepsilon_{c3} - \varepsilon_{c2}| = \varepsilon_{cu}\left(\frac{d - x_u}{x_u}\right)$$

Thus the total strain in tendon at failure, ε_{pu} is:

$$\varepsilon_{pu} = \varepsilon_{p1} + \varepsilon_{p2} + \varepsilon_{p3}$$

$$\varepsilon_{pu} = \frac{P_e}{A_p E_s} + \frac{1}{E_c}\left(\frac{P_e}{A} + \frac{P_e e_s^2}{I}\right) + \varepsilon_{cu}\left(\frac{d - x_u}{x_u}\right) \tag{5.23}$$

- *Material stress–strain curves:* The design stress–strain curves for concrete in compression and prestressing steel in tension discussed in Chapter 3 are used for analysis.

  After finding ε_{pu}, the ultimate tensile stress in tendon (f_{pu}) is determined using the design stress–strain curve for tendon.

- *Procedure for finding M_{uR}:* The ultimate moment of resistance M_{uR} depends on x_u and f_{pu} (Eq. (5.22)), which in turn depends on the ultimate tensile strain in tendon ε_{pu}. But ε_{pu} is governed by the depth of neutral axis x_u (Eq. (5.23)). Hence M_{uR} can be computed only by a iterative numerical procedure by assuming trial value of x_u until convergence is reached.

  The steps for the iterative procedure are:

  (i) Assume a trial value of x_u, say 0.4 times the overall depth.

  (ii) Find ε_{pu} using Eq. (5.23).

  (iii) Using the design stress–strain curve for prestressing steel, find f_{pu} for the strain ε_{pu}.

  (iv) Compute x_u using Eq. (5.21).

  (v) Compare this value of x_u with the trial value of x_u. If the difference between these values is acceptably small, accept the value of x_u obtained in step (iv). Else repeat the steps from (ii) to (v) with improved value of x_u till convergence is reached.

  (vi) Using the converged value of x_u, determine M_{uR} using Eq. (5.22).

Simplified Method to Find M_{uR}

As an alternative to the iterative strain compatibility method, the Code (Appendix B of the Code) recommends a simplified method to find M_{uR}. The code specifies x_u and f_{pu} (Table 5.2) based on the amount of prestressing steel. The latter is expressed as a prestressed reinforcement index ω_p.

$$\omega_p = \frac{A_p f_p}{bd\, f_{ck}} \tag{5.24}$$

where f_p is the characteristic strength of the prestressing steel.

The Code recommendations are applicable only if the effective prestress f_{pe} is not less than 0.45 f_p. After finding x_u and f_{pu}, M_{uR} can be determined using Eq. (5.22).

Table 5.2 x_u and f_{pu} for pre-tensioned and post-tensioned rectangular beams (Table 11 of the code)

$\omega_p = \frac{A_p f_p}{bd f_{ck}}$	$f_{pu}/0.87 f_p$		x_u/d	
	Pre-tensioned	**Bonded Post-tensioned**	**Pre-tensioned**	**Bonded Post-tensioned**
0.025	1.0	1.0	0.054	0.054
0.05	1.0	1.0	0.109	0.109
0.10	1.0	1.0	0.217	0.217
0.15	1.0	1.0	0.326	0.316
0.20	1.0	0.95	0.435	0.414
0.25	1.0	0.90	0.542	0.488
0.30	1.0	0.85	0.655	0.558
0.40	0.9	0.75	0.783	0.653

From Table 5.2, it can be observed that upto ω_p= 0.3 for pre-tensioned beams and ω_p = 0.15 for post-tensioned beams, $f_{pu}/0.87 f_p$ = 1. These limits for ω_p correspond to under-reinforced sections. If the amount of steel is greater than these limits, then the section becomes over-reinforced.

Safety at Ultimate Limit State in Flexure

The bending moment at a section at the ultimate limit state due to factored loads is known as the factored moment M_u. A beam section is considered to be safe if M_{uR} is greater than or equal to M_u.

Example 5.14

Find the ultimate moment of resistance of the pre-tensioned beam section of width 300 mm and effective depth 500 mm for the following data.

$$f_{ck} = 40 \text{ N/mm}^2,\ f_p = 1800 \text{ N/mm}^2,\ A_p = 500 \text{ mm}^2,\ f_{pe} = 0.6 f_p$$

Solution:

Since f_{pe} is greater than 0.45 f_p, the simplified Code procedure can be used to find M_{uR}

Step 1: *Calculation of* ω_p

$$\omega_p = \frac{A_p f_p}{bd\, f_{ck}} = \frac{500 \times 1800}{300 \times 500 \times 40} = 0.15$$

Step 2: *Finding* f_{pu} and *using Table 11 of the Code*

- Ultimate tensile stress in tendon f_{pu}

  For $\omega_p = 0.15; f_{pu}/0.87 f_p = 1$

  $\Rightarrow f_{pu} = 0.87 f_p = 0.87 \times 1800 = 1566 \text{ N/mm}^2$

- Depth of neutral axis x_u

  For $\omega_p = 0.15; x_u/d = 0.326$

  $\Rightarrow x_u = 0.326\, d = 0.326 \times 500 = 163 \text{ mm}$

Step 3: *Calculation of* M_{uR}

$$\begin{aligned} M_{uR} &= f_{pu} A_p (d - 0.42\, x_u) \\ &= 1566 \times 500 \times [500 - (0.42 \times 163)]/10^6 \\ &= 337.9 \text{ kN-m} \end{aligned}$$

Comments

Using the Code provisions, the total compression (C_u) and tension (T_{up}) forces are:

$$\begin{aligned} C_u &= 0.36 f_{ck}\, b\, x_u \\ &= \frac{0.36 \times 40 \times 300 \times 163}{1000} = 704.16 \text{ kN} \\ T_{up} &= f_{pu} A_p \\ &= \frac{1566 \times 500}{1000} = 783 \text{ kN} \end{aligned}$$

Hence, $C_u \neq T_{up}$.

From this example, it can be seen that the simplified Code approach fails to satisfy the force equilibrium condition.

Another shortcoming of the Code provision is the inability to account for the actual value of effective prestress. From Eq. (5.23), it is clear that the effective prestress has a significant effect in M_{uR}. Even though the Code distinguishes between pre-tensioned and post-tensioned members, the effective prestress is not used for finding M_{uR}.

Interested reader can refer the work of Paul and Menon[1], where improvements to Code recommendations are suggested to overcome the above limitations.

---

[1] Paul, S. and Menon, D., "Improvement of IS 1343 Procedure for Estimation of Ultimate Moment Capacity of PSC Sections with Bonded Tendons", *The Indian Concrete Journal*, May 2008, pp. 16–24.

5.8.2 Analysis of Post-tensioned Rectangular Beams Having Unbonded Tendons

In the case of unbonded post-tensioned beams, there is no strain compatibility between the tendon and concrete at any section. Since the tendon is anchored at the beam ends, the total deformation of the tendon (Δ_p) will be equal to the deformation of concrete at the level of tendon (Δ_{cp}). Hence the compatibility condition for this case is:

$$\Delta_p = \Delta_{cp} \tag{5.25}$$

It is very difficult to find f_{pu} using the above compatibility condition. As an alternative, the Code recommendations (Table 5.3) can be used to determine x_u and f_{pu} for a given value of ω_p and effective span (l) to effective depth (d) ratio.

Table 5.3 x_u and f_{pu} for unbonded post-tensioned rectangular beams (Table 12 of the code)

$\omega_p = \frac{A_p f_p}{bd f_{ck}}$	f_{pu}/f_p for Values of l/d			x_u/d for Values of l/d		
	30	20	10	30	20	10
0.025	1.23	1.34	1.45	0.10	0.10	0.10
0.05	1.21	1.32	1.45	0.16	0.16	0.18
0.10	1.18	1.26	1.45	0.30	0.32	0.36
0.15	1.14	1.20	1.36	0.44	0.46	0.52
0.20	1.11	1.16	1.27	0.56	0.58	0.64

Example 5.15

Find the ultimate moment of resistance of unbonded post-tensioned beam section of width 300 mm and effective depth 500 mm for the following data.

$f_{ck} = 40$ N/mm², $f_p = 1800$ N/mm², $A_p = 500$ mm², $f_{pe} = 0.5 f_p$, $l = 10$ m

Solution:

Step 1: *Calculation of ω_p and l/d*

$$\omega_p = \frac{A_p f_p}{bd\, f_{ck}} = \frac{500 \times 1800}{300 \times 500 \times 40} = 0.15$$

$$\frac{l}{d} = \frac{10 \times 10^3}{500} = 20$$

Step 2: *Finding f_{pu} and x_u using Table 12 of the Code*

- Ultimate tensile stress f_{pu}

  Effective prestress $f_{pe} = 0.5 f_p = 0.5 \times 1800 = 900$ N/mm²

  For $\omega_p = 0.15$ and $l/d = 20$; $f_{pu}/f_{pe} = 1.2$

  $\Rightarrow f_{pu} = 1.2 f_{pe} = 1.2 \times 900 = 1080$ N/mm²

- Depth of neutral axis x_u

  For $\omega_p = 0.15$ and $l/d = 20$; $x_u/d = 0.46$

  $\Rightarrow x_u = 0.46\, d = 0.46 \times 500 = 230$ mm

Step 3: *Calculation of M_{uR}*

$$M_{uR} = f_{pu} A_p (d - 0.42\, x_u)$$
$$= 1080 \times 500 \times [500 - (0.42 \times 230)]/10^6$$
$$= 217.84 \text{ kN-m}$$

Comments

Total compression (C_u) and tension (T_{up}) forces are:

$$C_u = 0.36 f_{ck}\, b\, x_u$$
$$= \frac{0.36 \times 40 \times 300 \times 230}{1000} = 993.6 \text{ kN}$$
$$T_{up} = f_{pu} A_p$$
$$= \frac{1080 \times 500}{1000} = 540 \text{ kN}$$

Also in this example, the Code provisions do not satisfy the force equilibrium condition ($C_u \neq T_{up}$).

5.8.3 Analysis of Flanged Sections

The analysis of flanged section depends on whether the neutral axis lies within the flange region (Fig. 5.35(a)) or in the web region (Fig. 5.35(b)).

In Fig. 5.35, b_f is the effective width of the flange, which can be determined using Cl. 23.1.2 of IS 456:2000. D_f is the depth of the flange and b_w is the width of the web.

If the neutral axis lies within the flange ($x_u \leq D_f$), then the analysis can be done by treating the section as a rectangular section of width b_f and effective depth d. But when the neutral axis falls in the web region ($x_u > D_f$), the section should be treated as a flanged section.

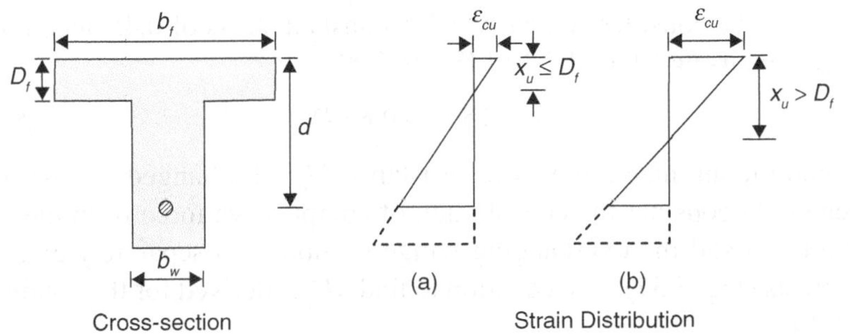

Fig. 5.35 Strain distribution in a flanged section at ultimate limit state (a) neutral axis in flange and (b) neutral axis in web

Two situations arise when neutral axis lies in the web. In the first case (Fig. 5.36(a)), the depth of the flange D_f is small ($D_f \leq 3/7\ x_u$), so that a constant compressive stress $0.45 f_{ck}$ acts over the entire depth D_f. In the next case (Fig. 5.36(b)), D_f is greater than $3/7\ x_u$. Hence the compressive stress block acting on the flange consists of a rectangular area plus a truncated parabolic area. In such cases, to simplify the analysis, an equivalent depth of

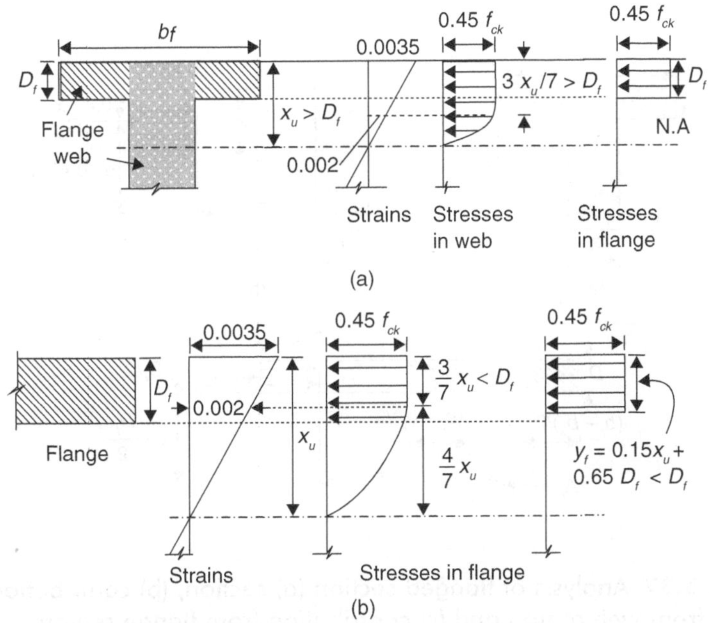

Fig. 5.36 Stress distribution in flange when $X_u > D_f$ (a) neutral axis (N.A) outside flange, $X_u > D_f$, but $3\ X_u/7 > D_f$ and (b) concept of equivalent flange thickness- y_f for ($D_f < X_u < 7/3\ D_f$)

flange y_f can be considered over which a constant stress of 0.45 acts. y_f can be calculated using Cl. G-2.2.1 of IS 456:2000.

$$y_f = 0.15\, x_u + 0.65\, D_f \tag{5.26}$$

To find the ultimate moment of resistance M_{uR} of a flanged section, it is convenient to consider the contribution of compressive force from the web portion (C_{uw}) and the overhanging flange portion (C_{uf}) separately and add their effects (Fig. 5.37). The equation to find M_{uR} is derived for the condition $D_f \leq 3/7\, x_u$.

In Fig. 5.37, T_{uw} is the component of the total tension force T_{up} which balances C_{uw} and T_{uf} is the force which balances C_{uf}. The compression and tension forces are:

$$C_{uw} = 0.36 f_{ck}\, b_w\, x_u \tag{5.27a}$$

$$C_{uf} = 0.45\, \text{fck}\, (b_f - b_w)\, D_f \tag{5.27b}$$

$$T_{uw} = f_{pu}\, A_{pw} \tag{5.27c}$$

$$T_{uf} = f_{pu}\, A_{pf} \tag{5.27d}$$

where A_{pw} is the steel area that contributes to T_{uw} and A_{pf} is the part of the total steel area A_p that contributes to T_{uf}.

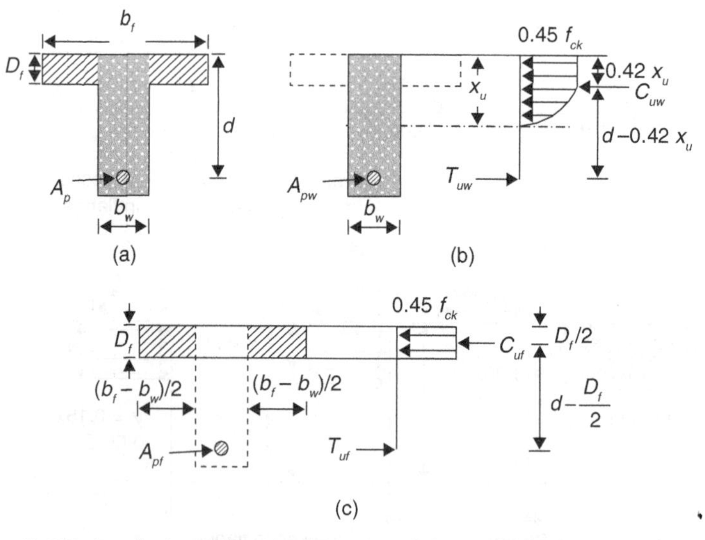

Fig. 5.37 Analysis of flanged section (a) section, (b) contribution from web region and (c) contribution from flange region

The force equilibrium in the horizontal direction gives:

Total tensile force T_{up} = Total compression force C_u

where,

$$T_{up} = T_{uw} + T_{uf}$$

$$C_u = C_{uw} + C_{uf}$$

Also,

$$T_{uf} = C_{uf}$$

Substituting the values of T_{uf} and C_{uf} gives:

$$f_{pu} A_{pf} = 0.45 f_{ck} (b_f - b_w) D_f$$

$$\Rightarrow \qquad A_{pf} = \frac{0.45 f_{ck} (b_f - b_w) D_f}{f_{pu}} \qquad (5.28)$$

But, the total area of prestressing steel A_p is:

$$A_p = A_{pw} + A_{pf}$$

Hence,

$$A_{pw} = A_p - A_{pf} \qquad (5.29)$$

Using the moment equation, M_{uR} is obtained as:

$$M_{uR} = T_{uw}(d - 0.42x_u) + T_{uf}(d - 0.5D_f)$$

$$M_{uR} = f_{pu} A_{pw}(d - 0.42x_u) + f_{pu} A_{pf}(d - 0.5D_f) \qquad (5.30)$$

By replacing D_f with y_f, the above equations can be used for the situation $D_f > 3/7\, x_u$.

M_{uR} for flanged section having bonded tendons can be determined using the strain compatibility method. Alternatively the simplified code procedure recommended for rectangular cross-sections can be used to determine M_{uR}. In this case, the reinforcement index (ω_{pw}) is calculated based on A_{pw}, i.e.,

$$\omega_{pw} = \frac{A_{pw} f_p}{b_w\, d\, f_{ck}} \qquad (5.31)$$

But A_{pw} depends on A_{pf}, which in turn depends on f_{pu}. f_{pu} being a unknown at the initial stage necessitates a iterative procedure for finding M_{uR} using Code recommendations. The steps required for finding M_{uR} are:

1. Assume a trial value of f_{pu}; say $f_{pu} = 0.87 f_p$.
2. Find A_{pf} and A_{pw} using Eqs. (5.28) and (5.29).
3. Calculate ω_{pw} using Eq. (5.31).
4. For ω_{pw} determine f_{pu} and x_u using either Table 5.2 or 5.3 depending on whether the tendon is bonded or not.

5. Compare f_{pu} with the trial value. If the difference between these values are acceptably small, accept the value of f_{pu} obtained using step (4). Else repeat the steps from (2) to (4) with improved values of f_{pu} till convergence.
6. Using the converged value of f_{pu} and x_u, calculate M_{uR} using Eq. (5.30).

Example 5.16

Determine the ultimate moment of resistance of the post-tensioned box section having bonded tendons (Fig. 5.38(a)) for the following data.

$$f_{ck} = 40 \text{ N/mm}^2, f_p = 1800 \text{ N/mm}^2, f_{pe} = 0.7 f_p$$

The total cross-sectional area of the tendons is 1000 mm² with the centroid at a distance of 100 mm from the bottom fibre.

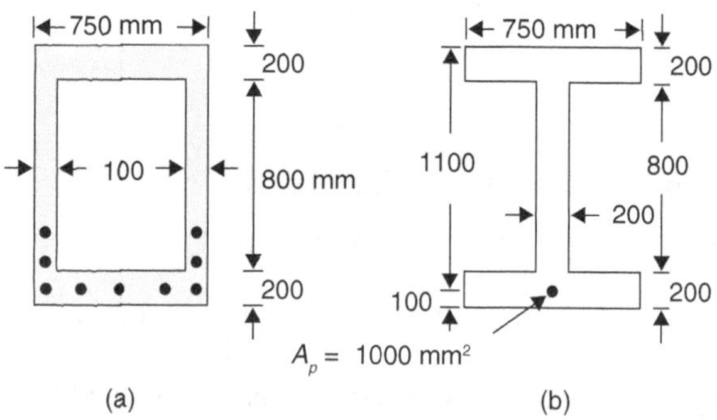

Fig. 5.38 Cross-section of prestressed concrete beam (a) box section and (b) equivalent flanged section

Solution:

The box section is replaced by a equivalent flanged section (Fig. 5.38(b)).

Step 1: *Calculation of* ω_p

Assuming the neutral axis lies in the flange ($b = b_f$),

$$\omega_p = \frac{A_p f_p}{bd\, f_{ck}} = \frac{1000 \times 1800}{750 \times 1100 \times 40} = 0.055$$

Step 2: *Finding* x_u and f_{pu} *using Table 11 of the Code*

- Depth of neutral axis x_u

  for $\omega_p = 0.055; x_u/d = 0.12$

  $\Rightarrow x_u = 0.12d = 0.12 \times 1100 = 132$ mm ($x_u < D_f = 200$ mm)

Since $x_u < D_f$, neutral axis lies in the flange. Hence M_{uR} can be determined for a rectangular section of width 750 mm and effective depth 1100 mm.

- Ultimate tensile stress in tendon f_{pu}

  For $\omega_p = 0.055; f_{pu}/0.87 f_p = 1$

  $\Rightarrow f_{pu} = 0.87 f_p = 0.87 \times 1800 = 1566 \text{ N/mm}^2$

Step 3: *Calculation of M_{uR}*

$M_{uR} = f_{pu} A_p (d - 0.42\, x_u)$

$= 1566 \times 1000 \times [1100 - (0.42 \times 132)] / 10^6$

$= 1635.78$ kN-m

Example 5.17

Compute the ultimate moment of resistance of the box section in Example 5.16 if the area of the tendon is 2500 mm^2. The centroid of the tendon is 100 mm from the bottom fibre.

Solution:

Step 1: *Calculation of ω_p*

Assuming the neutral axis lies in the flange ($b = b_f$),

$$\omega_p = \frac{A_p f_p}{bd\, f_{ck}} = \frac{2500 \times 1800}{750 \times 1100 \times 40} = 0.136$$

Step 2: *Finding x_u using Table 11 of the Code*

For $\omega_p = 0.136; x_u/d = 0.288$

$\Rightarrow x_u = 0.288\, d = 0.288 \times 1100 = 316.8$ mm ($x_u > D_f = 200$ mm)

Since the neutral axis lies outside the flange, the section should be analyzed as a flanged section.

Step 3: *Calculation of f_{pu}*

- Trial 1

  Assume $f_{pu} = 0.87 f_p = 0.87 \times 1800 = 1566 \text{ N/mm}^2$

$$A_{pf} = \frac{0.45 f_{ck}(b_f - b_w)D_f}{f_{pu}}$$

$$= \frac{0.45 \times 40 \times (750 - 200) \times 200}{1566} = \frac{1980000}{1566}$$

$= 1264.37 \text{ mm}^2$

$$A_{pw} = A_p - A_{pf} = 2500 - 126.37 = 1235.63 \text{ mm}^2$$

$$\omega_{pw} = \frac{A_p f_p}{b_w d f_{ck}} = \frac{1235.63 \times 1800}{200 \times 1100 \times 40} = \frac{1235.63 \times 9}{44000} = 0.25$$

from Table 11 of the Code, for $\omega_{pw} = 0.25; f_{pu}/0.87 f_p = 0.9$
$\Rightarrow f_{pu} = 0.9 \times 0.87 f_p = 0.9 \times 1566 = 1409.4 \text{ N/mm}^2$

- Trial 2

$$f_{pu} = 1409.4 \text{ N/mm}^2$$

$$A_{pf} = \frac{1980000}{1409.4} = 1404.85 \text{ mm}^2$$

$$A_{pw} = 2500 - 1404.85 = 1095.15 \text{ mm}^2$$

$$\omega_{pw} = \frac{1095.15 \times 9}{44000} = 0.22$$

for $\omega_{pw} = 0.22; f_{pu}/0.87 f_p = 0.93$
$\Rightarrow f_{pu} = 0.93 \times 1566 = 1456.38 \text{ N/mm}^2$

- Trial 3

$$f_{pu} = 1456.38 \text{ N/mm}^2$$

$$A_{pf} = \frac{1980000}{1456.38} = 1359.54 \text{ mm}^2$$

$$A_{pw} = 2500 - 1359.54 = 1140.47 \text{ mm}^2$$

$$\omega_{pw} = \frac{1140.47 \times 9}{44000} = 0.23$$

for $\omega_{pw} = 0.23; f_{pu}/0.87 f_p = 0.92$
$\Rightarrow f_{pu} = 0.92 \times 1566 = 1440.72 \text{ N/mm}^2$

- Trial 4

$$f_{pu} = 1440.72 \text{ N/mm}^2$$

$$A_{pf} = \frac{1980000}{1440.72} = 1374.313 \text{ mm}^2$$

$$A_{pw} = 2500 - 1374.313 = 1125.69 \text{ mm}^2$$

$$\omega_{pw} = \frac{1125.69 \times 9}{44000} = 0.23$$

for $\omega_{pw} = 0.23; f_{pu}/0.87 f_p = 0.92$
$\Rightarrow f_{pu} = 0.92 \times 1566 = 1440.72 \text{ N/mm}^2$

Hence the converged values are:

$$f_{pu} = 1440.72 \text{ N/mm}^2$$
$$A_{pf} = 1374.31 \text{ mm}^2$$
$$A_{pw} = 1125.69 \text{ mm}^2$$
$$\omega_{pw} = 0.23$$

Step 4: *Calculation of M_{uR}*

- Moment of resistance due to contribution from web

$$M_{uR,\,w} = f_{pu} A_{pw} (d - 0.42\, x_u)$$
$$= 1440.72 \times 1125.69 \times [1100 - (0.42 \times 503.8)] / 10^6$$
$$= 1440.82 \text{ kN-m}$$

- Moment of resistance due to contribution from flange

$$M_{uR,\,f} = f_{pu} A_{pf} (d - 0.5\, D_f)$$
$$= 1440.72 \times 1374.31 \times [1100 - (0.5 \times 200)] / 10^6$$
$$= 1980 \text{ kN-m}$$

- Ultimate moment of resistance of the section M_{uR}

$$M_{uR} = M_{uR,\,w} + M_{uR,\,f}$$
$$= 1440.82 + 1980 = 3420.82 \text{ kN-m}$$

Problems

5.1 Fig. 5.39 shows the cross-section of a simply supported type 1 post-tensioned beam prestressed using straight tendon. Determine the maximum span of the beam based on stresses at transfer. The beam is subjected to a live load of 8 kN/m. Find the extreme fibre stresses at the mid-span during service condition using

(i) Combined load approach

(ii) Internal couple approach and

(iii) Equivalent load approach.

Data:

$$f_{ck} = 40 \text{ N/mm}^2, f_{ci} = 25 \text{ N/mm}^2, f_p = 2000 \text{ N/mm}^2, \beta = 0.9, \eta = 0.8,$$
$$A_p = 1800 \text{ mm}^2$$

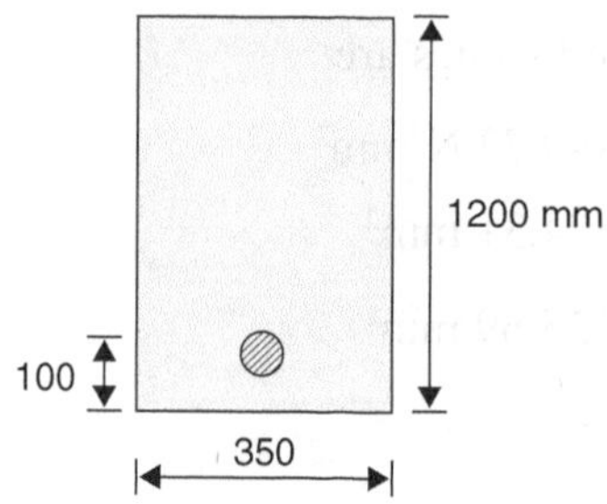

Fig. 5.39

5.2 Fig. 5.40 shows the cross-section of a cantilever post-tensioned (type 2) beam of span 6 m. The beam is prestressed using straight tendon. Find the maximum value of live load that can be applied on the beam.

Data:

$f_{ck} = 40$ N/mm^2, $f_{ci} = 20$ N/mm^2, $f_p = 2000$ N/mm^2, $\beta = 0.9$, $\eta = 0.75$, $A_p = 1800$ mm^2.

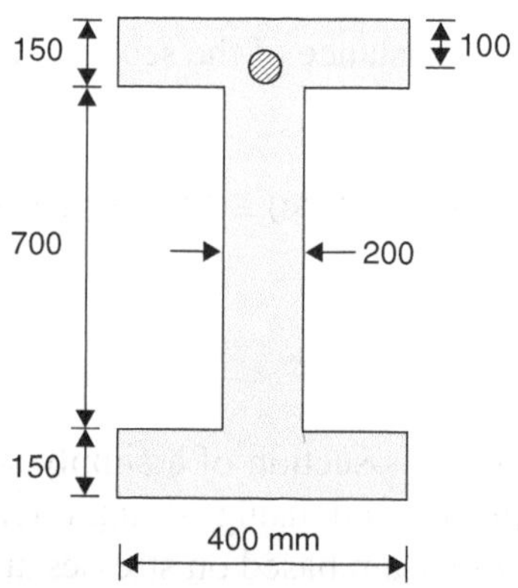

Fig. 5.40

5.3 Use strain compatibility method to find the ultimate moment of resistance of the beam given in Example 5.14. Assume that the beam is prestressed using strands.

chapter 6

Shear and Torsion

6.1 Introduction

Using the concept of mechanics, it can be proved that the shear force (V) is related to bending moment (M) by the equation:

$$V = \frac{dM}{dx}$$

Hence bending moment (varying) is always accompanied by transverse shear force and some times by torsion. This chapter deals with the ultimate limit state in shear and torsion.

The shear capacity of prestressed concrete beam depends on whether the section in the region of maximum shear has cracked by flexure or not. The mode of failure is different for these two cases. The failure by shear is sudden and is very difficult to predict. This failure is caused by the diagonal tension developed in the web of the beam due to shear and longitudinal stresses.

Torsion is the moment about the longitudinal axis of the member. In this chapter, torsion is discussed together with shear because torsion also causes shear stress in the member resulting in the development of diagonal tension. Hence the resisting and failure mechanism in torsion are similar to that of shear.

Shear and torsion are resisted by concrete and by the shear and torsion reinforcement provided in the member. The reinforcement in the form of transverse stirrups is usually provided to take the tension developed in the web of beam after diagonal cracking. This reinforcement should be provided in sufficient quantities such that the member fails by flexure rather than by shear or torsion. Failure by flexure is preferred because it can be predicted with sufficient accuracy and the failure will be ductile in nature (for under-reinforced flexural members).

The web reinforcement (stirrups) becomes effective only after the shear capacity of concrete is exhausted. Hence design for shear and torsion is done at the ultimate limit state.

6.2 Design for Shear

6.2.1 Effect of Prestress in Shear Strength

The effect of prestress in reducing the diagonal tension and thereby increasing the shear strength of concrete can be appreciated by considering the state of stress in element 1 located at the centroidal axis of reinforced and prestressed concrete beams shown in Fig. 6.1. The principal stresses and their directions are obtained from the Mohr circle. When the principal tensile stress f_2 exceeds the tensile strength of concrete, the beam cracks. The cracks occur in the direction of principal compressive stress f_1 (i.e., perpendicular to f_2).

From Fig. 6.1(a), it can be seen that the magnitude of the principal tensile stress f_2 for element 1 in a reinforced concrete beam is equal to the shear stress t, and the diagonal cracking occurs at an angle 45° with the horizontal. In the case of prestressed concrete beam (Fig. 6.1(b)), due to prestress, the magnitude of f_2 is reduced considerably, and the diagonal cracking occurs at a smaller angle ($\alpha < 45°$). Hence the cracks intercepts (or crosses) more number of stirrups there by improving the effectiveness of the stirrups.

Further, in the case of inclined tendons, the vertical component of the prestressing force P_v will be usually in the opposite direction of the external shear force V (Fig. 6.2), thereby reducing the shear force at the section. This beneficial effect of inclined tendon is accounted by including P_v in the shear capacity of concrete V_c.

6.2.2 Identification of Zones for Shear Design

For the purpose of shear design, four design zones (zones A, B, C and D) are identified in a prestressed concrete beam (Fig. 6.3). The important properties of these zones are given below.

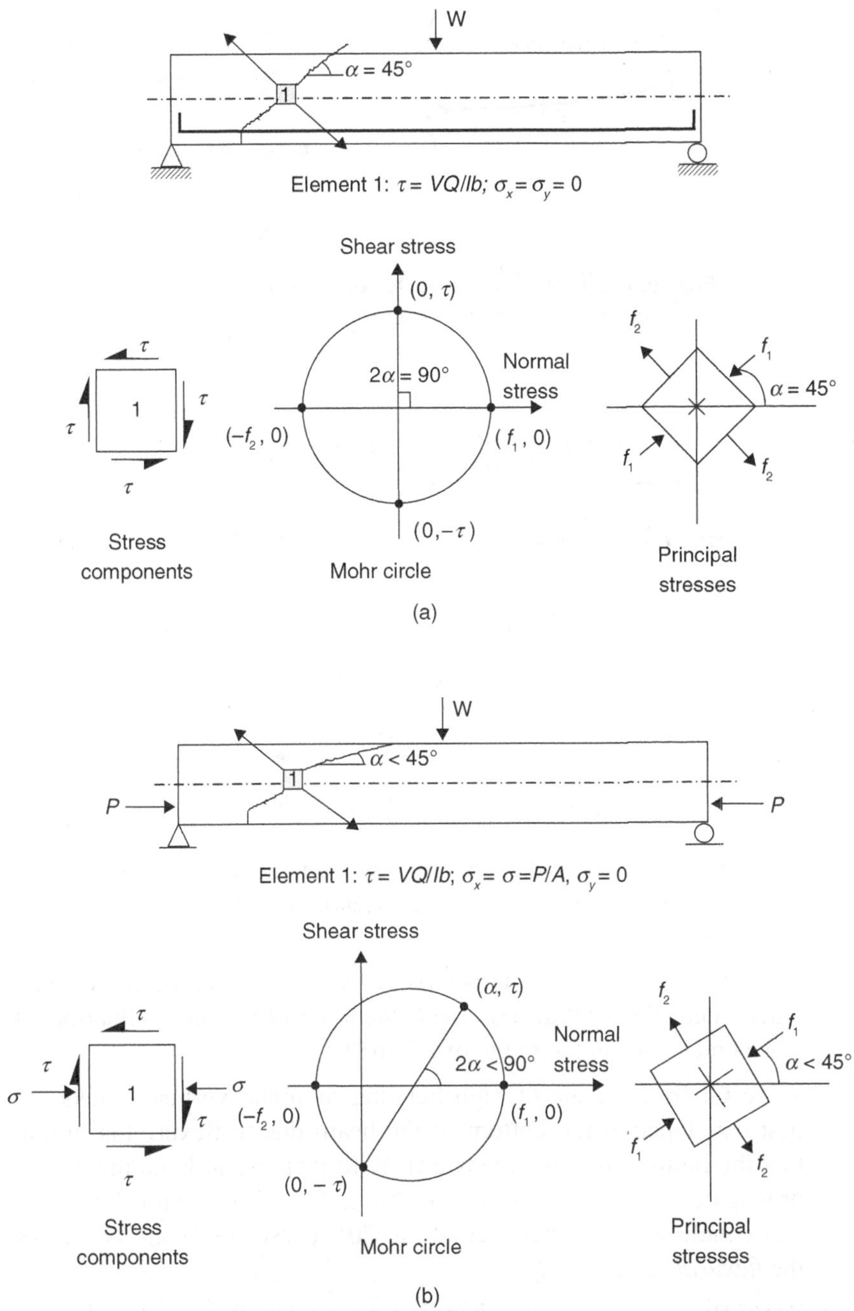

Fig. 6.1 Effect of prestress on diagonal tension and cracking (a) reinforced concrete beam and (b) prestressed concrete beam

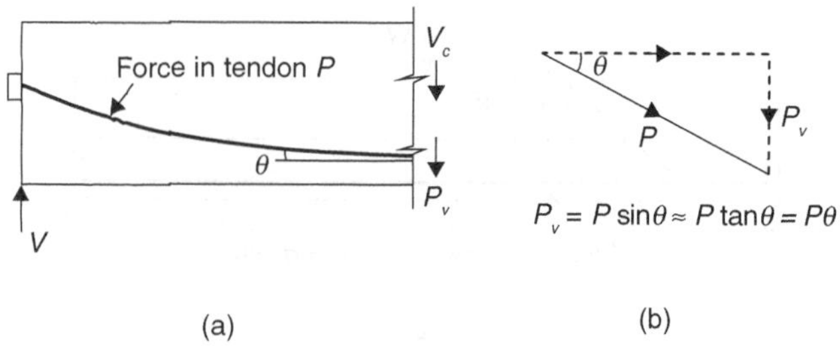

Fig. 6.2 Effect of inclined tendon (a) free body diagram and (b) force components of P

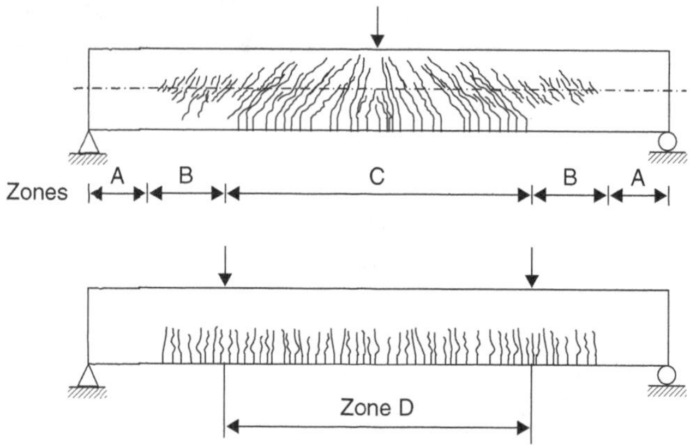

Fig. 6.3 Zones for shear design

- **Zone A:** This zone in the beam remains uncracked till failure. For a simply supported beam, this zone extends to a distance of effective depth from the supports.
- **Zone B:** This zone is characterized by the occurrence of inclined cracks near the neutral axis (*web-shear cracks*). The inclination of these cracks lies in the range of 20° to 30°.
- **Zone C:** In this zone of high bending moment, vertical cracks are first developed at the bottom of the beam due to flexure (assuming that the bottom fibre is in tension). With increase in load, the cracks propagate in an inclined direction due to both flexure and shear. The crack inclination lies between 40° to 70°. These cracks are known as the *flexural-shear cracks*.
- **Zone D:** In this zone, the bending moment is dominant and shear force is negligible. Hence only vertical cracks (flexural cracks) are formed, and shear failure does not occur in this region.

Zones A and B are combined together and is treated as the *zone of web-shear cracking*. Zone C is known as the *zone of flexural-shear cracking*. In the Code, the shear capacity of concrete in the zones of web-shear and flexural-shear cracking are denoted as V_{co} and V_{cr} respectively. The ultimate shear capacity of concrete V_{uc} is taken as the minimum of V_{co} and V_{cr}.

6.2.3 Ultimate Shear Resistance

The ultimate shear resistance V_{uR} at any section of the beam is given by:

$$V_{uR} = V_{uc} + V_{us} \tag{6.1}$$

where V_{uc} and V_{us} are the ultimate shear resistance of concrete and shear reinforcement respectively.

For the beam to be safe against shear mode of failure, at any section,

$$V_{uR} \geq V_{u} \tag{6.2}$$

where V_u is the factored shear force at the section.

6.2.4 Ultimate Shear Resistance of Concrete V_{uc}

The ultimate shear capacity of concrete at *sections uncracked in flexure* (i.e., section in zone of web-shear cracking) and *cracked in flexure* (i.e., section in zone of flexural-shear cracking) are determined, and the minimum capacity is taken as the ultimate shear resistance of concrete.

Ultimate Shear Resistance of Section Uncracked in Flexure V_{co}

V_{co} is the shear force at which the principal tensile stress (f_2) at the centroid of the section just reaches the tensile strength of concrete (f_t). To find V_{co}, f_t is taken as $0.24\sqrt{f_{ck}}$ (Cl. 22.4.1 of the Code). The principal stress f_2 is:

$$f_2 = -f_t = \left(\frac{\sigma_x + \sigma_y}{2}\right) - \sqrt{\left(\frac{\sigma_x - \sigma_y}{2}\right)^2 + \tau^2} \tag{6.3}$$

where σ_x is the compressive stress at the centroid of the section, which is taken as $0.8\,f_{cp}$. f_{cp} is the compressive stress at the centroidal axis due to prestress ($f_{cp} = P_e/A$). σ_y is the normal stress in the vertical direction, which can be taken as zero. τ is the shear stress developed due to bending which is given by:

$$\tau = \frac{VQ}{Ib} \tag{6.4}$$

Q is the first moment of area (area of section above the layer at which τ is required) is considered with respect to the centroid.

At $V = V_{co}$ and for a rectangular section of width b and depth D, τ at the centroid using Eq. (6.4) is:

$$\tau = \frac{3}{2}\frac{V_{co}}{bD}$$

By substituting the above values, Eq. (6.3) can be written as:

$$-f_t = \frac{0.8 f_{cp}}{2} - \sqrt{\left(\frac{0.8 f_{cp}}{2}\right)^2 + \left(\frac{3}{2}\frac{V_{co}}{bD}\right)^2}$$

$$\Rightarrow \qquad \frac{V_{co}}{bD} = \frac{2}{3}\sqrt{f_t^2 + 0.8 f_{cp} f_t}$$

Hence,

$$V_{co} = 0.67bD\sqrt{f_t^2 + 0.8 f_{cp} f_t} \tag{6.5}$$

When inclined tendons are provided and if the vertical component of prestressing force P_v reduces the shear force at the section, then P_v is added to V_{co}. Thus,

$$V_{co} = 0.67bD\sqrt{f_t^2 + 0.8 f_{cp} f_t} + P_v \tag{6.6}$$

But when P_v is in the same direction of applied shear force, then P_v should be deducted from V_{co}.

For flanged sections having centroid in the flange, the principal tensile stress is calculated at the web-flange junction. f_2 is assumed to reach f_t at this section and σ_x is taken as 0.8 times the stress due to prestress at this intersection.

Comments

- For flanged sections having centroid in the web, the quantity Ib/Q at the centroid will be greater than $0.67bD$. Hence using Eq. (6.5) for such sections will lead to underestimation of V_{co}. But for flanged sections, maximum principal stress occurs at the web-flange junction. Thus checking the principal stress at the centroid will be on the unsafe side. These two effects compensate each other, and hence Eq. (6.5) can be also used for flanged sections by replacing b by b_w, the width of web.

- The tensile strength f_t of concrete lies between $0.3\sqrt{f_{ck}}$ and $0.4\sqrt{f_{ck}}$. For design purpose, f_t is taken as:

$$f_t = 0.3\sqrt{\frac{f_{ck}}{\gamma_m}} = \frac{1}{\sqrt{\gamma_m}} \times 0.3\sqrt{f_{ck}} = 0.8 \times 0.3\sqrt{f_{ck}}$$

$$= 0.24\sqrt{f_{ck}} \ (\gamma_m \text{ for concrete} = 1.5)$$

Thus a safety factor of 0.8 is applied to f_t. The Code recommends to use the same safety factor to f_{cp}. Hence σ_x is taken as $0.8 f_{cp}$.

Ultimate Shear Resistance of Section Cracked in Flexure V_{cr}

The Code (Cl. 22.4.2) recommends an empirical formula to find V_{cr} which is given by:

$$V_{cr} = \left(1 - 0.55\frac{f_{pe}}{f_p}\right)\tau_c bd + M_o \frac{V_u}{M_u} \tag{6.7}$$

where f_p is the characteristic strength of prestressing steel and f_{pe} is the effective prestress. For the calculation of V_{cr}, f_{pe} should not exceed $0.6 f_p$. τ_c is the design shear strength of concrete. b and d are the width and effective depth of the beam (effective depth is the distance between the extreme compression fibre and the centroid of the tendon). V_u and M_u are the ultimate shear force and bending moment at the section considered due to applied loads. M_o is the moment required to produce zero stress in concrete at the level of prestressing steel.

The effect of vertical component of prestressing P_v should not be considered for finding V_{cr}. The Code also suggests the minimum value of V_{cr} as $0.1\, bd\sqrt{f_{ck}}$.

Calculation of M_o

The compressive stress in concrete due to prestress at the level of tendon (f_{pt}) is:

$$f_{pt} = \frac{P_e}{A} + \frac{(P_e e_s)e_s}{I} \tag{6.8}$$

The design stress in concrete due to prestress $f_{pt,\,d}$ is obtained by applying a safety factor 0.8 to f_{pt}. Hence, $f_{pt,\,d} = 0.8\,f_{pt}$. To obtain zero stress in concrete at the level of tendon, the tensile stress due to moment M_o (assumed as sagging moment) at the same level i.e., $\frac{M_o e_s}{I}$ should be equal to $f_{pt,\,d}$. Hence,

$$\frac{M_o e_s}{I} = f_{pt,\,d} = 0.8\,f_{pt}$$

$$\Rightarrow \qquad M_o = \frac{0.8 f_{pt} I}{e_s} \qquad (6.9)$$

Design Shear Strength of Concrete τ_c

The design shear strength of concrete depends on the compressive strength of concrete and the percentage of prestressing steel $p_p = 100\,A_p/(bD)$. The recommended values of tc given in the Code (Table 6 in Code) are based on the following empirical equation[1]:

$$\tau_c = 0.85\sqrt{0.8 f_{ck}}\,(\sqrt{1+5\beta}-1)/(6\beta)$$

where $\beta = \begin{cases} 0.8 f_{ck}/(6.89 p_p) \\ 1 \end{cases}$ which ever is greater.

τ_c for different grades of concrete and percentage of prestressing steel are given in Table 6.1.

Table 6.1 Design shear strength of concrete τ_c (N/mm²)

$100\frac{A_p}{bd}$	Grade of Concrete		
	M 30	**M 35**	**≥ M 40**
0.2	0.334	0.337	0.339
0.4	0.452	0.457	0.461
0.6	0.536	0.543	0.549
0.8	0.601	0.611	0.618
1.0	0.656	0.667	0.677
1.2	0.703	0.716	0.727
1.4	0.744	0.759	0.771
1.6	0.780	0.797	0.811

[1] Explanatory Handbook on *Indian Standard Code of Practice for Plain and Reinforced Concrete* (IS 456:1978), Special Publication SP:24, Bureau of Indian Standards, New Delhi, 1983.

2.0	0.843	0.864	0.880
2.2	0.871	0.893	0.911
2.4	0.896	0.919	0.939
2.6	0.919	0.944	0.965
2.8	0.941	0.968	0.990
3.0	0.962	0.989	1.013

6.2.5 Design of Shear Reinforcement

If the ultimate shear force V_u at a section is less than V_{uc}, then only nominal (minimum) shear reinforcement is required. Further if V_u is less than or equal to 0.5 V_{uc}, no shear reinforcement is required. This situation occurs in slabs, pile caps and walls.

The shear reinforcement has to be designed only when V_u exceeds V_{uc}, and it should be designed to carry the excess shear force ($V_u - V_{uc}$). For the purpose of design, it is necessary to find the shear carrying capacity of shear reinforcement.

The ultimate shear resistance provided by shear reinforcement V_{us} is obtained by considering the free body diagram of the beam cracked in shear (Fig. 6.4).

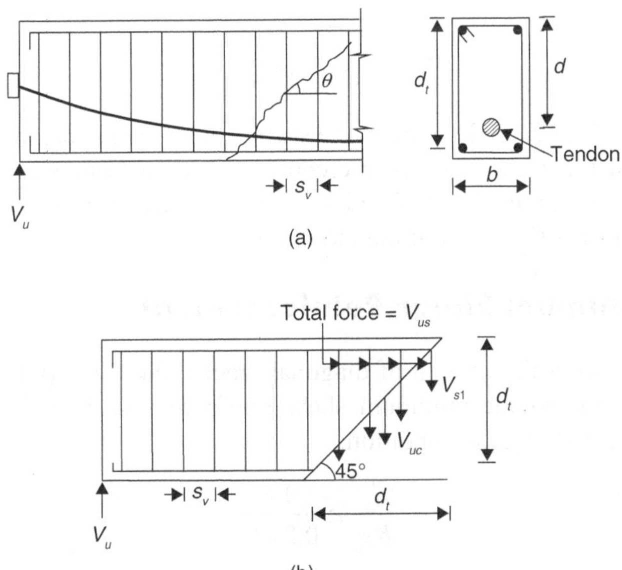

Fig. 6.4 Ultimate shear resistance of vertical stirrups (a) prestressed concrete beam with vertical stirrups (θ assumed as 45°) and (b) free body diagram of beam segment cracked in shear

Since the effect of vertical prestressing force P_v is already included in V_{uc}, the prestressing force is not shown in the free body diagram (Fig. 6.4(b)). In Fig. 6.4, s_v is the spacing of stirrups, θ is the inclination of shear crack and d_t is the distance from the extreme compression fibre to the longitudinal bar (untensioned reinforcement) or to the centroid of tendon, which ever is greater.

V_{s1} is the shear resistance provided by a single stirrup given by:

$$V_{s1} = 0.87 f_v A_{sv}$$

where f_v is the characteristic strength of stirrup reinforcement which should not be more than 415 N/mm^2. A_{sv} is the total cross-sectional area of stirrup legs effective in shear.

If n is the number of stirrups intercepted by the shear crack, then the shear resistance of web reinforcement V_{us} is:

$$V_{us} = nV_{s1} = \frac{d_t}{s_v} V_{s1} = \frac{d_t}{s_v}(0.87 f_v A_{sv})$$

But $V_{us} \geq (V_u - V_{uc})$. Hence,

$$\frac{d_t}{s_v}(0.87 f_v A_{sv}) \geq (V_u - V_{uc})$$

Thus the required stirrup spacing s_v is:

$$s_v \leq \frac{(0.87 f_v A_{sv}) d_t}{(V_u - V_{uc})} \tag{6.10}$$

The Code (Cl. 22.4.3.2) restricts the stirrup spacing to a maximum value ($s_{v,\,m}$), so that the shear crack is intercepted by at least one stirrup. $s_{v,\,m}$ is taken as the minimum of 0.75 d_t and 4 b_w (b_w is the width of web). When V_u is greater than 1.8 V_{uc}, $s_{v,\,m}$ is limited to 0.5 d_t.

6.2.6 Minimum Shear Reinforcement

In order to restrict the growth of diagonal cracks, the Code (Cl. 22.4.3.1) recommends to provide minimum shear reinforcement in regions where $V_u \Rightarrow 0.5\, V_{uc}$ satisfying the condition:

$$\frac{A_{sv}}{b s_v} \geq \frac{0.4}{0.87 f_v}$$

$$\Rightarrow \qquad s_v \leq \frac{2.175 f_v A_{sv}}{b} \tag{6.11}$$

However s_v should not exceed $s_{v,\,m}$.

The Code (Cl. 18.6.3.2) also recommends to provide a minimum transverse reinforcement of 0.1 percent of the sectional area of the web in plan for beams having depth of web more than four times the web thickness and not subjected to dynamic loading.

6.2.7 Maximum Ultimate Shear Force ($V_{u,\,max}$)

To make sure that the stirrup yields at failure, the Code (Cl. 22.4.4) has imposed a upper limit for the shear force ($V_{u,\,max}$) that can act at a section.

$$V_{u,\,max} = \tau_{c,\,max} bd \tag{6.12}$$

where the values of maximum shear stress $\tau_{c,\,max}$ ($\approx 0.64\sqrt{f_{ck}}$) in N/mm^2 are 3.5, 3.7, 4.0, 4.3, 4.6 and 4.8 for concrete grades M 30, M 35, M 40, M 45, M 50 and M 55 respectively (Table 7 of the Code).

6.2.8 Steps for Shear Design

The steps to be followed for the shear design of a prestressed concrete member are listed below:

(i) Determine V_u at the required section. The critical section for shear design can be taken at the face of supports.

(ii) Check whether V_u is less than or equal to $V_{u,\,max}$. If V_u exceeds $V_{u,\,max}$, increase the cross-sectional dimensions of the member such that $V_u \leq V_{u,\,max}$.

(iii) Calculate V_{uc}, which is taken as the minimum of V_{co} and V_{cr}.

(iv) Compare V_{uc} with V_u.

(v) If $V_{uc} > V_u$ and $V_u \geq 0.5\ V_{uc}$, provide minimum shear reinforcement (Eq. (6.11)). When $V_u < 0.5\ V_{uc}$, no shear reinforcement is required.

(vi) Find the required spacing of stirrups when V_u exceeds V_{uc} (Eq. (6.10)).

(vii) The stirrup spacing obtained in steps (v) or (vi) should not exceed $s_{v,\,m}$ and must satisfy the recommendations for minimum transverse reinforcement as per Cl. 18.6.3.2 of the Code.

Example 6.1

Determine the required shear reinforcement at the critical section of the prestressed concrete beam shown in Fig. 6.5(a). The material properties are f_p = 1800 N/mm^2, f_{pe} = 1200 N/mm^2 and f_{ck} = 40 N/mm^2. The applied live load and super imposed dead loads are 12 kN/m and 3 kN/m. Assume that the width of column is 500 mm.

Solution:

Step 1: *Calculation of V_u and M_u at critical section*

- Self-weight of the beam w_{SW}

  Cross-sectional area $A = 2(0.5 \times 0.15) + (0.6 \times 0.18) = 0.258\ \text{m}^2$

  $w_{SW} = \gamma_c A = 24 \times 0.258 = 6.19$ kN/m

- Factored load w_u

  Superimposed dead load $w_{SIDL} = 3$ kN/m

  Dead load $w_{DL} = w_{SW} + w_{SIDL} = 6.19 + 3 = 9.19$ kN/m

  Live load $w_{LL} = 12$ kN/m

  Factored load $w_u = 1.5\,(w_{DL} + w_{LL}) = 31.79$ kN/m

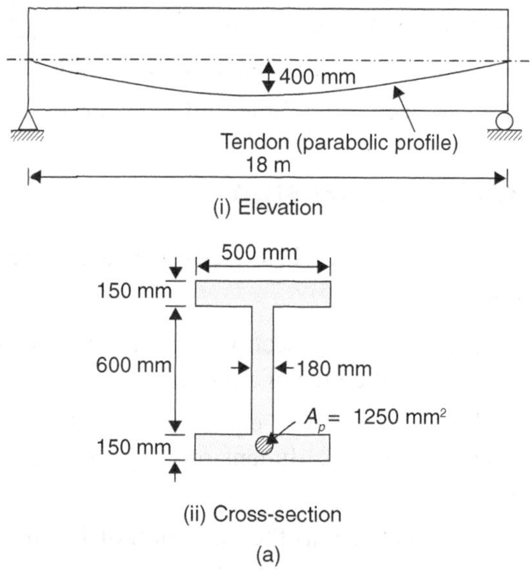

Fig. 6.5 (a) Prestressed concrete beam

- Factored shear force at critical section V_u

  The critical section is taken at the face of support, i.e., at a distance of half of bearing width ($0.5\,w_b$) from the centerline of support.

  Width of bearing w_b = width of column = 500 mm

  Distance of critical section from support $x_c = 0.5\,w_b = 250$ mm

  Support reaction $R = R = \dfrac{w_u l}{2} = 31.79 \times \dfrac{18}{2} = 286.11$ kN

  Shear force at critical section $V_u = R - w_u x_c = 286.11 - (31.79 \times 0.25)$
  $= 278.16$ kN

- Factored bending moment at critical section M_u

  $M_u = R\,x_c - w_u x_c^{\,2}/2 = (286.11 \times 0.25) - (31.79 \times 0.25^2/2)$
  $= 70.53$ kN-m

Step 2: *Check for maximum shear force at critical section*

- Eccentricity of tendon (e_s) at critical section

  The equation of the cable profile (y) gives the variation of e_s.

$$y = \frac{4h}{l^2}(lx - x^2)$$

At critical section: $x = x_c = 0.25$ m, $h = 0.4$ m, $l = 18$ m

$$e_s = y \text{ (at } x = 0.25\text{m)}$$

$$= \frac{4 \times 0.4}{18^2}\ [(18 \times 0.25) - 0.25^2]$$

$$= 0.02191 \text{ m} = 21.91 \text{ mm}$$

- Effective depth d at critical section

  Depth of beam $D = 900$ mm.

$$\text{Effective depth } d = \frac{D}{2} + e_s = \left(\frac{900}{2}\right) + 21.91 = 471.91 \text{ mm}$$

- Maximum ultimate shear force $V_{u,\,max}$

  $b = b_w = 180$mm

  $\tau_{c,\,max}$ $4 =$ N/mm^2 (for M 40 grade concrete, Table 7 of Code)

$$V_{u,\max} = \tau_{c,\max} bd = \frac{4 \times 180 \times 471.91}{1000} = 339.78 \text{ kN} = 339.78 \text{ kN} (> V_u, \text{OK})$$

  Since V_u is less than $V_{u,\,max}$, there is no need to revise the section.

Step 3: *Calculation of ultimate shear resistance of section uncracked in flexure* V_{co}

- Stress at the centroid due to prestress f_{cp}

  Area of tendon $A_p = 1250$ mm^2

$$\text{Effective prestressing force } P_e = A_p . f_{pe} = \frac{1250 \times 1200}{1000} = 1500 \text{ kN}$$

$$f_{cp} = \frac{P_e}{A} = \frac{1250 \times 10^3}{0.258 \times 10^6} = 5.81 \text{ N/mm}^2$$

- Tensile strength of concrete f_t

$$f_t = 0.25\sqrt{f_{ck}} = 0.25\sqrt{40} = 1.58 \text{ N/mm}^2$$

- Vertical component of prestressing force P_v

  Since the effect of P_v is to reduce the shear force, P_v is added to V_{co}.

$$\text{Slope of the cable at critical section } \theta = \frac{dy}{dx} = \frac{4h}{l^2}(l - 2x)$$

$$\theta = \frac{4\times 0.4}{18^2}\times[18-(2\times 0.25)] = 0.086 \text{ radians}$$

$P_v = P_e\,\theta = 1500 \times 0.086 = 129$ kN

- Shear capacity of section uncracked in flexure V_{co}

$$V_{co} = 0.67bD\sqrt{f_t^2 + 0.8 f_{cp} f_t} + P_v$$

$$= \left[\frac{0.67\times 180\times 900\sqrt{1.58^2 + (0.8\times 5.81\times 1.58)}}{1000}\right] + 129$$

$$= 469.48 \text{ kN}$$

Step 4: *Calculation of ultimate shear resistance of section cracked in flexure* V_{cr}

- Moment of inertia of the cross-section I

$$I = 2\left[\left(\frac{500\times 150^3}{12}\right) + (500\times 150\times 375^2)\right] + \frac{180\times 600^3}{12}$$

$$= 2.4615 \times 10^{10} \text{ mm}^4$$

- Stress in concrete at the level of tendon due to prestress (f_{pt})

$$f_{pt} = \frac{P_e}{A} + \frac{(P_e e_s)e_s}{I}$$

$$= \left[\frac{1500\times 10^3}{0.258\times 10^6}\right] + \left[\frac{1500\times 10^3\times 21.91^2}{2.4615\times 10^{10}}\right] = 5.84 \text{ N/mm}^2$$

- Moment M_o

$$M_o = \frac{0.8 f_{pt} I}{e_s} = \frac{0.8\times 5.84\times 2.4615\times 10^{10}}{21.91\times 10^6} = 5248.8 \text{ kN-m}$$

- Design strength of concrete τ_c

  Percentage of prestressing steel $p_p = \dfrac{100A_p}{bd} = \dfrac{100\times 1250}{180\times 471.91} = 1.47$

  For 1.47 percentage of prestressing steel,
  $\tau_c = 0.784$ N/mm² (Table 6 of the Code)

- Ratio of effective prestress to characteristic strength (f_{pe}/f_p)

$$\frac{f_{pe}}{f_p} = \frac{1200}{1800} = 0.67$$

For calculating V_{cr}, f_{pe}/f_p should not exceed 0.6. Hence, $\dfrac{f_{pe}}{f_p} = 0.6$

- Minimum value of V_{cr}

$$(V_{cr})_{min} = 0.1bd\sqrt{f_{ck}}$$

$$= \frac{0.1 \times 180 \times 471.91 \times \sqrt{40}}{1000} = 53.72 \text{ kN}$$

- Shear capacity of section cracked in flexure V_{cr}

$$V_{cr} = \left(1 - 0.55\frac{f_{pe}}{f_p}\right)\tau_c bd + M_o \frac{V_u}{M_u}$$

$$= \left[\frac{(1-0.55\times0.6)\times0.784\times180\times471.91}{1000}\right] + \left[\frac{5248.8\times278.16}{70.53}\right]$$

$$= 20745.12 \text{ kN } (>(V_{cr})_{min})$$

Step 5: *Finding the ultimate shear resistance of concrete V_{uc}*

V_{uc} is taken as the minimum of V_{co} and V_{cr}. Hence in this case,
$V_{uc} = V_{co} = 469.48$ kN

Step 6: *Calculation of shear reinforcement*

Since $V_{uc} > V_u$ and $V_u \geq 0.5\ V_{uc}$, provide minimum shear reinforcement.

- Spacing of stirrups s_v for minimum shear reinforcement

$$s_v \leq \frac{2.175 f_y A_{sv}}{b}$$

Assuming f_y = 415 N/mm^2 and 2 legged stirrups of 8 mm diameter (ϕ_s = 8mm).

$$A_{sv} = 2 \times \frac{\pi}{4} \times 8^2 = 100.53 \text{ mm}^2$$

$$s_v \leq \frac{2.175 \times 415 \times 100.53}{180} = 504.12 \text{ mm}$$

- Maximum stirrup spacing $s_{v,m}$

  Assuming a clear cover (c) of 30 mm and diameter of longitudinal reinforcement (ϕ_l) as 10 mm,

  $$d_t = D - c - \phi_s - 0.5\ \phi_l$$
  $$= 900 - 30 - 8 - (0.5 \times 10) = 857 \text{ mm}$$

  $s_{v,m}$ is the minimum of $0.75d_t$ and 4 b_w. Thus,

  $s_{v,m}$ = Minimum of 642.75 mm and 720 mm = 642.75 mm

- Spacing based on minimum transverse reinforcement (Cl. 18.6.3.2 of Code)

  Fig. 6.5(b) shows the elevation and horizontal section of the web of the beam.

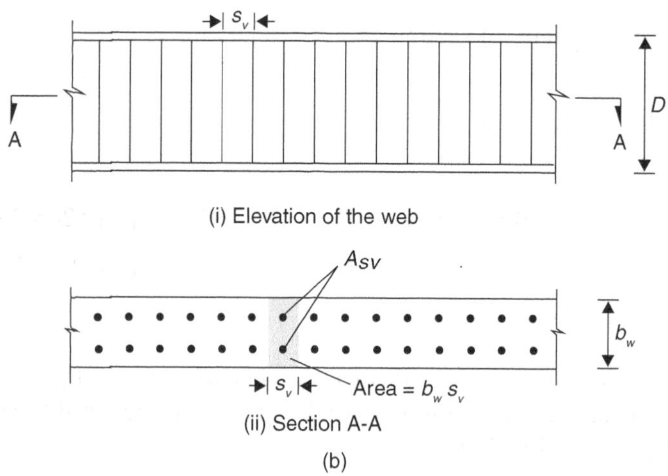

Fig. 6.5 (b) Elevation and horizontal section of the web

For the given beam, $\dfrac{D}{b_w} = \dfrac{900}{180} = 5$

Since $D/b_w > 4$, a minimum transverse reinforcement of 0.1 percentage of the sectional area of the web (in plan) should be provided.

For the web area (in plan) of $b_w\ s_v$ (Fig. 6.5(b)), the area of the transverse steel is A_{sv}. Hence,

$$100\frac{A_{sv}}{b_w s_v} \geq 0.1$$

$$\Rightarrow s_v \leq 1000\frac{A_{sv}}{b_w}$$

$$s_v \leq \frac{1000 \times 100.53}{180} = 558.5 \text{ mm}$$

s_v should be taken as the least of 504.12 mm, 642.75 mm and 538.5 mm. Provide 2 legged stirrups of 8 mm diameter at 500 mm c/c spacing.

6.3 Design for Torsion

Torsion in concrete structures usually occurs in combination with flexure and shear. Due to inherent heterogeneous and nonlinear behaviour of concrete, predicting the combined behaviour of flexure, shear and torsion is a difficult task. Hence the codes of practice usually recommend highly simplified design procedures based on theoretical and experimental studies.

6.3.1 Equilibrium Torsion and Compatibility Torsion

From a design point of view, there are two types of torsion. The first type known as *equilibrium torsion* is developed to maintain equilibrium of the structure. In this case, the structure does not have any other mechanism (or load path) to carry the load without developing torsion. The torsion is determined by using equilibrium equations alone, and the torsion developed does not depend on the torsional stiffness of the member. For this case, the member must be designed to resist torsion. An eccentrically loaded beam shown in Fig. 6.6(a) is a good example for equilibrium torsion.

The second type of torsion known as *compatibility torsion* occurs in indeterminate structure having rigidly connected members. This type of torsion

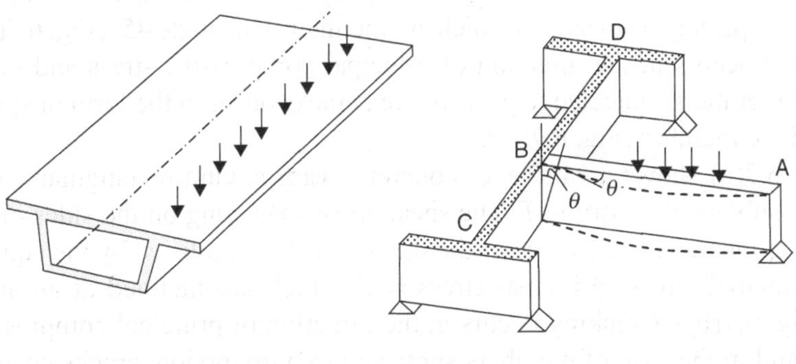

Fig. 6.6 Different types of torsion (a) equilibrium torsion and (b) compatibility torsion

is statically indeterminate and it depends on the torsional stiffness of the member. Both equilibrium and compatibility conditions are required to find the torsional moment.

The compatibility torsion is developed to maintain the deformation compatibility of the members meeting at a joint. For example, in Fig. 6.6(b), the beam AB rests on the member CD having torsional restraints at the ends. The bending of the member AB causes a rotation θ at the end B. But the member AB is rigidly connected to member CD. Hence to maintain compatibility of deformation at joint B, the member CD must also rotate by the same amount θ. Since the member CD is torsionally restrained at the ends, torsional moments are developed in CD due to this rotation.

In the case of compatibility torsion, the structure will have alternate load paths by which the load can be transferred without developing torsion. For the structure in Fig. 6.6(b), if the designer neglects torsional stiffness of member CD (*torsion release*), the beam AB can be designed by assuming a hinge support at B. The resulting structure will be stable and can be designed to carry the applied loads. But this design results in the formation of large cracks at joint B. However, these cracks can be controlled by providing adequate reinforcement. Due to these reasons, compatibility torsion is usually ignored in design practice.

6.3.2 Failure of Concrete Member Due to Torsion

Concrete Member Subjected to Pure Torsion

Fig. 6.7(a) shows a cantilever concrete beam having rectangular cross-section subjected to torque T. The shear stress (τ) acting on the sides of the elements located on each face of the member can be replaced by the equivalent principal stress (principal stress = τ) which are inclined at an angle 45° (Fig. 6.7(b)). Cracking occurs in the direction of principal compressive stress and in the case of members subjected to pure torsion, cracks occur in the form of spirals around the member (Fig. 6.7(c)).

Fig. 6.7(a) shows a cantilever concrete beam having rectangular cross-section subjected to torque T. The shear stress (τ) acting on the sides of the elements located on each face of the member can be replaced by the equivalent principal stress (principal stress = τ) which are inclined at an angle 45° (Fig. 6.7(b)). Cracking occurs in the direction of principal compressive stress and in the case of members subjected to pure torsion, cracks occur in the form of spirals around the member (Fig. 6.7(c)).

Ideally reinforcements in the form of spirals are required for the control of cracks. But due to practical considerations, equivalent longitudinal

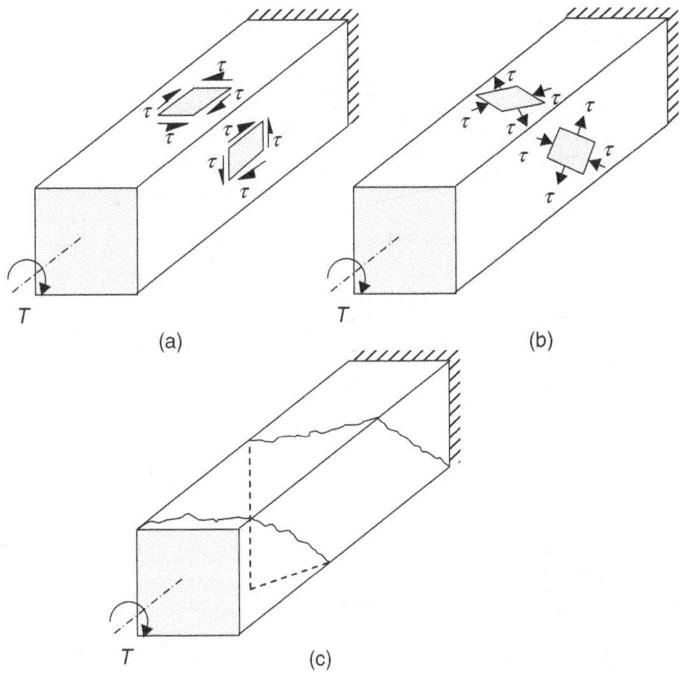

Fig. 6.7 Concrete beam subjected to pure torsion (a) shear stresses, (b) principal stresses and (c) cracks due to torsion

and transverse reinforcements are provided as torsion reinforcement. Due to spiral cracks, *closed stirrups* are required as transverse reinforcement.

Concrete Member Subjected to Flexure, Shear and Torsion

If the cantilever beam is subjected to both torque T and shear force V (Fig. 6.8), the shear stresses due to T and V gets added on one face (face ABCD) and counteract each other on the other (face EFGH). In this case, the cracks start from the face where shear stress gets added up (crack 1-2) and propagates on the flexural tensile face of the beam (crack 2-3). Since the beam is cantilever, the flexural tensile face is at the top of the beam. If the bending moment is large, it will cause the crack to extend in the vertical direction (crack 3-4). The flexural compressive zone at the bottom of the beam prevents the crack to reach the bottom of the beam on both faces ABCD and EFGH.

Prestressed Concrete Members Subjected to Torsion

Similar to members subjected to shear, prestressing the member delays the formation of cracks and changes the direction of crack propagation.

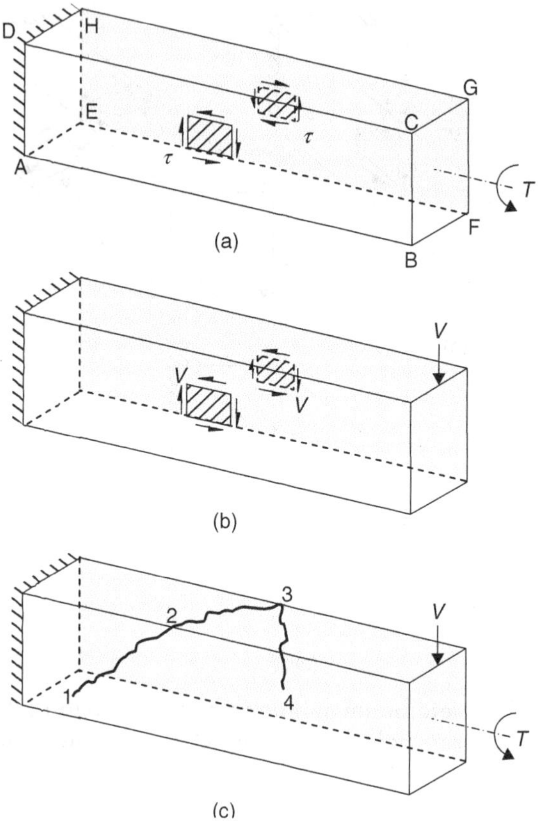

Fig. 6.8 Concrete beam subjected to flexure, shear and torsion
(a) shear stresses due to torsion T,
(b) shear stresses due to shear V and (c) crack pattern

6.3.3 Design Methods for Torsion

The two common methods used for torsion design are the *plastic space truss method* and the *skew bending theory*.

Plastic Space Truss Design Method

In the space truss method developed by Lampert and Thurlimann[2], the cracked concrete beam (solid or hollow beam) is replaced by a three dimensional truss (Fig. 6.9). This truss can represent the behaviour of concrete beam subjected

---

[2] Lampert, P. and Thurlimann, B., *Ultimate Strength and Design of Reinforced Concrete Beams in Torsion and Bending*, International Association for Bridge and Structural Engineering, 1971, Vol. 311, pp. 107–131.

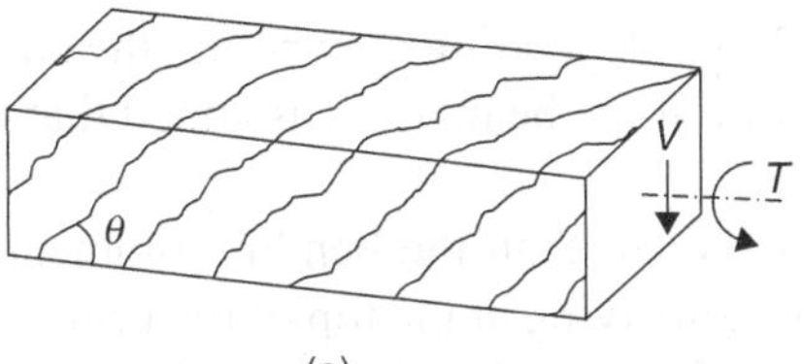

(a)

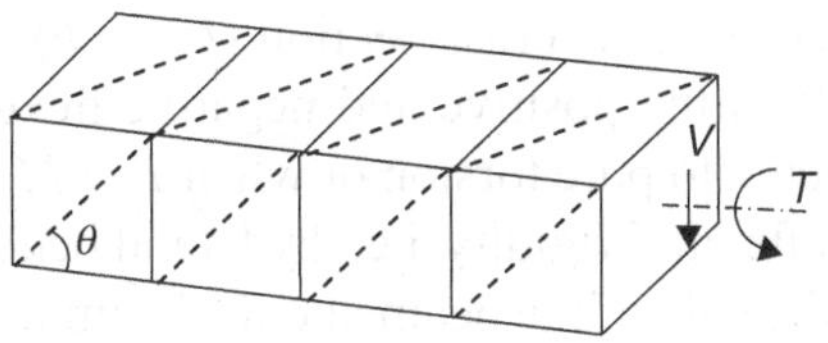

(b)

Fig. 6.9 Plastic space truss model for concrete beam (a) cracked beam and (b) space truss model

to flexure, shear and torsion. The dotted lines in the truss are the compression members, which represent the concrete between the inclined cracks. The solid lines in the truss are the tension members representing the steel reinforcement. The truss thus developed is analysed for the applied loads. Once the force is truss members are determined, they are used to check the stresses in compression members and to find the area of steel required for tension members.

Skew Bending Theory

The skew bending theory[3, 4] developed by Lessig forms the basis of torsion design recommendations given in the Code. This theory assumes that torsion and shear are partly resisted by concrete and the remaining by longitudinal and transverse reinforcements. The failure mode as per this theory is due to bending acting on a skewed surface formed by spiral cracks on three faces of the beam (Fig. 6.8 (c)).

For a beam in pure bending, the moment vector will be perpendicular to the beam axis, and the failure surface will be at right angles to the axis. When torsion is also present, the resultant moment-torque vector will be inclined to the beam axis (Fig. 6.10(a)). Hence the failure surface will be inclined to the axis and is also warped.

Basically three modes of failures are identified for beams subjected to combined torsion and bending (Figs. 6.10(b)–(d)). The failure modes are

---

[3] Rangan, B.V., and Hall, A.S., "Strength of Rectangular Prestressed Concrete Beams in Combined Torsion, Bending and Shear", *ACI Journal*, April 1973, pp. 270–278.

[4] Rangan, B.V. and Hall, A.S., "Design of Prestressed Concrete Beams Subjected to Combined Bending, Shear and Torsion", *ACI Journal*, March 1975, pp. 89–93.

classified as Mode 1, 2 or 3 depending on whether the compression zone lies at the top, side or bottom of the beam (it is assumed that bending moment is a sagging moment).

When the torsion T_u is less than the bending moment M_u, the beam still fails with compression zone lying at the top of the beam. The beam is said to have failed in Mode 1 or by a *modified bending failure* (Fig. 6.10(b)).

For beams with very narrow cross-section ($D >> b$), the lateral flexural rigidity will be less than the positive and negative flexural rigidity. When such beams are subjected to pure torsion or when $T_u > M_u$, the beam will fail about the axis of least flexural rigidity, i.e., by lateral bending. The compression zone will lie on the side of the beam and the beam fails in Mode 2 or by *lateral bending failure* (Fig. 6.10(c)).

In beams having nearly square cross-section, with top steel less than the bottom steel, the negative flexural strength will be less and hence governs the failure mode. When such beams are subjected to pure torsion or when $T_u > M_u$, the beam will fail with compression zone lying at the bottom of the beam. Even though the applied bending moment is positive (sagging), the torsion has resulted in a negative bending failure. The beam is said to have failed in Mode 3 or by a *negative bending failure* (Fig. 6.10(d)).

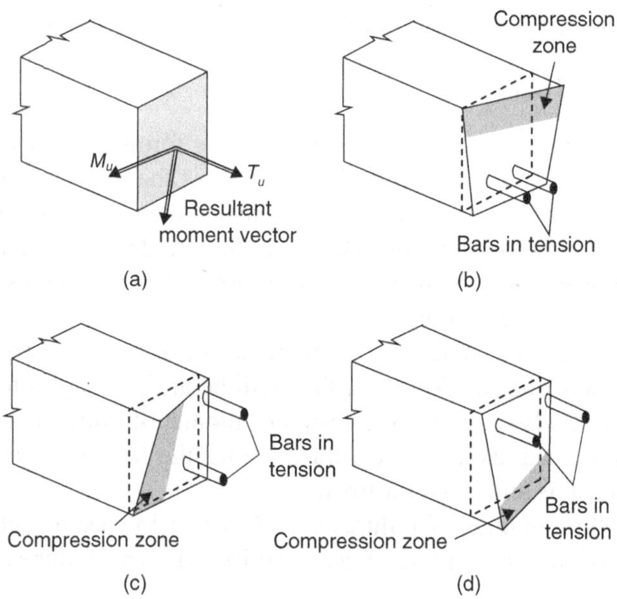

Fig. 6.10 Failure modes for combined flexure and torsion (a) rectangular beam section subjected to bending and torsion, (b) mode 1 (modified bending failure, (c) mode 2 (lateral bending failure) and (d) mode 3 (negative bending failure)

The ultimate torsional capacity of beams for the three failure modes can be determined using the expressions given in Rangan and Hall[3, 4]. The lowest among them will be the torsional strength of the beam.

The presence of shear affects the bending-torsion failure. A small shear will change only the shape of the failure surface without significantly affecting the capacity of the section. If the shear is large, the shear stress in the compression zone may lead to premature failure of the member (known as *shear compression failure*).

6.3.4 Design Provisions for Torsion as per IS:1345–1980

The Code provisions (Cl. 22.5) are applicable for beams having the following cross-sections:

- Solid rectangular sections with $D > b$ (D and b are the depth and width of the section).
- Hollow rectangular sections with $D > b$ and $t \geq b/4$ (t is the wall thickness).
- Flanged sections ($b = b_w$)

In all these cases, f_{cp} should not exceed $0.3 f_{ck}$.

Design of Longitudinal Reinforcement

The Code (Cl. 22.5.3) recommends simplified procedure based on skew bending theory for the design of longitudinal reinforcement in beams subjected to flexure and torsion. The factored torsional moment T_u is converted into an effective bending moment M_t defined by the equation:

$$M_t = T_u \sqrt{1 + \left(\frac{2D}{b} \right)} \tag{6.13}$$

M_t thus obtained is added with the factored bending moment M_u (for Mode 1 and Mode 3) to get the equivalent ultimate bending moment M_e. The longitudinal reinforcement is provided to resist M_e.

Equivalent Ultimate Bending Moment for Mode 1 Failure (M_{e1})

For Mode 1 failure,

$$M_{el} = M_u + \mathrm{M}_t \tag{6.14}$$

The longitudinal reinforcement designed for M_{e1} is provided in the flexural tension zone of the beam.

Equivalent Ultimate Bending Moment for Mode 3 Failure (M_{e2})

When M_u is less than M_t, the beam is designed for M_{e2} given by:

$$M_{e2} = M_t - M_u \tag{6.15}$$

The longitudinal reinforcement for M_{e2} is provided at the flexural compression zone of the beam.

Equivalent Ultimate Bending Moment for Mode 2 Failure (M_{e3})

When $M_u \leq M_t$, the beam is designed to withstand a equivalent transverse moment M_{e3} given by:

$$M_{e3} = M_t\left(1+\frac{x_1}{2e}\right)^2\left[\frac{1+\dfrac{2b}{D}}{1+\dfrac{2D}{b}}\right] \tag{6.16}$$

when x_1 is the smaller dimension of the closed stirrup and e is the ratio of T_u and M_u ($e = T_u/M_u$).

The moment M_{e3} acts in a direction at right angles to M_u.

Design of Transverse Reinforcement

The torsional moment at failure (T_u) can be written as the sum of torsion resisted by concrete (T_{uc1}) and stirrups (T_{us}). Similarly, shear force at failure (V_u) is the sum of shear force taken by concrete (V_{uc1}) and stirrups (V_{us}). Hence,

$$T_u = T_{uc1} + T_{us} \tag{6.17a}$$

$$V_u = V_{uc1} + V_{us} \tag{6.17b}$$

Evaluation of T_{uc1} and V_{uc1}

In the presence of flexural shear, the torsional capacity of concrete reduces from T_{uc} to T_{uc1}. Similarly, due to torsion, the flexural shear capacity of concrete decreases from V_{uc} to V_{uc1}. To find T_{uc1} and V_{uc1}, a linear interaction is assumed, i.e.,

$$\frac{T_{uc1}}{T_{uc}} + \frac{V_{uc1}}{V_{uc}} = 1 \tag{6.18}$$

where,

T_{uc} = torsional resisted by concrete when $V_u = M_u = 0$, i.e., T_{uc} is the capacity of concrete under pure torsion.

V_{uc} = shear resisted by concrete when $T_u = 0$ ($M_u \neq 0$). V_{uc} is the ultimate shear resistance of concrete, which is taken as the minimum of V_{co} and V_{cr} (see Section 6.2.4).

For a given beam, the ratio of T_u and V_u (e) is known. It is assumed that the ratio of T_{uc1} and V_{uc1} is also equal to e. Hence,

$$e = \frac{T_u}{V_u} = \frac{T_{uc1}}{V_{uc1}}$$

$$\Rightarrow \qquad V_{uc1} = \frac{T_{uc1}}{e}$$

Substituting V_{uc1} in Eq. (6.18) gives:

$$T_{uc1} = T_{uc}\left(\frac{e}{e + T_{uc}/V_{uc}}\right)$$

Introducing the notation e_c as T_{uc} / V_{uc} gives:

$$T_{uc1} = T_{uc}\left(\frac{e}{e + e_c}\right) \tag{6.19}$$

From Eq. (6.18),

$$\frac{V_{uc1}}{V_{uc}} = 1 - \frac{T_{uc1}}{T_{uc}} = 1 - \left(\frac{e}{e + e_c}\right) = \frac{e_c}{e + e_c}$$

$$\Rightarrow \qquad V_{uc1} = V_{uc}\left(\frac{e_c}{e + e_c}\right) \tag{6.20}$$

Capacity of Concrete Under Pure Torsion (T_{uc})

T_{uc} can be determined using the *sand-heap analogy*. If sand is poured over the cross-section, a heap is formed with a constant slope on all faces. According to sand-heap analogy, twice the volume of sand-heap is the ultimate torque, and the slope represents the shear stress due to torsion (τ_t).

Consider a rectangular section of width b and depth D in plastic condition. The sand-heap for this section is shown in Fig. 6.11.

Volume of sand heap $V = \frac{1}{2}bH(D - b) + \frac{1}{3}b^2H = \frac{bH}{2}[D - (b/3)]$

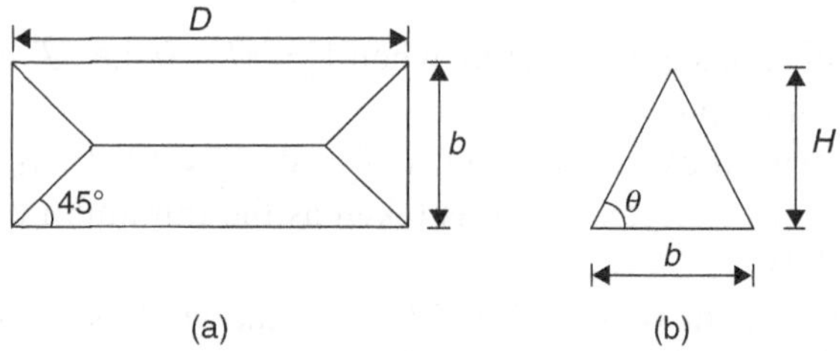

Fig. 6.11 Sand-heap for rectangular section (a) plan and (b) end view

Slope
$$\theta = \tau_t = \frac{2H}{b}$$
$$\Rightarrow H = b\tau_t/2$$

Substituting H in the equation for volume gives:

$$V = \frac{\tau_t b^2}{4}[D - (b/3)]$$

Since the ultimate torque T_{uc} is equal to twice the volume,

$$T_{uc} = 2V = \frac{\tau_t b^2}{2}[D - (b/3)] \tag{6.21}$$

At the time of failure, τ_t is equal to f_t, the tensile strength of concrete ($f_t = 0.24\sqrt{f_{ck}}$ N/mm^2). But for a material like concrete, the assumption of full plastification of the section used in sand-heap analogy cannot be justified and will cause a over estimation of T_{uc}. To avoid this, a reduced value for τ_t is to be used to find T_{uc}. τ_c can be taken as $0.2\sqrt{f_{ck}}$ N/mm^2. Hence,

$$T_{uc} = 0.2\sqrt{f_{ck}}\frac{b^2}{2}[D - (b/3)]$$
$$= 0.1b^2D\left[1 - \left(\frac{b}{3D}\right)\right]\sqrt{f_{ck}} \tag{6.22}$$

A flanged section can be considered to be made by combining small rectangular sections (Fig. 6.12). T_{uc} for such sections can be determined by adding the torque capacity of individual rectangular sections.

$$T_{uc} = \sum 0.1b^2D\left[1 - \left(\frac{b}{3D}\right)\right]\sqrt{f_{ck}} \tag{6.23}$$

In the above equation, b and D are lengths of the short side and long side of the individual rectangle.

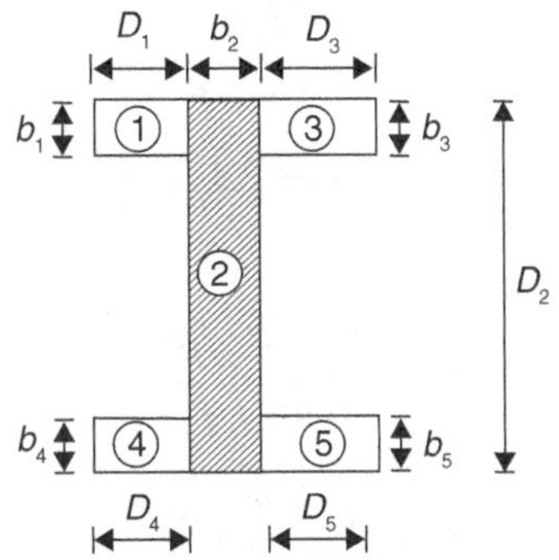

Fig. 6.12 Flanged section

Prestressing concrete increases the torsional resistance. If f_{cp} is the average prestress, it can be shown that[5] T_{uc} is increased by a factor λ_p given by:

$$\lambda_p = \sqrt{1+\left(\frac{12f_{cp}}{f_{ck}}\right)} \tag{6.24}$$

Hence the capacity of prestressed concrete section to resist pure torsion is:

$$T_{uc} = \sum 0.1b^2D\left[1-\left(\frac{b}{3D}\right)\right]\lambda_p\sqrt{f_{ck}} \tag{6.25}$$

Comments

The equation to find T_{uc} recommended by the Code (Cl. 22.5.4.1) is:

$$T_{uc} = \sum 1.5b^2D\left[1-\left(\frac{b}{30}\right)\right]\lambda_p\sqrt{f_{ck}}$$

The coefficient 1.5 appearing in the equation can be used if imperial units are followed and not applicable for SI unit system[6]. Similarly the factor $\left(1-\frac{b}{30}\right)$ should be replaced by $\left(1-\frac{b}{3D}\right)$. Due to the above reasons, Eq. (6.25) can be used to find T_{uc}, instead of the Code recommendations.

Transverse Reinforcement (Cl. 22.5.4.3 of the Code)

The area of cross-section (A_{sv}) of the closed rectangular stirrup is taken as the larger of the values given by Eqs. (6.26) and (6.27).

---

[5] Hsu, T.T.C., "Torsion of Structural Concrete – Uniformly Prestressed Rectangular Members without Web Reinforcement", *PCI Journal*, April 1968, pp. 34–44.

[6] Pandit, G.S. and Gupta, S.P., *Prestressed Concrete,* CBS Publishers, 2008.

$$A_{sv} = \frac{M_t s_v}{1.5 b_1 d_1 f_v} \qquad (6.26)$$

$$A_{sv} = A_v + 2A_T \qquad (6.27)$$

where,

$$A_v = \frac{(V_u - V_{uc1}) s_v}{0.87 d_1 f_v} \qquad (6.28)$$

$$A_T = \frac{(T_u - T_{uc1}) s_v}{0.87 b_1 d_1 f_v} \qquad (6.29)$$

In the above equations, b_1 and d_1 are the centre to centre distance between corner bars in the direction of width and depth of the section.

Based on experimental studies[3], it is found that:

$$M_t = \Omega\, A_{sv}\, f_v \frac{b_1 d_1}{s_v}$$

where Ω is a factor which vary between 1 and 2. Eq. (6.26) is obtained by taking a average value of 1.5 for Ω.

Eq. (6.27) is obtained based on the concept of total shear, i.e., the reinforcement provided for torsion must be combined with that required for shear. For a two-legged stirrup, A_{sv} is the area of stirrup (both legs) for resisting both shear and torsion. A_v is the area of stirrup (both legs) for resisting shear and A_T is the area of stirrup (one leg) for resisting torsion (refer addition of shear stress due to V and T at face ABCD of the beam in Fig. 6.8). Equations for finding A_v and A_T are derived using the truss model for concrete beam.

Minimum Transverse Reinforcement

The Code (Cl. 22.5.4.4) recommends to provide a minimum amount of area of stirrup A_{sv} such that:

$$\frac{A_{sv}}{b s_v} = \frac{0.4}{0.87 f_v} \qquad (6.30)$$

6.3.5 Detailing of Torsion Reinforcement

The detailing recommendations given in the Code (Cl. 22.5.5) are listed below:

- There should be at least one longitudinal bar of diameter not less than 12mm at each corner of the stirrup.

- The web reinforcement should be in the form of closed stirrups perpendicular to the axis of the member.
- The spacing of the stirrups should not exceed the least of $(x_1 + y_1)/4$ and 200 mm. Where x_1 and y_1 are the short and long dimensions of stirrup.
- Torsional reinforcement should be continued till a distance of $(D + b)$ beyond the theoretical cut off point.

6.3.6 Steps for Torsion Design

The design steps to be followed are given below:

(i) Determine M_u, V_u and T_u at the required section.

(ii) Check whether the Code provisions can be used for the given section.

Design for Longitudinal Reinforcement

(iii) Calculate M_t and hence M_{e1}.

(iv) Find the area of steel to resist M_{e1} and provide in the flexural tension zone.

(v) If $M_t > M_u$, then the beam can also fail by Mode 2 or by Mode 3. Hence find M_{e2}, and provide longitudinal steel for M_{e2} in the flexural compression zone. Also check whether the lateral flexural strength satisfies M_{e3}. Else provide adequate side face reinforcement.

Design for Transverse Reinforcement

(vi) Calculate T_{uc}, V_{uc}, T_{uc1} and V_{uc1}.

(vii) Find A_{sv} and check whether it is greater than the requirement for minimum transverse reinforcement.

(viii) Find the spacing of stirrups.

(ix) Check whether the detailing requirements are satisfied.

Example 6.2

A post-tensioned bonded prestressed concrete beam of rectangular cross-section 400 mm wide and 550 mm deep is subjected to service load bending moment of 250 kN-m, shear force of 100 kN and torsional moment of 40 kN-m. An effective prestress of 700 kN is applied at an eccentricity of 150 mm. The area of prestressing steel is 700 mm^2 with an ultimate strength of 1840 N/mm^2. The compressive strength of concrete is 40 N/mm^2. Design suitable longitudinal and transverse reinforcement.

Solution:

Step 1: *Calculation of M_u, V_u and T_u*

$M_u = 1.5\ M = 1.5 \times 250 = 375$ kN-m

$V_u = 1.5\ V = 1.5 \times 100 = 150$ kN-m

$T_u = 1.5\ T = 1.5 \times 40 = 60$ kN-m

Step 2: *Checking the applicability of Code recommendations*

- Average intensity of prestress f_{cp}

  Effective prestressing force $P_e = 700\ kN$

  Cross-sectional area of beam $A = bD = 400 \times 550 = 220000$ mm²

  $$f_{cp} = \frac{P_e}{A} = \frac{700 \times 10^3}{220000} = 3.18 \text{ N/mm}^2$$

  Maximum value of $f_{cp} = 0.3 f_{ck} = 0.3 \times 40 = 12$ N/mm²

  Since f_{pe} is less than 12 N/mm², and the depth of the beam is greater than the width, the Code recommendations can be used for the design of the beam.

Step 3: *Calculation of M_t and M_{e1}*

- Effective bending moment M_t

  $$M_t = T_u \sqrt{1 + \left(\frac{2D}{b}\right)}$$

  $$= 60\sqrt{1 + \left(\frac{2 \times 550}{400}\right)} = 116.19 \text{ kN-m}$$

- Equivalent ultimate bending moment for Mode 1 failure (M_{e1})

  $$M_{e1} = M_u + M_t = 375 + 116.19 = 491.19 \text{ kN-m}$$

Step 4: *Calculation of area of steel for M_{e1}*

- Ultimate moment of resistance of section M_{uR}

  Area of prestressing steel $A_p = 700$ mm²

  Tendon eccentricity $e_s = 150$ mm

  Effective depth $d = D/2 + e_s = 425$mm

  $$\text{Effective prestress } f_{pe} = \frac{P_e}{A_p} = \frac{700 \times 10^3}{700} = 1000 \text{ N/mm}^2$$

  $$\frac{f_{pe}}{f_p} = \frac{1000}{1840} = 0.54 \ (> 0.45, \text{OK})$$

$$\omega_p = \frac{A_p f_p}{bd\, f_{ck}} = \frac{700 \times 1840}{400 \times 425 \times 40} = 0.19$$

From Table 11 of the Code,

For $\omega_p = 0.19$; $\dfrac{f_{pu}}{0.87 f_p} = 0.96$, $\dfrac{x_u}{d} = 0.394$

$$f_{pu} = 0.96 \times 0.87 f_p = 0.96 \times 0.87 \times 1840 = 1536.77 \text{ N/mm}^2$$

$$x_u = 0.394\, d = 0.394 \times 425 = 167.45 \text{ mm}$$

$$M_{uR} = f_{pu} A_p (d - 0.42\, x_u)$$

$$= 1536.77 \times 700 \times [425 - (0.42 \times 167.45)]/10^6$$

$$= 381.53 \text{ kN-m } (< M_{e1})$$

Hence additional steel reinforcement is necessary to resist M_{e1}.

- Balance moment $M_{u,\,b}$

$$M_{u,\,b} = M_{e1} - M_{uR} = 491.19 - 381.53 = 109.66 \text{ kN-m}$$

This balance moment is resisted by providing untensioned steel reinforcement.

- Area of steel reinforcement A_{st} for $M_{u,\,b}$

$$M_{ub} = 0.87 f_y A_{st} d_t \left[1 - \left(\frac{A_{st} f_y}{b d_t f_{ck}}\right)\right]$$

Assume $f_y = 415$ N/mm^2 and $d_t = 500$ mm. Hence,

$$109.66 \times 10^6 = 0.87 \times 415 \times A_{st} \times 500 \left[1 - \left(\frac{A_{st} \times 415}{400 \times 500 \times 40}\right)\right]$$

Solving the equation gives:

$$A_{st} = 627.47 \text{ mm}^2$$

Provide two reinforcing bars of diameter 20 mm ($A_{st} = 628$ mm^2). Also provide two corner bars of 20 mm diameter at the top of the beam.

Step 5: *Check for Mode 2 and Mode 3 failure*

Since $M_t < M_{e1}$, the beam will not fail in Mode 2 or Mode 3.

Step 6: *Calculation of T_{uc}*

$$\lambda_p = \sqrt{1 + \left(\frac{12 f_{cp}}{f_{ck}}\right)} = \sqrt{1 + \left(\frac{12 \times 3.18}{40}\right)} = 1.4$$

$$T_{uc} = 0.1b^2D\left[1-\left(\frac{b}{3D}\right)\right]\lambda_p\sqrt{f_{ck}}$$

$$= 0.1\times 400^2 \times 550\times\left[1-\left(\frac{400}{3\times 550}\right)\right]\times 1.4\times\sqrt{40}/10^6$$

$$= 59 \text{ kN-m}$$

Step 7: *Calculation of V_{uc}*

- Shear capacity of concrete uncracked in flexure V_{co}

  Tensile strength of concrete $f_t = 0.24\sqrt{f_{ck}} = 0.24\sqrt{40} = 1.52$ N/mm²

$$V_{co} = 0.67bD\sqrt{f_t^2 + 0.8f_{cp}f_t}$$

$$= 0.67\times 400\times 550\times\sqrt{1.52^2 + (0.8\times 3.18\times 1.52)}/10^3$$

$$= 366.35 \text{ kN}$$

- Shear capacity of concrete cracked in flexure V_{cr}

  Moment of inertia of the cross-section $I = \dfrac{bD^3}{12}$

$$I = \frac{400\times 550^3}{12} = 55.458\times 10^8 \text{ mm}^4$$

Stress in concrete at the level of tendon due to prestress f_{pt}:

$$f_{pt} = \frac{P_e}{A} + \frac{(P_e e_s)e_s}{I}$$

$$= \left(\frac{700\times 10^3}{220000}\right) + \left(\frac{700\times 10^3\times 150\times 150}{55.458\times 10^8}\right) = 6.02 \text{ N/mm}^2$$

$$M_o = \frac{0.8f_{pt}I}{e_s} = \frac{0.8\times 6.02\times 55.458\times 10^8}{150\times 10^6} = 178.06 \text{ kN-m}$$

Since longitudinal steel reinforcement consists of both prestressed and nonprestressed reinforcement, an equivalent area of prestressing steel ($A_{p,e}$) is determined to find the percentage of prestressing steel.

$$A_{p,e} = A_p + A_{st}(f_y/f_p)$$

$$= 700 + \left[\frac{628\times 415}{1840}\right] = 841.64 \text{ mm}^2$$

Percentage of prestressing steel $p_p = 100\dfrac{A_{p,e}}{bd} = \dfrac{100 \times 841.64}{400 \times 425} = 0.5$

From Table 6 of the Code, for M 40 grade concrete and p_p = 0.5, $\tau_c = 0.51 \text{N/mm}^2$

$$\frac{f_{pe}}{f_p} = \frac{1000}{1840} = 0.54 \ (< 0.6, \text{OK})$$

Minimum value of $V_{cr} = (V_{cr})_{min} = 0.1\,bd\sqrt{f_{ck}}$

$$(V_{cr})_{\min} = \frac{0.1 \times 400 \times 425 \times \sqrt{40}}{1000} = 107.52 \text{ kN}$$

$$V_{cr} = \left(1 - 0.55\frac{f_{pe}}{f_p}\right)\tau_c bd + M_o \frac{V_u}{M_u}$$

$$= \left[\frac{(1 - 0.55 \times 0.54) \times 0.51 \times 400 \times 425}{1000}\right] + \left[\frac{178.06 \times 150}{375}\right]$$

$$= 132.17 \text{ kN } (>(V_{cr})_{min})$$

- Ultimate shear capacity of concrete V_{uc}

  V_{uc} is the minimum of V_{co} and V_{cr}.
  Hence, V_{uc} = 132.17 kN

Step 8: *Calculation of T_{ucl} and V_{ucl}*

$$e = \frac{T_u}{V_u} = \frac{60}{150} = 0.4$$

$$e_c = \frac{T_{uc}}{V_{uc}} = \frac{59}{132.17} = 0.45$$

$$T_{ucl} = T_{uc}\frac{e}{(e + e_c)} = 59 \times \left(\frac{0.4}{0.4 + 0.45}\right) = 27.76 \text{ kN-m}$$

$$V_{ucl} = V_{uc}\frac{e_c}{(e + e_c)} = 132.17 \times \left(\frac{0.45}{0.4 + 0.45}\right) = 69.97 \text{kN-m}$$

Step 9: *Calculation of transverse reinforcement*

- Calculation of A_{sv} / s_v

  Assuming $f_v = 415 \text{ N/mm}^2$ and effective cover to corner reinforcing bars as 50 mm (Fig. 6.13).

$$\frac{A_{sv}}{s_v} = \frac{M_t}{1.5b_1d_1f_v} = \frac{116.19\times10^6}{1.5\times300\times450\times415} = 1.38 \text{ mm}^2/\text{mm} \tag{6.30}$$

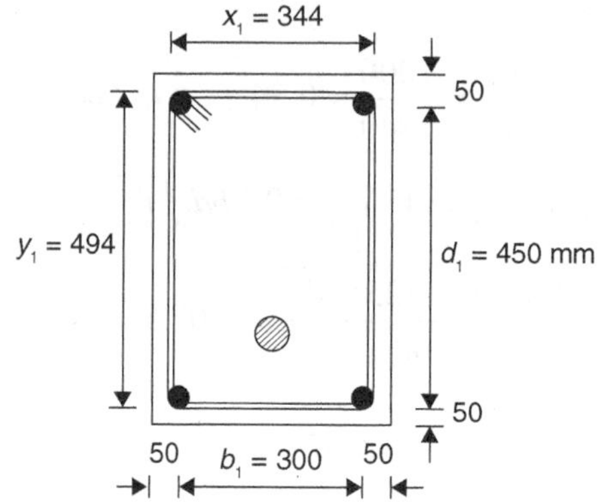

Fig. 6.13 Cross-section of prestressed concrete beam

$$\frac{A_v}{s_v} = \frac{(V_u - V_{uc1})}{0.87d_1f_v} = \frac{(150-69.97)\times10^3}{0.87\times450\times415} = 0.49 \text{ mm}^2/\text{mm}$$

$$\frac{A_T}{s_v} = \frac{(T_u - T_{uc1})}{0.87bd_1f_v} = \frac{(60-27.76)\times10^6}{0.87\times300\times450\times415} = 0.66 \text{ mm}^2/\text{mm}$$

$$\frac{A_{sv}}{s_v} = \frac{A_v}{s_v} + 2\frac{A_T}{s_v} = 0.49 + (2\times0.66) = 1.81 \text{ mm}^2/\text{mm} \tag{6.31}$$

Using minimum transverse reinforcement requirement, $\frac{A_{sv}}{s_v} = \frac{0.4b}{0.87f_v}$

$$\Rightarrow \qquad \frac{A_{sv}}{s_v} = \frac{0.4\times400}{0.87\times415} = 0.44 \text{ mm}^2/\text{mm} \tag{6.32}$$

$\frac{A_{sv}}{s_v}$ is taken as the maximum of values given in Eqs. (6.30–6.32). Hence,

$$\frac{A_{sv}}{s_v} = 1.81 \text{ mm}^2/\text{mm}$$

- Calculation of required spacing of stirrups s_v

  Try 2 legged stirrups of 12 mm diameter.

$$A_{sv} = 2\times\frac{\pi}{4}\times12^2 = 226 \text{ mm}^2$$

Required spacing of stirrups $s_v = \frac{A_{sv}}{1.81} = \frac{226}{1.81} = 124.86 \text{ mm}^2$

- Maximum spacing of stirrups $s_{v,m}$

  Diameter of longitudinal bar $\phi_l = 20$ mm

  Diameter of stirrups $\phi_s = 12$ mm

  From Fig. 6.13,

$$x_1 = b_1 + \phi_l + 2\phi_s = 300 + 20 + (2 \times 12) = 344 \text{ mm}$$
$$y_1 = d_1 + \phi_l + 2\phi_s = 450 + 20 + (2 \times 12) = 494 \text{ mm}$$
$$(x_1 + y_1)/4 = (344 + 494)/4 = 209.5 \text{ mm}$$

  $s_{v,m}$ is the minimum of $(x_1 + y_1)/4$ and 200 mm

  Hence $s_{v,m} = 200$ mm

  Provide 2 legged stirrups of 12 mm diameter at 120 mm c/c spacing.

Problems

6.1 A beam of rectangular cross-section 250 × 500 mm is prestressed with an effective force of 350 kN at a eccentricity of 150 mm. Design the shear reinforcement at a section uncracked in flexure to resist a ultimate shear force of 180 kN. The grade of concrete is M 50.

6.2 Design a rectangular section to carry the following ultimate loads: $M_u = 300$ kN-m, $V_u = 150$ kN and $T_u = 90$ kN-m.

Data:

$$f_{ck} = 40 \text{ N/mm}^2, f_{pe} = 0.6 f_p, f_p = 1800 \text{ N/mm}^2.$$

6.3 The cross-section of a simply supported prestressed concrete beam is shown in Fig. 6.14. The beam is subjected to a factored shear force of 500 kN, factored bending moment of 80 kN-m and a factored tosional moment of 30 kN-m at the critical section. Design the required reinforcements to resist the shear force and torsion.

Data:

$$f_{ck} = 50 \text{ N/mm}^2, f_{pe} = 960 \text{ N/mm}^2, f_p = 1860 \text{ N/mm}^2, A_p = 850 \text{ mm}^2$$

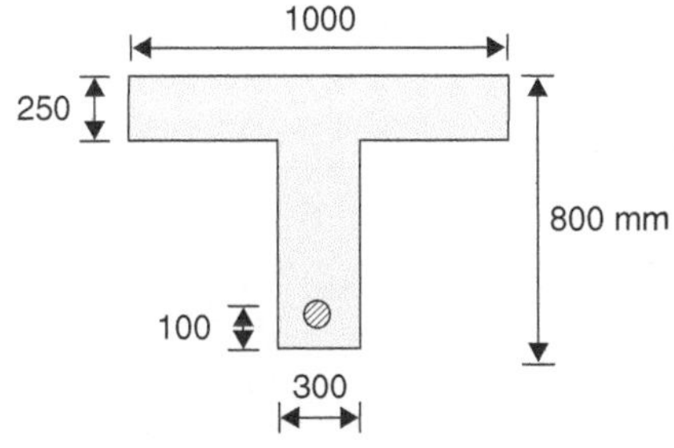

Fig. 6.14

chapter 7

Anchorage Zones

7.1 Introduction

The force in the prestressing steel is transferred to concrete either by bond in the case of pre-tensioned members or through bearing-type anchorages in post-tensioned members. In either case, the compression force is transferred to concrete in a concentrated manner (usually at the ends of the member).

Due to these concentrated forces, a very complex state of stress is developed, and in these regions the assumption of plane section remaining plane after loading will not be valid. Hence the stress distribution cannot be determined using the concepts of Strength of Materials (Sections 1 and 2 in Fig. 7.1(a)). These regions where complex stress distributions are developed due to transfer of prestressing force are known as *anchorage zones*.

But beyond a certain distance from the concentrated force, the complex stress distribution vanishes and the stress can be determined using the Strength of Materials approach (Section 3 in Fig. 7.1(a)). This aspect can be used to define the extent of anchorage zone.

In the anchorage zones, transverse tensile stress are developed (Fig. 7.1(c)), which causes longitudinal cracks in the member. Design and detailing of reinforcement to control there cracks is a very important part in the design of prestressed concrete members.

Due to a gradual transfer of prestress by bond, the anchorage zone in a pre-tensioned member will be less critical than in a post-tensioned member. Also in pre-tensioned member, usually the tendons are relatively smaller and well distributed over the cross-section. Further, the large bearing stresses

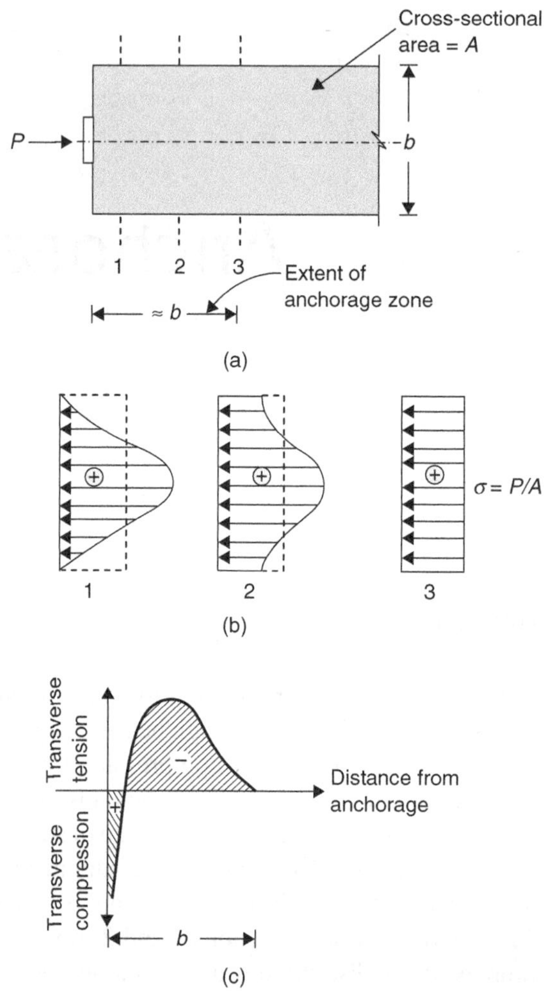

Fig. 7.1 Stress distribution near concentrated loads (a) member subjected to concentrated load P, (b) longitudinal stress distribution at different vertical sections and (c) transverse stress distribution (Section along centroid axis)

behind the anchorage or bearing plates in post-tensioned member are absent in pre-tensioned member.

7.2 Anchorage Zones in Pre-tensioned Members

In a pre-tensioned member, the tendon is initially tensioned between the end anchors and the concrete is cast around the tendon. Once concrete has gained sufficient strength, the end anchorages are released and the prestressing force

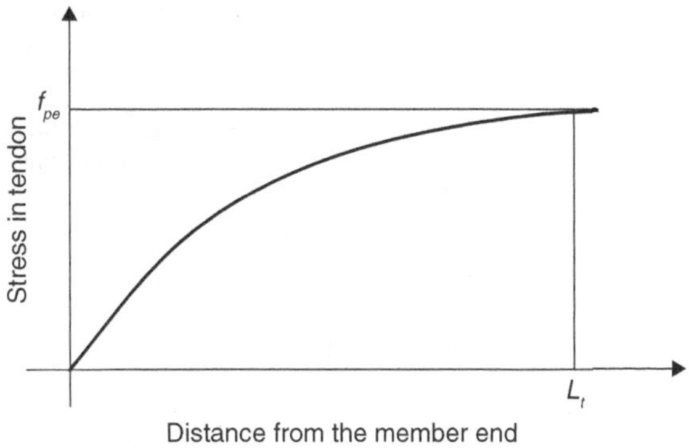

Fig. 7.2 Variation of stress in tendon along transmission length

is transferred to concrete through the bond between concrete and steel. The transfer of prestress occurs usually at the end of the member with stress in the steel varying from zero at the end to the full effective prestress (f_{pe}) at some distance away from the end (Fig. 7.2). This distance over which the transfer of prestress occurs is known as the *transmission length* (L_t).

7.2.1 Bond Mechanisms

There are three bond mechanisms responsible for transfer of prestress from the tendon to concrete, namely—chemical adhesion, Hoyer effect (or Wedge action) and mechanical interlock. These bond mechanisms are briefly discussed below.

Chemical Adhesion

Adhesion refers to the bond between two different materials, in this case concrete and steel. It should be noted that adhesion does not contribute to bond once the slip between steel and concrete takes place. But at transfer, the tendon slips relative to concrete, and hence the contribution of this mechanism is usually neglected.

Hoyer Effect (Wedge Action)

Due to Poisson's effect, tensioning the tendon causes a reduction in its diameter. When the tendon is released, the stress in tendon at the member ends is zero. Hence at these locations, the tendon regains its original diameter. But at some distance from the member end, since the prestress is not equal to zero, the tendon can not fully recover its original diameter. Thus within the transmission length, there is a variation of tendon diameter, and there is also radial pressure exerted by the surrounding concrete. This radial pressure causes development of friction force, which helps to

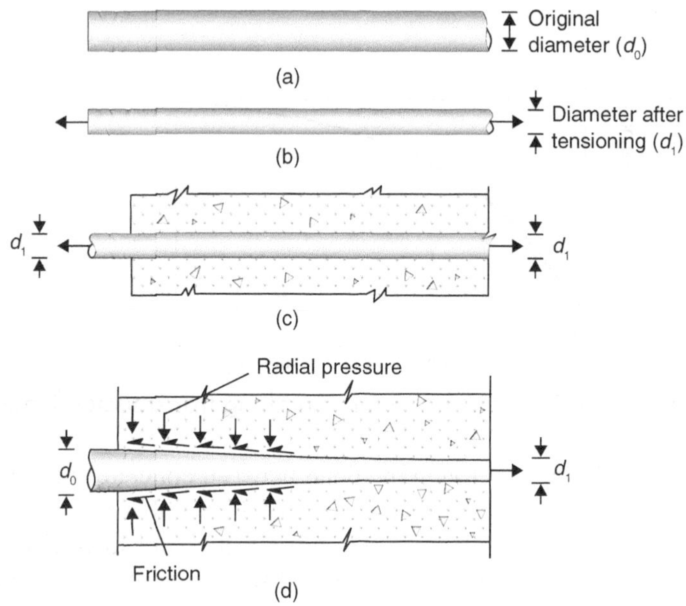

Fig. 7.3 Hoyer effect (a) unstressed tendon, (b) stretching the tendon, (c) casting of concrete and (d) at transfer stage

transfer the force from steel to concrete. This wedge effect is known as Hoyer effect. The different stages in the development of Hoyer effect are depicted in Fig. 7.3.

Mechanical Interlock

This mechanism is similar to the pullout resistance offered by deformed reinforced bars due to the surface protrusions in these bars.

7.2.2 Transmission Length

Transmission length (L_t) can be defined as the length of bond required to develop effective prestress (f_{pe}) in the tendon. Alternatively, the transmission length can be defined as the length of bond required to transfer the effective prestress from the tendon to concrete.

The transmission length depends on a number of factors like the compressive strength of concrete at transfer, size and type of tendon, surface condition of tendon, degree of concrete compaction in the anchorage zone, method of release of prestressing force etc. In the absence of test data, the Code recommendations (Cl. 18.6.1.1) can be used to find L_t (Table 7.1). These recommendations can be used in the case of well compacted concrete having compressive strength more than 35 N/mm^2 at transfer, and when the

Table 7.1 Transmission length as per IS: 1343–1980

Type of Tendon	Lt
Plain and indented wires	100 ϕ
Crimped wires	65 ϕ
Strands	30 ϕ

tendons are gradually released. Further, the recommended values for L_t are valid for wires of diameter up to 5 mm and strands of diameter not exceeding 18 mm. In the table, ϕ is the nominal diameter of tendon.

In a prestressed concrete member, the sections having high moments should be located beyond transmission length so that at these sections, the stress in the tendon is at least equal to f_{pe}. To avoid transmission length in the clear span of the beam, the Code (Cl. 18.6.1.1) suggests the following guidance for beams having different support conditions.

i) In the case of simply supported beam, the overhang beyond the support should be at least equal to $L_t/2$ (Fig. 7.4(a)).

ii) At fixed supports, the length of fixity should be at least equal to L_t (Fig. 7.4(b)).

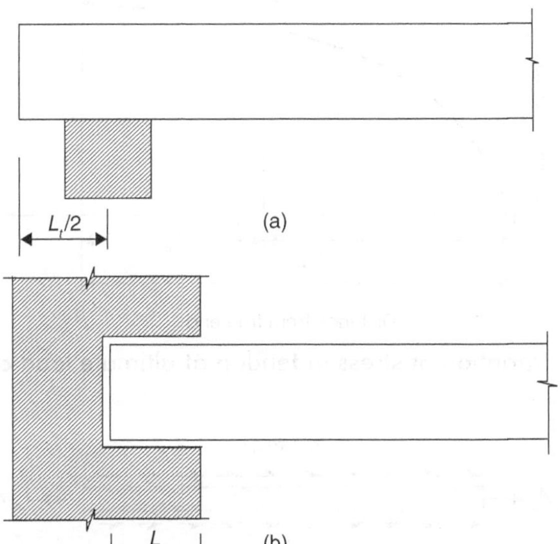

Fig. 7.4 End conditions for pre-tensioned beam (a) simply supported ends and (b) fixed ends

7.2.3 Flexural Bond Length

The flexural bond length (L_b) is the length of bond over which the stress in the tendon can increase from the effective prestress (f_{pe}) to ultimate prestress (f_{pu}). The variation of prestress in tendon at the ultimate load condition is shown in Fig. 7.5.

An equation to find L_b can be derived by considering the free body diagram of the tendon in the flexural bond length (Fig. 7.6).

In the figure, τ_{bd} is the average design bond stress, A_p is the cross-sectional area of tendon of diameter $\phi\left(A_p = \frac{\pi}{4}\phi^2\right)$. τ_{bd} can be obtained from Cl. 26.2.1.1 of IS 456:2000*.

From the free body diagram, the force equilibrium in the horizontal direction gives:

$$(f_{pu} - f_{pe})\frac{\pi}{4}\phi^2 = \tau_{bd}\,\pi\phi L_b$$

$$\Rightarrow \qquad L_b = \frac{(f_{pu} - f_{pe})\phi}{4\tau_{bd}} \tag{7.1}$$

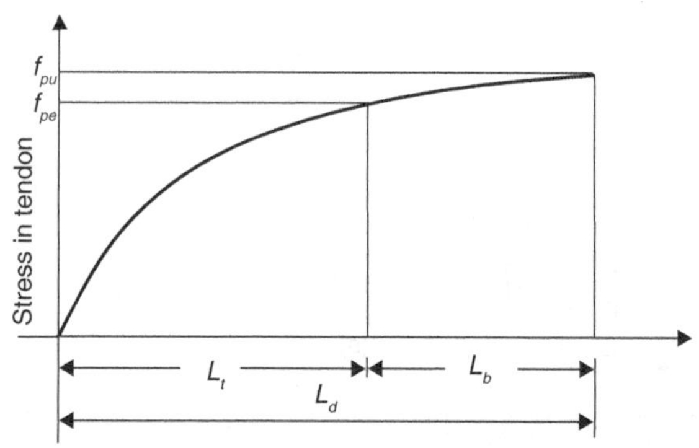

Fig. 7.5 Variation of stress in tendon at ultimate load condition

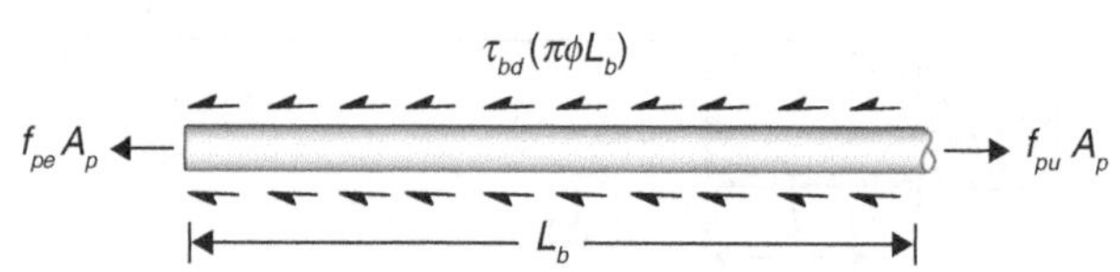

Fig. 7.6 Free body diagram of tendon

* Indian standard, plain and reinforced concrete code of practice.

7.2.4 Development Length

Development length (L_d) is the length of bond required to develop ultimate prestress (f_{pu}) in tendon. L_d is the sum of L_t and L_b (Fig. 7.5).

$$L_d = L_t + L_b \tag{7.2}$$

In order to prevent bond failure, the available bond length at the critical section (usually the section of maximum bending moment) should be greater than or equal to L_d.

7.2.5 End Zone Reinforcement

To control longitudinal cracks developed due to transverse tensile stresses (Fig.7.1(c)), transverse reinforcement in the form of stirrups are provided at the ends of the member. These reinforcements are known as the end zone reinforcement.

Based on experimental studies, Marshall and Mattock[1] developed an empirical equation to find the total amount of transverse reinforcement A_{st} (in the form of stirrups) as:

$$A_{st} = 0.021\frac{D}{L_t}\frac{P_o}{f_s} \tag{7.3}$$

where D is the depth of member, P_o is the prestressing force after immediate losses and f_s is the permissible stress in steel which can be taken as 150 N/mm^2. The transverse reinforcement should be equally distributed within a distance of 0.2D from the end face of the member.

Example 7.1

Design the end zone reinforcement for the pre-tensioned beam (Fig. 7.7) if the prestressing force after immediate losses is 1500 kN. The diameter of prestressing wires is 5 mm.

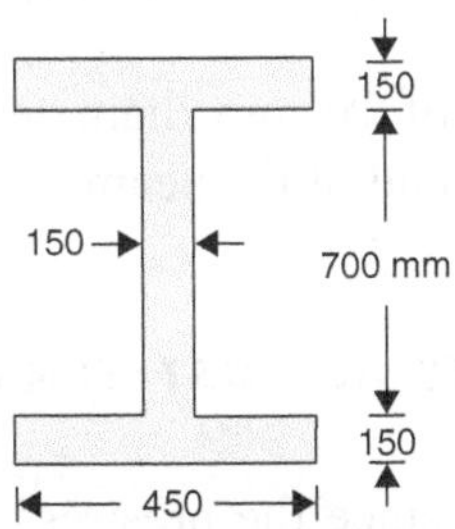

Fig. 7.7 Cross-section of beam

[1] Marshall, W.T. and Mattock, A.H., "Control of Horizontal Cracking in the Ends of Pre-tensioned Prestressed Concrete Girders", *PCI Journal*, October 1962, pp. 56–74.

Solution:

Step 1: *Calculation of transmission length (L_t)*

Diameter of tendon $\phi = 5$ mm

$L_t = 100\ \phi = 100 \times 5 = 500$ mm

Step 2: *Calculation of transverse reinforcement*

Depth of the member $D = 1000$ mm

Prestressing force after immediate losses $P_o = 1500$ kN

Permissible stress in steel $f_s = 150$ N/mm^2

- Total area of transverse reinforcement A_{st}

$$A_{st} = 0.021 \frac{D}{L_t} \frac{P_o}{f_s}$$

$$= \frac{0.021 \times 1000 \times 1500 \times 10^3}{500 \times 150} = 420 \text{ mm}^2$$

- Total cross-sectional area of stirrup legs A_{sv}

  Assuming 2 legged 10 mm stirrups,

$$A_{sv} = 2 \times \frac{\pi}{4} \times 10^2 = 157.08 \text{ mm}^2$$

- Number of stirrups (n)

$$n = \frac{A_{st}}{A_{sv}} = \frac{420}{157.08} = 2.7 \text{ say 3 stirrups}$$

- Spacing of stirrups (s_v)

  The stirrups should be kept in a distance of 0.2D from the beam end.

$$s_v = \frac{0.2D}{(n-1)} = \frac{0.2 \times 1000}{(3-1)} = 100 \text{ mm}$$

Provide two legged stirrups of 10 mm diameter at 100 mm c/c within a distance of 200 mm from the end of the beam.

7.3 Anchorage Zones in Post-tensioned Members

In a post-tensioned member, since the prestress is transferred by bearing, the stress in the tendon attains the full prestress at the beginning of the anchorage zone itself. Hence there is no need for transmission length and development length in the case of post-tensioned members. Due to this full

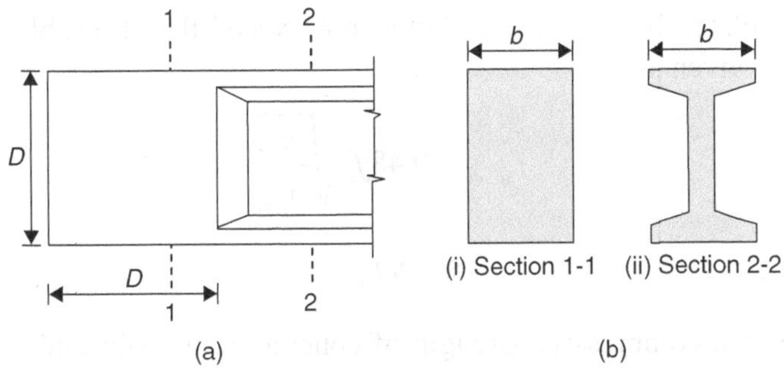

Fig. 7.8 End blocks in post-tensioned beam (a) elevation of beam and (b) cross-section

prestressing force, large stresses are developed in the anchorage zone. To take care of these stresses, the end regions of the flanged beams are usually enlarged to get more area. The cross-section of the beam at these regions will be rectangular in shape (Fig. 7.8). These enlarged regions are known as *end blocks*, which extend to about one member depth from the ends of the member.

The failure of anchorage zones can be due to the large compressive stresses (bearing stresses) developed behind the anchorage or bearing plates, and also due to the transverse tensile forces (bursting force) developed in these regions (Fig. 7.1 (c)). The bearing failure can be avoided by providing bearing plates of adequate dimensions and by increasing the compressive strength of concrete in these regions by using high strength concrete or by confining the concrete (usually done by providing spiral reinforcement). The longitudinal cracks due to bursting force can be controlled by providing sufficient and well distributed transverse reinforcement.

Anchorage zone failure is the most common mode of failure of post-tensioned members during the construction stage. These types of failure are very difficult to repair and usually require the replacement of the member itself.

7.3.1 Bearing Stresses Behind Anchorage

As per Cl. 18.6.2.1 of Code, the bearing stress (f_{br}) is calculated as:

$$f_{br} = \frac{P_e}{A_{pun}} \tag{7.4}$$

where P_e is the effective prestressing force and A_{pun} is the punching area (i.e., the loaded area).

The calculated bearing stress should not exceed the allowable bearing stress ($f_{br,\,all}$) given as:

$$f_{br,all} = 0.48 f_{ci} \sqrt{\frac{A_{br}}{A_{pun}}}$$

$$\leq 0.8 f_{ci} \qquad (7.5)$$

where f_{ci} is the compressive strength of concrete at transfer and A_{br} is the bearing area. A_{br} is taken as the maximum area which is geometrically similar to A_{pun}, with the same centroid and which is possible to inscribe in the cross-sectional area A (Fig. 7.9).

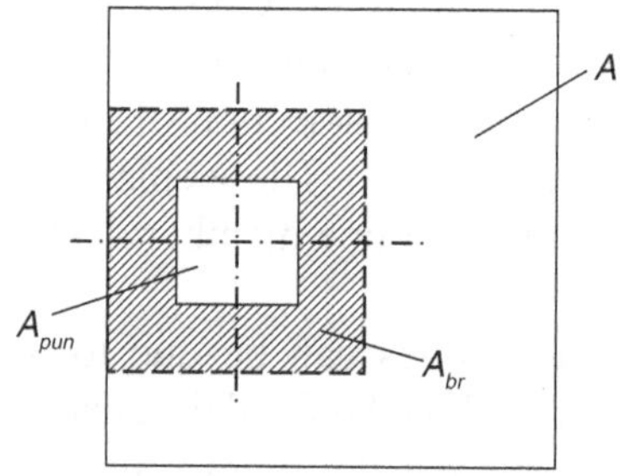

Fig. 7.9 Determining the bearing area A_{br}

Comments

- To determine A_{pun} for circular bearing plates, they are replaced by equivalent square plates having same area.
- In anchorage zones having several anchorages, A_{br} of each individual anchorage plate is determined such that the bearing areas do not overlap.

7.3.2 Bursting Forces in Anchorage Zones

In a post-tensioned member, the prestressing force is transferred through the anchorage as concentrated force (P_k). This force causes development of transverse tensile stresses (Fig. 7.10). The resultant of the tensile stresses is known as the bursting force (F_{bst}), and transverse reinforcement should be provided to restrict the longitudinal cracks caused due to F_{bst}.

Though large numbers of theoretical and experimental studies were done to determine the bursting force, only the Strut-and-Tie method and the Code recommendations are discussed below.

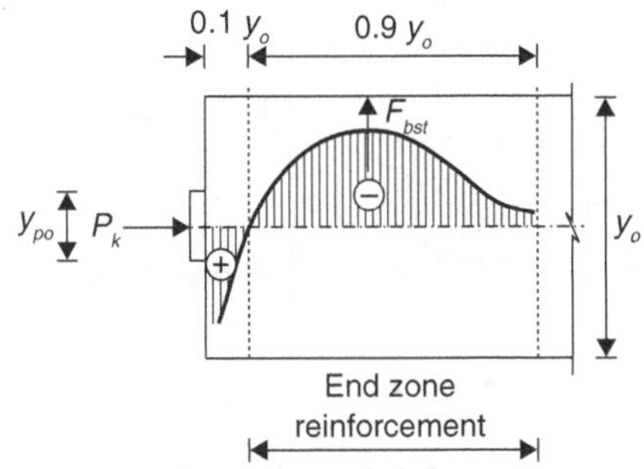

Fig. 7.10 Transverse stress distribution in the end zone

Strut-and-Tie Method for Anchorage Zone

In the Strut-and-Tie method[2,3] (STM), the concrete structure is replaced by a hypothetical truss that can resist the applied load. STM is based on the lower bound theorem of plasticity and hence the design using STM will be on the safer side. Strut-and-Tie models for the end zone of post-tensioned members are given in Fig. 7.11.

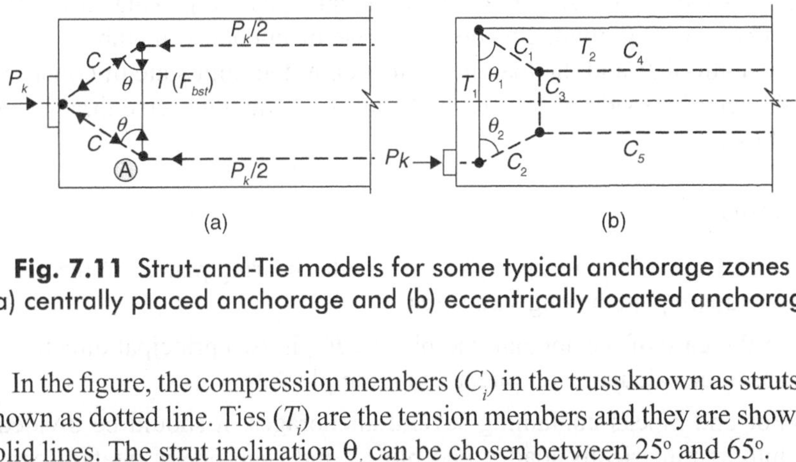

Fig. 7.11 Strut-and-Tie models for some typical anchorage zones (a) centrally placed anchorage and (b) eccentrically located anchorage

In the figure, the compression members (C_i) in the truss known as struts are shown as dotted line. Ties (T_i) are the tension members and they are shown as solid lines. The strut inclination θ_i can be chosen between 25° and 65°.

Once the truss model is developed, the truss can be analyzed to get the force in the ties (T_i). After finding T_i, the area of steel to resist T_i can be obtained by dividing the force by the permissible stress of steel.

---

[2] Schlaich, J., Schafer, K. and Jennewein, M., "Toward a Consistent Design of Structural Concrete", *PCI Journal*, June 1987, pp. 74–150.

[3] Schlaich, J. and Schafer, K., "Design and Detailing of Structural Concrete using Strut-and Tie Models", *The Structural Engineer*, March 1991, pp. 113–125.

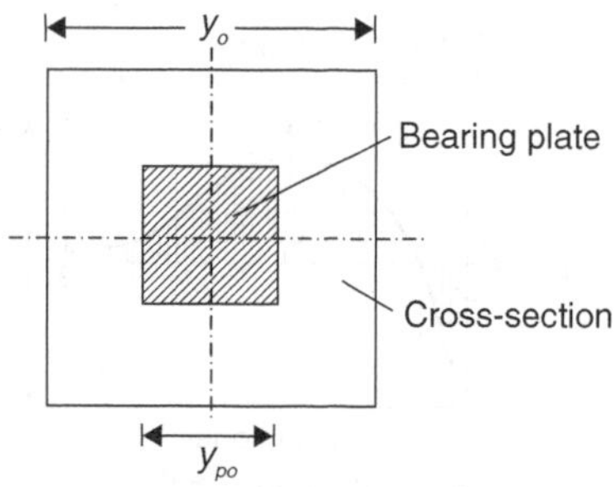

Fig. 7.12 End block

Code Recommendations for Finding F_{bst}

According to the Code (Cl. 18.6.2.2), the bursting force F_{bst} in a individual square end block loaded by a symmetrically placed square bearing plate (Fig. 7.12) can be found as:

$$F_{bst} = P_k \left[0.32 - 0.3 \frac{y_{po}}{y_o} \right] \tag{7.6}$$

where P_k is the prestressing force which is taken as the jacking force (P_i) for members having bonded tendons. In case of unbonded members, P_k is the maximum of P_i and the load in the tendon at the limit state of collapse. y_{po} is the length of side of bearing plate and y_o is the transverse dimension of end block.

Comments

- To determine F_{bst} using Eq. (7.6), circular bearing plates are replaced by square plate having same area.
- In the case of rectangular end blocks, F_{bst} in two principal directions are to be calculated (explained in Example 7.2).
- For end blocks containing several anchorages, it should be divided into symmetrically loaded end blocks and each one is treated separately. Additional reinforcement must be provided by considering the anchorages as a whole (see Example 7.3).

7.3.3 End Zone Reinforcement

Transverse reinforcement to control cracking due to F_{bst} are provided and they are distributed within a length from 0.1 y_o to y_o from the ends of the member (Fig. 7.10). The required area of transverse reinforcement (A_{st}) is:

$$A_{st} = \frac{F_{bst}}{f_s} \tag{7.7}$$

where f_s is the stress in steel reinforcement, which can be taken as $0.87 f_y$, f_y being the yield strength of steel. But when the cover to reinforcement is less than 50 mm, f_s should be limited to the value corresponding to a strain of 0.001 (Cl. 18.6.2.2 of the Code).

Example 7.2

Design the bearing plate and the end zone reinforcement for the end block of bonded post-tensioned beam of dimension 350 mm × 500 mm. The effective prestressing force is 900 kN and tendon is centrally placed at the ends. The strength of M50 grade concrete at transfer is 40 N/mm². Use the Code recommendations and compare the results using STM.

Solution:

Step 1: *Design of bearing plate*

Try a bearing plate of dimension 200 mm × 200 mm (Fig. 7.13).

- Punching area A_{pun} and bearing area A_{br}

  $A_{pun} = 200 \times 200 = 40000 \text{ mm}^2$

  $A_{br} = 350 \times 350 = 122500 \text{ mm}^2$

- Bearing stress f_{br}

$$f_{br} = \frac{P_e}{A_{pun}} = \frac{900 \times 10^3}{40000} = 22.5 \text{ N/mm}^2$$

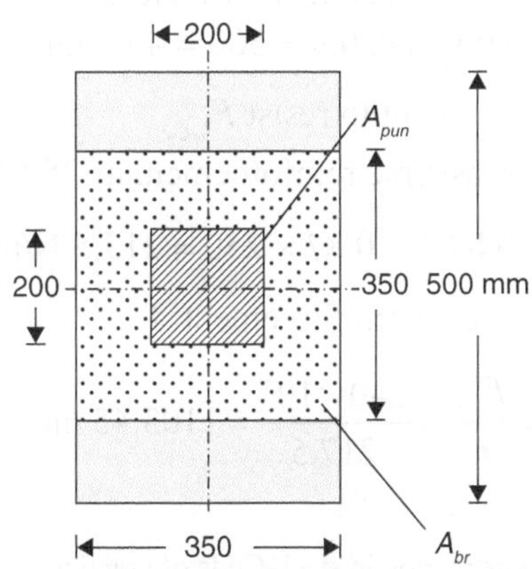

Fig. 7.13 A_{br} and A_{pun}

- Allowable bearing stress ($f_{br,\,all}$)

$$f_{br,all} = 0.48 f_{ci}\sqrt{\frac{A_{br}}{A_{pun}}} = 0.48 \times 40 \times \sqrt{\frac{122500}{40000}} = 33.6 \text{ N/mm}^2$$

$$\leq 0.8 f_{ci} = 0.8 \times 40 = 32 \text{ N/mm}^2$$

$$\therefore f_{br,\,all} = 32 \text{ N/mm}^2 \ (> f_{br}, \text{ OK})$$

The thickness of the bearing plate can be determined using the per missible stress in bending given in IS 800: 2007[4]

Step 2: *Calculation of bursting force and transverse reinforcement*

- Prestressing force P_k
  For bonded post-tensioned beam, $P_k = P_i$
  Assuming loss ratio $\eta = 0.75$;

$$P_i = \frac{P_e}{\eta} = \frac{900}{0.75} = 1200 \text{ kN}$$

- Bursting force in vertical direction $F_{bst,\,v}$

$$y_{po} = 200 \text{ mm}$$
$$y_o = 500 \text{ mm}$$

$$F_{bst,\,v} = P_k\, F_{bst,\,v} = P_k\left[0.32 - 0.3\frac{y_{po}}{y_o}\right]$$

$$= 1200\left[0.32 - \left(0.3 \times \frac{200}{500}\right)\right] = 240 \text{ kN}$$

The 240 kN is to be resisted using transverse reinforcement provided over a length of $0.9\, y_o$, i.e. $0.9 \times 500 = 450$ mm.

- Transverse reinforcement to resist $F_{bst,\,v}$
  Assuming f_y for transverse reinforcement as 250 N/mm²,

$$f_s = 0.87 f_y = 0.87 \times 250 = 217.5 \text{ N/mm}^2$$

Total area of transverse reinforcement A_{st}:

$$A_{st} = \frac{F_{bst}}{f_s} = \frac{240 \times 10^3}{217.5} = 1103.45 \text{ mm}^2$$

---

[4] Indian standard, general construction in steel–Code of practice.

Providing 12 mm diameter, two legged stirrups,

Number of required stirrups $n = \dfrac{1103.45}{2 \times \dfrac{\pi}{4} \times 12^2} = 4.9$, say 5 stirrups

Spacing of stirrups $s_v = \dfrac{0.9y_o}{(n-1)} = \dfrac{450}{(5-1)} = 112.5\,\text{mm}$, say 110 mm

- Bursting force in horizontal direction ($F_{bst,\,h}$)

  $y_{po} = 200$ mm

  $y_o = 350$ mm

  $$F_{bst,\,h} = P_k \left[0.32 - 0.3\frac{y_{po}}{y_o}\right]$$

  $$= 1200\left[0.32 - \left(0.3 \times \frac{200}{350}\right)\right] = 178.29\ \text{kN}$$

  This force is to be resisted within a length of $0.9 \times 350 = 315$ mm

- Transverse reinforcement to resist $F_{bst,\,h}$

  $$A_{st} = \frac{F_{bst}}{f_s} = \frac{178.29 \times 10^3}{217.5} = 819.72\ \text{mm}^2$$

  Number of required stirrups $n = \dfrac{819.72}{2 \times \dfrac{\pi}{4} \times 12^2} = 3.6$, say 4 stirrups

  Spacing of stirrups $s_v = \dfrac{315}{(4-1)} = 105\,\text{mm}$, say 100 mm

  Provide 2 legged 12 mm diameter stirrups at 100 mm c/c over a length of 450 mm.

Note: The transverse reinforcement should be calculated based on the maximum demand caused either due to $F_{bst,\,v}$ or $F_{bst,\,h}$. Here providing 2 legged 12 mm diameter stirrups at 100 mm c/c over a length of 450 mm satisfies both requirements.

Step 3: *Finding bursting force using STM*

The force in the tie can be obtained by considering the free body diagram (Fig. 7.14) of joint A of the truss shown in Fig. 7.11(a).

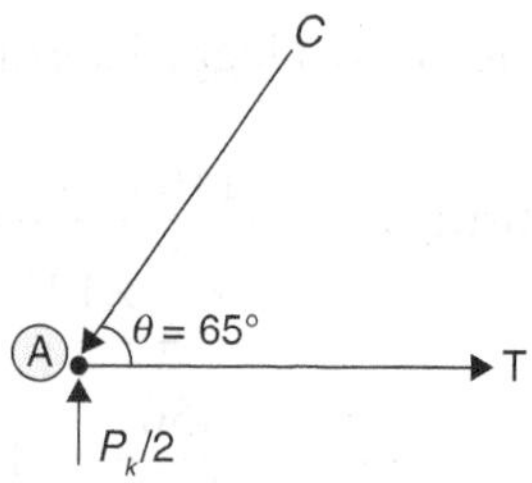

Fig. 7.14 Free body diagram of joint A

From Fig. 7.14, the force in the tie T is :

$$T = \frac{P_k/2}{\tan\theta} = \frac{(1200/2)}{\tan 65^0} = 279.78 \text{ kN}$$

The bursting force calculated using STM is slightly higher than that obtained using the Code provisions. This could be due to the lower bound nature of STM.

Example 7.3

Design the end zone reinforcement for the end block of the post-tensioned beam of cross-section 350 mm × 500 mm (Fig. 7.15). The end block has four anchorages with 100 mm square bearing plates, and an initial prestressing force of 400 kN is applied to each anchorage. The compressive strength of concrete at transfer is 40 N/mm^2. Use Fe 415 grade steel having modulus of elasticity 2×10^5 N/mm^2.

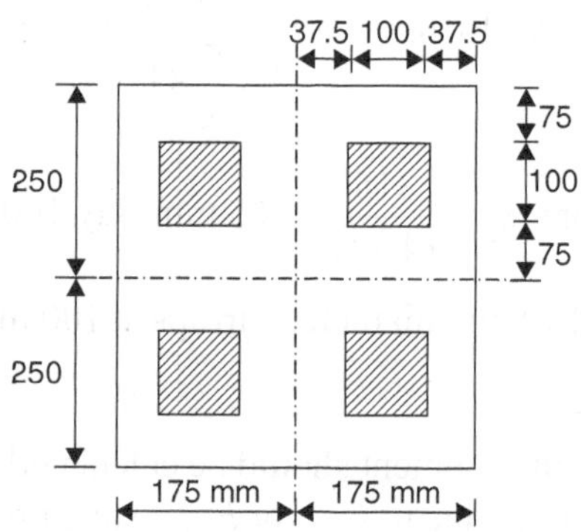

Fig. 7.15 End block

Solution:

The overall end block is divided into four separate effective end blocks of dimensions 175 mm × 250 mm.

Step 1: *Calculation of bearing stress behind bearing plate*

- Effective prestressing force P_e

  $P_e = \eta\, P_i = 0.75 \times 400 = 300$ kN (assuming $\eta = 0.75$)

- Punching area A_{pun}

  $A_{pun} = 100 \times 100 = 10000\ \text{mm}^2$

- Bearing area A_{br}

  $A_{br} = 175 \times 175 = 30625\ \text{mm}^2$

- Allowable bearing stress ($f_{br,\,all}$)

$$f_{br,all} = 0.48 f_{ci}\sqrt{\frac{A_{br}}{A_{pun}}} = 0.48 \times 40\sqrt{\frac{30625}{1000}} = 33.6\ \text{N/mm}^2$$

$$\leq 0.8 f_{ci} = 0.8 \times 40 = 32\ \text{N/mm}^2$$

$$\therefore f_{br,\,all} = 32\ \text{N/mm}^2$$

- Bearing stress f_{br}

$$f_{br} = \frac{P_e}{A_{pun}} = \frac{300 \times 10^3}{10000} = 30\ \text{N/mm}^2\ (< f_{br,\,all},\ \text{OK})$$

Step 2: *Calculation of bursting force and transverse reinforcement for individual anchorage*

- Bursting force in vertical direction ($F_{bst,\,v}$)

$$P_k = P_i = 400\ \text{kN}$$

$$y_{po} = 100\ \text{mm}$$

$$y_o = 250\ \text{mm}$$

$$F_{bst,\,v} = P_k\left[0.32 - 0.3\frac{y_{po}}{y_o}\right]$$

$$= 400\left[0.32 - \left(0.3 \times \frac{100}{250}\right)\right] = 80\ \text{kN}$$

  The 80 kN force is to be resisted using stirrups provided over a length of $0.9 \times 250 = 225$ mm.

- Transverse reinforcement to resist $F_{bst,\,v}$

  Yield stress of steel $f_y = 415\ \text{N/mm}^2$

  Assuming the cover to reinforcement is less than 50 mm,

  $f_s = 0.87 f_y = 0.87 \times 415 = 361.05\text{N/mm}^2$

$\leq 0.001\ E_s = 0.001 \times 2 \times 10^5 = 200\ \text{N/mm}^2$
$\therefore f_s = 200\ \text{N/mm}^2$

Total area of transverse reinforcement A_{st}:

$$A_{st} = \frac{F_{bst}}{f_s} = \frac{80 \times 10^3}{200} = 400\ \text{mm}^2$$

Providing 10 mm diameter, two legged stirrups,

$$\text{Number of required stirrups } n = \frac{400}{2 \times \frac{\pi}{4} \times 10^2} = 2.55, \text{ say 3 stirrups}$$

$$\text{Spacing of stirrups } s_v = \frac{255}{(3-1)} = 112.5\ \text{mm.}$$

- Bursting force in horizontal direction ($F_{bst,\,h}$)

$$y_{po} = 100\ \text{mm}$$
$$y_o = 175\ \text{mm}$$
$$F_{bst,\,h} = P_k \left[0.32 - 0.3\frac{y_{po}}{y_o}\right]$$
$$= 400\left[0.32 - \left(0.3 \times \frac{100}{175}\right)\right] = 59.43\ \text{kN}$$

The transverse reinforcement to resist this force should be provided within a length of 0.9 × 175 = 157.5 mm.

- Transverse reinforcement to resist $F_{bst,\,h}$

$$\text{Area of transverse reinforcement } A_{st} = \frac{59.43 \times 10^3}{200} = 297.15\ \text{mm}^2$$

$$\text{Number of 2 legged 10 mm stirrups } n = \frac{297.15}{2 \times \frac{\pi}{4} \times 10^2} = 1.9, \text{ say 2 stirrups}$$

$$\text{Spacing of stirrups } s_v = \frac{157.5}{(2-1)} = 157.5\ \text{mm}$$

Provide 2 legged 10 mm diameter stirrups at 110 mm c/c around each anchorage over a length of 225 mm.

Step 3: *Calculation of bursting force and transverse reinforcement for whole anchorage*

Total area of bearing plate = 4 × 100 × 100 = 40000 mm^2

- Dimensions of equivalent square bearing plate y_{po}

$$y_{po} = \sqrt{40000} = 200 \text{ mm}$$

- Prestressing force P_k

$$P_k = P_i = 4 \times 400 = 1600 \text{ kN}$$

- Bursting force in vertical direction ($F_{bst,\,v}$)

$$y_{po} = 200 \text{ mm}$$

$$y_o = 500 \text{ mm}$$

$$F_{bst,\,v} = P_k\left[0.32 - 0.3\frac{y_{po}}{y_o}\right]$$

$$= 1600\left[0.32 - \left(0.3 \times \frac{200}{500}\right)\right] = 320 \text{ kN}$$

The transverse reinforcement for this force should be provided in a length of 0.9 × 500 = 450 mm.

- Transverse reinforcement to resist $F_{bst,\,v}$

$$f_s = 200 \text{ N/mm}^2$$

$$A_{st} = \frac{320 \times 10^3}{200} = 1600 \text{ mm}^2$$

Number of 4 legged 10 mm stirrups $n = \dfrac{1600}{4 \times \dfrac{\pi}{4} \times 10^2} = 5$

Spacing of stirrups $s_v = \dfrac{450}{(5-1)} = 112.5$ mm.

- Bursting force in horizontal direction ($F_{bst,\,h}$)

$$y_{po} = 200 \text{ mm}$$

$$y_o = 350 \text{ mm}$$

$$F_{bst,\,h} = P_k\left[0.32 - 0.3\frac{y_{po}}{y_o}\right]$$

$$= 1600\left[0.32 - \left(0.3 \times \frac{200}{350}\right)\right] = 237.7 \text{ kN}$$

This force has to be resisted by providing stirrups within a length of 0.9 × 350 = 315 mm.

- Transverse reinforcement to resist $F_{bst,\,h}$

$$f_s = 200\ \text{N/mm}^2$$

$$A_{st} = \frac{237.7 \times 10^3}{200} = 118.5\ \text{mm}^2$$

Number of 4 legged 10 mm stirrups $n = \dfrac{1188.5}{4 \times \dfrac{\pi}{4} \times 10^2} = 3.8$, say 4 stirrups

Spacing of stirrups $s_v = \dfrac{315}{(4-1)} = 105\ \text{mm}$

Provide 10 mm diameter 4 legged stirrups at 100 mm c/c enclosing all the four anchorages over a length of 450 mm. The detailing of end block is shown in Fig. 7.16.

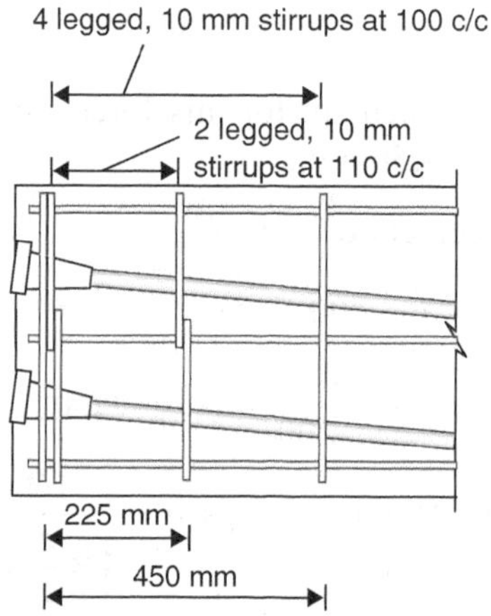

Fig. 7.16 Detailing in the end zone

PROBLEMS

7.1 A post-tensioned concrete beam 300 mm wide and 600 mm deep is prestressed by a effective prestressing force of 1000 kN at an eccentricity of 100 mm. The size of the bearing plate is 300 mm × 300 mm. Calculate the bursting force and design reinforcement to resist this force.

7.2 The end block of a post-tensioned beam has three anchorages with 300 mm square bearing plates (Fig. 7.17). A initial prestressing force of 700 kN is applied to each anchorage. Design the end zone reinforcement.

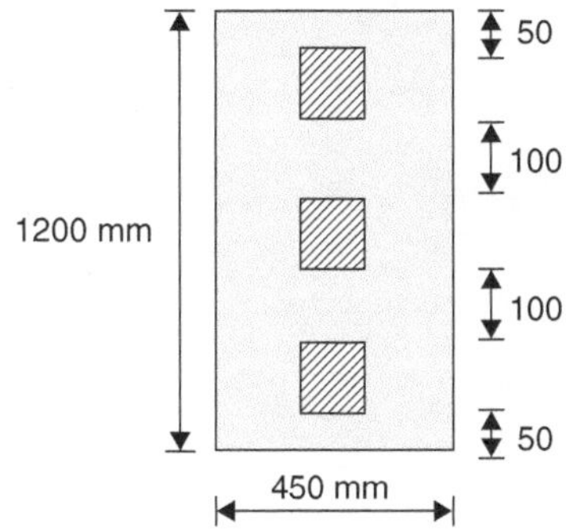

Fig. 7.17

7.3 Fig. 7.18 shows the anchorage zone of a T-beam. The initial pre-stressing force is 1500 kN and the gade of concrete is M 50. The size of the bearing plate is 250 mm × 250 mm. Design the end zone reinforcement.

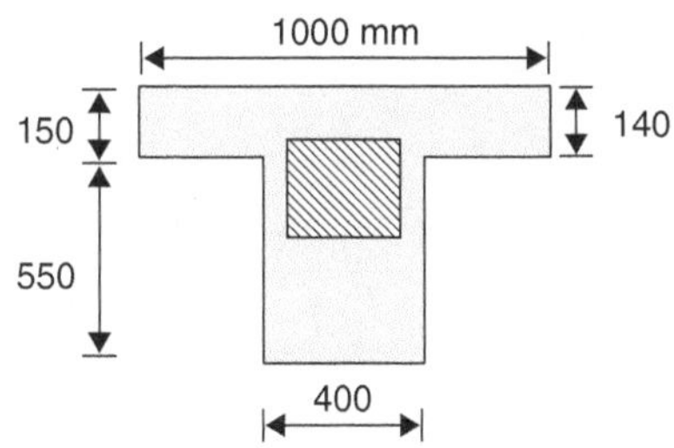

Fig. 7.18

chapter 8

Deflections

8.1 Introduction

Predicting the deflections is necessary to check the performance of flexural members at serviceability limit state. Since prestressed concrete members are slender than reinforced concrete members, check for deflection becomes more important. Deflection in a prestressed concrete member is usually checked at transfer (*short term deflection*) and at service condition (*long term deflection*).

At transfer, the member is subjected to prestressing force and the self-weight of member, usually resulting in an upward deflection (also known as *camber*). The amount of camber should be controlled in roof slabs to avoid drainage problems. Also in projects involving precast members, excessive camber will cause difficulty in the alignment of the member.

Generally, the prestressed concrete member will have a downward deflection in service condition. Excessive downward deflection will cause roof ponding, damages to partitions and finishes, uncomfortable feeling to occupants, etc.

Though the calculation of deflection is important, it should be kept in mind that deflection of prestressed concrete members cannot be predicted with high degree of accuracy. This is due to the nonlinear behaviour of concrete, change in prestressing force with time, difficulty in predicting the effects of creep and shrinkage, change in stiffness of the member due to cracking etc. Hence simplified methods are usually used to predict deflections with a reasonable degree of accuracy.

8.2 Deflections in Uncracked Beams

Type 1 and type 2 members are designed to remain uncracked under service loads. For calculating deflections in these members, the flexural rigidity is taken as $E_c I$, where I is the gross moment of inertia of the cross-section and E_c is the modulus of elasticity of concrete. Even though E_c changes with time (due to change in compressive strength), for calculating deflections using the simplified methods, it can be taken as $E_c = 5000\sqrt{f_{ck}}$ N/mm^2.

The reader may refer books on structural analysis for the various methods used to calculate deflection. Expressions to find mid-span deflection (Δ_m) of a simply supported beam of span l subjected to some typical loading cases are given in Table 8.1.

Table 8.1 Mid span deflection

No.	Type of Loading	Δ_m
1.	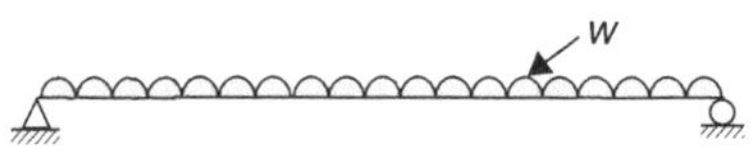	$\dfrac{5}{384}\dfrac{wl^4}{EI}$
2.	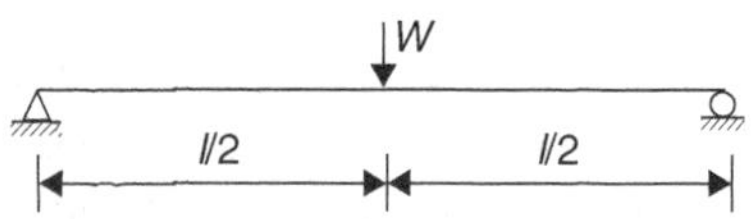	$\dfrac{Wl^3}{48EI}$
3.	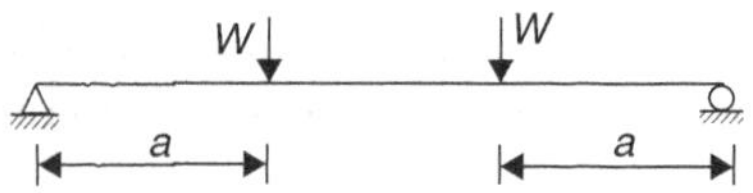	$\dfrac{Wa}{24EI}(3l^2 - 4a^2)$
4.	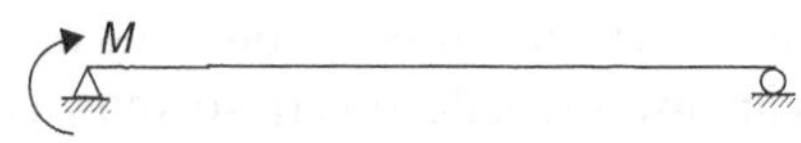	$\dfrac{Ml^2}{16EI}$
5.	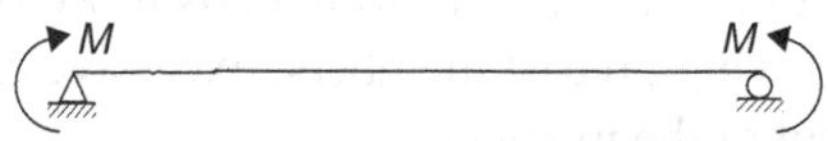	$\dfrac{Ml^2}{8EI}$

8.2.1 Short Term Deflection at Transfer Δ_{st}

At transfer, the deflection of the member (Fig. 5.6) is caused due to the prestressing force (P_o) and the self-weight of the member (w_{SW}). The deflection due to P_o (Δ_{po}) and self-weight (Δ_{SW}) are calculated separately and added to get the short term deflection. The short term deflection at transfer Δ_{st} is:

$$\Delta_{st} = \Delta_{po} + \Delta_{SW} \tag{8.1}$$

8.2.2 Long Term Deflection at Service Condition Δ_{ls}

The short term deflection discussed in the previous section was caused due to the elastic deformation of concrete. However, creep and shrinkage movements will also cause deflection in concrete members.

The effect of creep in deflection is calculated by using the effective modulus of elasticity $E_{c,\,eff}$ given by:

$$E_{c,eff} = \frac{E_c}{1+\theta} \tag{8.2}$$

where θ is the creep coefficient (Table 4.3).

If Δ_i is the instantaneous deflection due to sustained load W, then Δ_i can be written as:

$$\Delta_i = \frac{CW}{E_c I} \tag{8.3}$$

where the constant C depends on span, loading and support condition. The long term deflection Δ_l (including the effect of creep) due to W is:

$$\Delta_l = \frac{CW}{E_{c,eff} I} = \frac{CW}{E_c I}(1+\theta) = \Delta_i(1+\theta) \tag{8.4}$$

In the above equation, the deflection component $\Delta_i\theta$ represent the deflection due to creep.

At the service condition, the force in the tendon is P_e, and the applied loading on the member is the full dead and live load. The long term deflection at service condition (Δ_{ls}) can be determined using the *approximate time-steps method* recommended in ACI 435R–95[1] and is given by:

$$\Delta_{ls} = \Delta_{pe} + \left(\frac{\Delta_{pe} + \Delta_{po}}{2}\right)\theta + \Delta_{su}(1+\theta) + \Delta_{LL} \tag{8.5}$$

where Δ_{pe} is the deflection due to the prestressing force P_e, Δ_{su} is the deflection due to sustained load (usually the dead load*) and Δ_{LL} is the deflection due to live load.

In Eq. (8.5), it is assumed that creep occurs at a constant prestressing force equal to the average of P_o and P_e. Deflection due to shrinkage is not usually considered in prestressed concrete members.

---

[1] ACI Committee 435 Report (ACI 435R–95), *Control of Deflection in Concrete Structures,* American Concrete Institute, 2003.

* If a small fraction of live load is included as a part of sustained load, then the remaining live load is used to find Δ_{LL}.

8.3 Deflection of Type 3 Members

These types of members are expected to crack under service loads and the effective moment of inertia (which takes into account of tension stiffening effect) should be used for the calculation of deflection. Hence, the gross moment of inertia should not be used to find deflections in type 3 member. However, the Code (Cl. 22.6.2) recommends to use gross moment of inertia, when the permanent load is less than or equal to 25 percentage of the live load.

To avoid calculation of deflections, the Code recommends simplified checks for type 3 members subjected to permanent load greater than 25 percentage of live load. Such members can be assumed to satisfy the deflection criteria, if the effective span (l) to effective depth (d) ratio does not exceed the limits given in Table 8.2.

Table 8.2 Limits for l/d

No.	Type of Member	Limiting Value of l/d
1.	Cantilever	7
2.	Simply supported	20
3.	Continuous	26

The limits in Table 8.2 are applicable for $l \leq 10$m. For $l > 10$m, the limiting value of l/d should be multiplied by $10/l$ (l in metres), except for cantilevers where deflection calculations are necessary.

8.4 Deflection Limits

The Code (Cl. 19.3.1) prescribes the following limits for prestressed concrete flexural members:

- The final deflection due to all loads (including the effects of temperature, creep and shrinkage) should not exceed span/250. This limit is specified for the purpose of control of cracking and to prevent psychological discomfort to occupants.
- The deflection (including the effects of temperature, creep and shrinkage) that occurs after construction of partitions and finishes should not exceed the least of span/350 or 20 mm. The loads applied after construction of partitions and finishes will be in general the live load. This deflection limit is aimed to prevent damages to partitions and finishes.
- If finishes are applied, the total upward deflection should not exceed span/300.

Example 8.1

Determine the short term and long term deflections of the type1 prestressed concrete beam shown in Fig. 8.1(a). The beam is prestressed with a straight tendon at an eccentricity of 100 mm. The prestressing force after immediate and time dependent losses are 1100 kN and 950 kN respectively. The beam is subjected to a super imposed dead load of intensity (w_{SIDL}) 1.5 kN/m and live load of intensity 8 kN/m. The grade of concrete is M 50. Check whether the deflections satisfies the limits specified in IS:1343–1980.

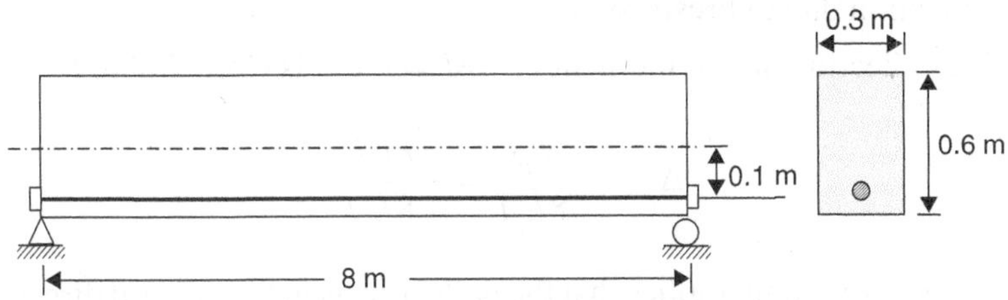

(a)

Fig. 8.1 (a) Prestressed concrete beam

Solution:

Step 1: *Calculation of flexural rigidity and self-weight of beam*

- Modulus of elasticity of concrete E_c

$$E_c = 5000\sqrt{f_{ck}} = 5000\sqrt{50} = 35355.34 \text{ N/mm}^2$$

- Gross moment of inertia I

$$I = \frac{300 \times 600^3}{12} = 54 \times 10^8 \text{ mm}^4$$

- Flexural rigidity $E_c I$

$$E_c I = 1.9092 \times 10^{14} \text{ N-mm}^2$$

- Self-weight w_{SW}

  Density of concrete $\gamma_c = 24$ kN/m^3

  Cross-sectional area of the beam $A = 0.3 \times 0.6 = 0.18$ m^2

  $w_{SW} = \gamma_c A = 24 \times 0.18 = 4.32$ kN/m

Step 2: *Calculation of equivalent loads and deflection due to prestress*

- Equivalent loads due to prestress

  The equivalent loads due to prestress are shown in Fig. 8.1(b).

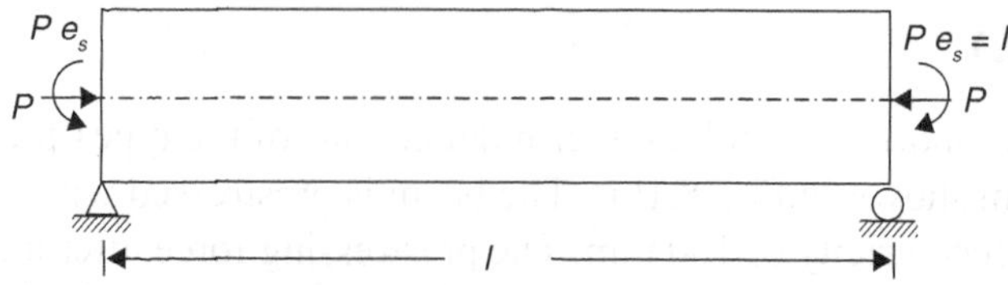

P = Prestressing force, e_s = 0.1 m, l = 8 m

(b)

Fig. 8.1 (b) Equivalent loads due to prestress

- Deflection due to prestress Δ_p

  The deflection due to moment M_p ($M_p = P\,e_s$) is (from Table 8.1):

$$\Delta_p = -\frac{M_p l^2}{8E_c I} = -\frac{Pe_s l^2}{8E_c I}$$

The negative sign implies that the deflection is in the upward direction.

Step 3: *Calculation of short term deflection at transfer* Δ_{st}

Deflection due to self-weight of the beam Δ_{SW}

- $w_{SW} = 4.32$ kN/m $= 4.32$ N/mm

$$\Delta_{SW} = \frac{5}{384}\frac{w_{SW} l^4}{E_c I} = \frac{5\times 4.32\times(8000)^4}{384\times 1.9092\times 10^{14}} = 1.21 \text{ mm}$$

Deflection due to prestressing force P_o (ΔP_o)

$$P_o = 1100 \text{ kN}$$

$$\Delta_{po} = -\frac{P_o e_s l^2}{8E_c I} = -\frac{1100\times 10^3\times 100\times 8000^2}{8\times 1.9092\times 10^{14}}$$

$$= -4.61 \text{ mm}$$

- Short term deflection at transfer

$$\Delta_{st} = \Delta_{po} + \Delta_{SW} = -4.61 + 1.21 = -3.4 \text{ mm (upward deflection)}$$

- Deflection limits as per Code

$$\text{Allowable upward deflection} = \frac{\text{span}}{300} = \frac{8\times 10^3}{300} = 26.67 \text{ mm}$$

Since the upward deflection (3.4 mm) is less than the allowable limit (26.67 mm), the short term deflection satisfies the deflection criteria of the Code.

Step 4: *Calculation of long term deflection at service condition* Δ_{ls}

- Creep coefficient θ

  Assuming the age of concrete at loading is 28 days,
  $\theta = 1.6$ (from Cl. 5.2.5.1 of the Code)

- Deflection due to P_e (ΔP_e)

$$P_e = 950 \text{ kN}$$

$$\Delta P_e = -\frac{P_e e_s l^2}{8E_c I} = -\frac{950\times 10^3 \times 100\times 8000^2}{8\times 1.9092\times 10^{14}} = -3.98 \text{ mm}$$

- Deflection due to sustained load (Δ_{su})

  Intensity of sustained load w_{su} = Dead load = $w_{SW} + w_{SIDL}$

$$w_{su} = 4.32 + 1.5 = 5.82 \text{ N/mm}$$

$$\Delta_{su} = \frac{5}{384}\frac{w_{su} l^4}{E_c I} = \frac{5\times 5.82\times (8000)^4}{384\times 1.9092\times 10^{14}} = 1.63 \text{ mm}$$

- Deflection due to live load (Δ_{LL})

  Live load intensity $w_{LL} = 8$ N/mm

$$\Delta_{LL} = \frac{5}{384}\frac{w_{LL} l^4}{E_c I} = \frac{5\times 8\times (8000)^4}{384\times 1.9092\times 10^{14}} = 2.24 \text{ mm}$$

- Long term deflection (Δ_{ls})

$$\Delta_{ls} = \Delta_{pe} + \left(\frac{\Delta_{po} + \Delta_{pe}}{2}\right)\theta + \Delta_{su}(1+\theta) + \Delta_{LL}$$

$$= -3.98 - \left[\frac{(4.61+3.98)}{2}\times 1.6\right] + \left[1.63\times(1+1.6)\right] + 2.24$$

$$= -4.61 \text{ mm}$$

  Δ_{ls} is less than 26.67 mm. Hence it satisfies the code requirements.

Example 8.2

Find the short term and long term deflections of the type 1 prestressed concrete beam in Fig. 8.2(a) and check the Code limits for deflection.

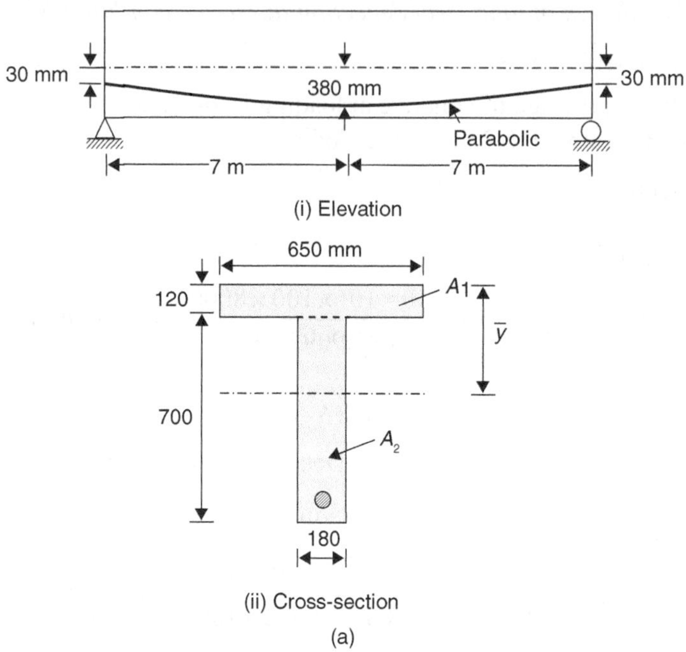

Fig. 8.2 (a) Prestressed concrete beam

Data:

$P_o = 900$ kN, $P_e = 600$ kN, $w_{LL} = 10$ kN/m, $w_{SIDL} = 3$ kN/m, $f_{ck} = 40$ N/mm^2

Solution:

Step 1: *Calculation of cross-sectional properties, flexural rigidity and self-weight of the beam*

- Area of cross-section (Fig. 8.2(a))

$$A_1 = 650 \times 120 = 78000 \text{ mm}^2$$
$$A_2 = 700 \times 180 = 126000 \text{ mm}^2$$
$$A = A_1 + A_2 = 204000 \text{ mm}^2$$

- Distance of centroid of the section from the top fibre ($\bar{y}$)

$$\bar{y} = \frac{(78000 \times 60) + (126000 \times 470)}{204000} = 313.24 \text{ mm}$$

- Moment of inertia about the centroid of the section I

$$I = \left[\frac{650 \times 120^3}{12}\right] + [78000 \times 253.24^2] + \left[\frac{180 \times 700^3}{12}\right] + [126000 \times 156.76^2]$$
$$= 1.3337 \times 10^{10} \text{ mm}^4$$

- Modulus of elasticity of concrete E_c

$$E_c = 5000\sqrt{f_{ck}} = 5000\sqrt{40} = 31622.78 \text{ N/mm}^2$$

- Flexural rigidity of the section $E_c I$

$$E_c I = 4.218 \times 10^{14} \text{ N-mm}^2$$

- Self-weight of the beam w_{SW}

$$w_{SW} = \gamma_c A = 24 \times 204000 \times 10^{-6} = 4.9 \text{ kN/m}$$

Step 2: *Calculation of equivalent loads and deflection due to prestress*

- Equivalent loads due to prestress
  The equivalent loads due to prestress are shown in Fig. 8.2(b).

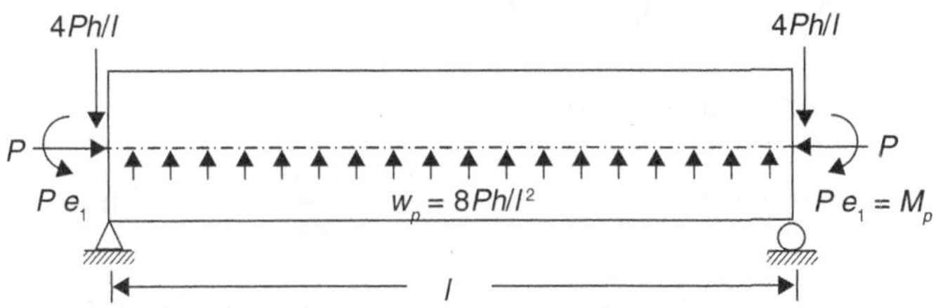

e_1 = 0.03 m, h = 0.38 – 0.03 = 0.35 m, l = 14 m

(b)

Fig. 8.2 (b) Equivalent loads due to prestress

- Deflection due to prestress Δ_p

Deflection due to $M_p = \Delta_{p1} = -\dfrac{M_p l^2}{8E_c I} = -\dfrac{Pe_1 l^2}{8E_c I}$

Deflection due to $w_p = \Delta_{p2} = -\dfrac{5}{384}\dfrac{w_p\, l^4}{E_c I}$

Deflection due to prestress $\Delta_p = \Delta_{p1} + \Delta_{p2}$

Step 3: *Calculation of short term deflection at transfer* Δ_{st}

- Deflection due to self-weight of the beam Δ_{SW}

$$w_{SW} = 4.9 \text{ kN/m}$$

$$\Delta_{SW} = \frac{5}{384}\frac{w_{SW} l^4}{E_c I} = \frac{5 \times 4.9 \times (14000)^4}{384 \times 4.218 \times 10^{14}} = 5.81 \text{ mm}$$

- Deflection due to prestressing force P_o (ΔP_o)

$$P_o = 900 \text{ kN}$$

$$\Delta_{po1} = -\frac{P_o e_1 l^2}{8 E_c I} = -\frac{900 \times 10^3 \times 30 \times (14000)^2}{8 \times 4.218 \times 10^{14}} = -1.57 \text{ mm}$$

$$w_{po} = \frac{8 P_o h}{l^2} = \frac{8 \times 900 \times 0.35}{14^2} = 12.86 \text{ kN/m} = 12.86 \text{ N/mm}$$

$$\Delta_{po2} = -\frac{5}{384} \frac{w_{po} l^4}{E_c I} = -\frac{5 \times 12.86 \times (14000)^4}{384 \times 4.218 \times 10^{14}} = -15.25 \text{ mm}$$

Deflection due to $\Delta_{po} = \Delta_{po1} + \Delta_{po2} = -1.57 - 15.25 = -16.82$ mm

- Short term deflection Δ_{st}

$$\Delta_{st} = \Delta_{po} + \Delta_{SW} = -16.82 + 5.81 = -11.01 \text{ mm}$$

- Deflection limits as per Code

Allowable upward deflection $= \frac{\text{span}}{300} = \frac{14000}{300} = 46.67$ mm

The upward deflection (11.01 mm) is less than the allowable limit 46.67 mm. Hence Δ_{st} satisfies Code recommendations.

Step 4: *Calculation of long term deflection at service condition Δ_{ls}*

- Creep coefficient θ (from Cl. 5.2.5.1 of the Code)

$\theta = 1.6$ (assuming age of concrete at service stage as 28 days)

- Deflection due to P_e (ΔP_e)

$$P_e = 600 \text{ kN}$$

$$\Delta P_{e1} = -\frac{P_e e_1 l^2}{8 E_c I} = -\frac{600 \times 10^3 \times 30 \times 14000^2}{8 \times 4.218 \times 10^{14}} = -1.05 \text{ mm}$$

$$w_{pe} = \frac{8 P_e h}{l^2} = \frac{8 \times 600 \times 0.35}{14^2} = 8.57 \text{ kN/m} = 8.57 \text{ N/mm}$$

$$\Delta_{pe2} = -\frac{5}{384} \frac{w_{pe} l^4}{E_c I} = -\frac{5 \times 8.57 \times (14000)^4}{384 \times 4.218 \times 10^{14}} = -10.16 \text{ mm}$$

Deflection due to $\Delta_{pe} = \Delta_{pe1} + \Delta_{pe2} = -1.05 - 10.16 = -11.21$ mm

- Deflection due to sustained load (Δ_{su})

Intensity of sustained load w_{su} = Dead load = $w_{SW} + w_{SIDL}$

$$w_{su} = 4.9 + 3 = 7.9 \text{ N/mm}$$

$$\Delta_{su} = \frac{5}{384}\frac{w_{su} l^4}{E_c I} = \frac{5 \times 7.9 \times (14000)^4}{384 \times 4.218 \times 10^{14}} = 9.37 \text{ mm}$$

- Deflection due to live load (Δ_{LL})

Live load intensity w_{LL} = 10 N/mm

$$\Delta_{LL} = \frac{5}{384}\frac{w_{LL} l^4}{E_c I} = \frac{5 \times 10 \times (14000)^4}{384 \times 4.218 \times 10^{14}} = 11.86 \text{ mm}$$

- Long term deflection (Δ_{ls})

$$\Delta_{ls} = \Delta_{pe} + \left(\frac{\Delta_{po} + \Delta_{pe}}{2}\right)\theta + \Delta_{su}(1+\theta) + \Delta_{LL}$$

$$= -11.21 - \left[\frac{(16.82 + 11.21)}{2} \times 1.6\right] + [9.37 \times (1 + 1.6)] + 11.86$$

$$= 2.59 \text{ mm}$$

- Deflection limits as per Code

Allowable downward deflection = $\frac{\text{span}}{250} = \frac{14000}{250}$ = 56 mm

Since the calculated downward deflection (2.59 mm) is less than allowable deflection, the long term deflection at service condition satisfies the deflection requirements of the Code.

Problems

8.1 Find the short term and long term deflections of the type 1 prestressed concrete beam shown in Fig. 8.3. The beam has a rectangular cross-section of 0.3 m × 0.6 m and a effective span of 16 m.

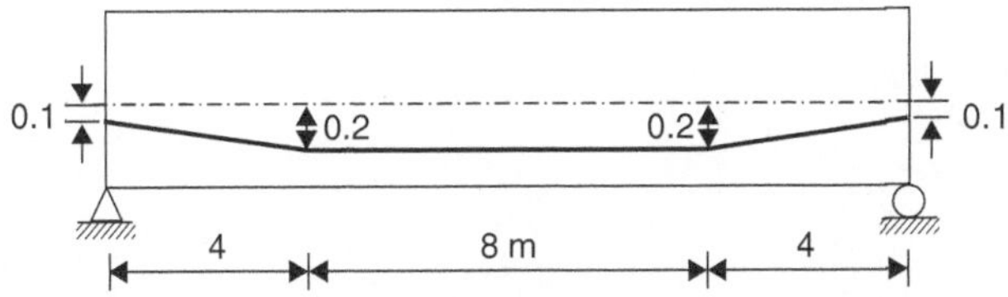

Fig. 8.3

Data:

$$f_{ck} = 50 \text{ N/mm}^2,\ P_o = 800 \text{ kN},\ P_e = 650 \text{ kN},\ w_{LL} = 10 \text{ kN/m},\ w_{SIDL} = 2 \text{ kN/m}$$

8.2 A pre-tensioned beam of rectangular cross-section 0.3 m × 0.6 m and of span 10 m is prestressed using straight tendon at a constant eccentricity of 150 mm. Determine (i) the camber and (ii) deflection due to uniformly distributed load causing decompression at the mid-span.

Data:

$$f_{ck} = 40 \text{ N/mm}^2,\ P_i = 900 \text{ kN},\ \beta = 0.9,\ \eta = 0.8$$

8.3 Find the long term deflection at the mid-span of the beam in Problem 8.2 if 85 percent of the uniformly distributed load is permanent.

chapter 9

Design of Members

9.1 Introduction

The analysis of section for a given prestressing force and applied loading is already discussed in Chapter 5. This chapter deals with the design of prestressed concrete members, which can be considered as the reverse of analysis. That is, given the applied loads and skeletal dimension of the member, it is required to arrive at suitable cross-sectional dimensions, prestressing force and tendon profile that would provide adequate safety and serviceability. Unlike an analysis problem which has a unique solution, the design problem will not have a single solution. This is because, different combinations of cross-sectional dimensions, prestressing force and cable profile will satisfy both serviceability and ultimate limit state requirements.

A reinforced concrete member is usually designed to satisfy the ultimate limit state and checked for its performance at serviceability limit state. But in the case of prestressed concrete members (type 1 and type 2), the serviceability limit state conditions are more critical than the ultimate limit state. Hence these members are designed to satisfy the serviceability limit state and checked whether the design satisfies the requirements at ultimate limit state.

Design of type 1 and type 2 members (i.e., uncracked prestressed concrete members in serviceability condition) are covered in this chapter.

9.2 Governing Inequalities

In order to derive the stress inequalities which govern the design at serviceability limit state, consider a simply supported beam shown in Fig. 9.1(a).

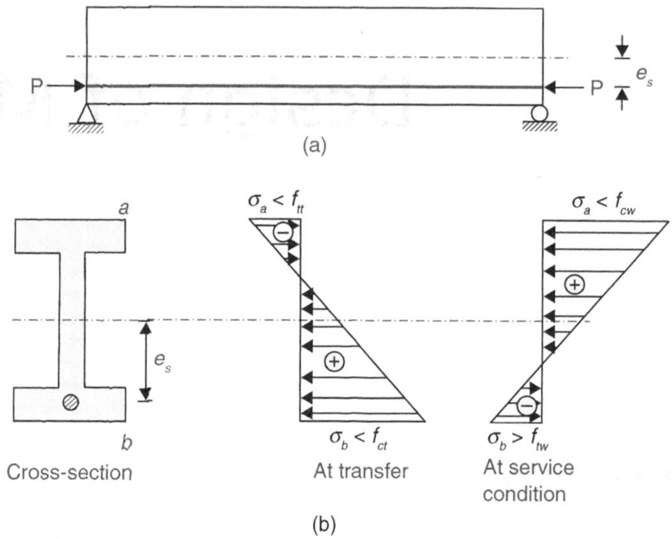

Fig. 9.1 Prestressed concrete beam and stress distribution (a) simply supported prestressed concrete beam and (b) longitudinal stress distribution at different loading stages

The total longitudinal stress at the top (σ_a) and the bottom (σ_b) fibres of the section are (Eq. (5.4)):

$$\sigma_a = \frac{P}{A} - \frac{Pe_s}{Z_a} + \frac{M}{Z_a} \tag{9.1}$$

$$\sigma_b = \frac{P}{A} + \frac{Pe_s}{Z_b} - \frac{M}{Z_b} \tag{9.2}$$

In the above equations, P and e_s are the prestressing force and its eccentricity. A is the area of cross-section of the beam. Z_a and Z_b are the elastic section moduli for the top and bottom fibres of the section. M is the bending moment due to applied loading (assumed as sagging).

If P_i is the initial prestressing force, β and η are the immediate and time dependent loss factors, then the prestressing force and the bending moment due to applied loading at the transfer loading stage are βP_i ($P_o = \beta P_i$) and M_o respectively. Similarly ηP_i ($P_e = \eta P_i$) and M_s denotes the prestressing force and bending moment at the service condition.

Due to the upward deflection of the beam at transfer (Fig. 5.6), the stresses at the top and bottom fibres of the beam are tensile and compressive in nature. But at the service condition, the bottom fibre is in tension and the top fibre in compression (Fig. 5.7).

If the maximum allowable compressive stress due to flexure in concrete at transfer and service condition are f_{ct} and f_{cw}, and the minimum permissible stress at transfer and service loading stages are f_{tt} and f_{tw} (f_{tt} and f_{tw} are tensile stresses and as per sign convention are negative stresses), then (Fig. 9.1(b)),

At Transfer:

$$\sigma_a \geq f_{tt} \tag{9.3a}$$

$$\sigma_b \leq f_{ct} \tag{9.3b}$$

At Service Condition:

$$\sigma_a \leq f_{cw} \tag{9.3c}$$

$$\sigma_b \geq f_{tw} \tag{9.3d}$$

In the above inequalities, it should be noted that the allowable tensile stress is a negative quantity. Hence the calculated stress should be greater than the allowable tensile stress (i.e., preferably the calculated stress should be a positive number indicating compressive stress). Similarly, the calculated stress must be less than the allowable compressive stress. The allowable stresses in concrete due to flexure are discussed in Section 3.7.

Using Eqs. (9.1–9.3), the governing stress inequalities can be written as:

$$\frac{\beta P_i}{A} - \frac{\beta P_i e_s}{Z_a} + \frac{M_o}{Z_a} \geq f_{tt} \tag{9.4a}$$

$$\frac{\beta P_i}{A} + \frac{\beta P_i e_s}{Z_b} - \frac{M_o}{Z_b} \leq f_{ct} \tag{9.4b}$$

$$\frac{\eta P_i}{A} - \frac{\eta P_i e_s}{Z_a} + \frac{M_s}{Z_a} \leq f_{cw} \tag{9.4c}$$

$$\frac{\eta P_i}{A} + \frac{\eta P_i e_s}{Z_b} - \frac{M_s}{Z_b} \geq f_{tw} \tag{9.4d}$$

9.3 Minimum Section Modulus

The governing stress inequalities considered at *critical section* (i.e., section where bending moment is maximum) are used to find the required minimum section modulus.

9.3.1 Minimum Section Modulus for the Top Fibre Z_a

The top fibre stresses should satisfy the stress criteria given by Eqs. (9.4a) and (9.4c). From Eq. (9.4a),

$$\frac{\beta P_i}{A} - \frac{\beta P_i e_s}{Z_a} + \frac{M_o}{Z_a} \geq f_{tt}$$

Multiplying both sides of the equation by η gives:

$$\beta\eta\frac{P_i}{A} - \beta\eta\frac{P_i e_s}{Z_a} + \frac{\eta M_o}{Z_a} \geq \eta f_{tt}$$

Multiplying the equation by – 1 and rearranging gives:

$$-\beta\eta\left[\frac{P_i}{A} - \frac{P_i e_s}{Z_a}\right] - \frac{\eta M_o}{Z_a} \leq -\eta f_{tt} \quad (9.5a)$$

Similarly from Eq. (9.4c):

$$\frac{\eta P_i}{A} - \frac{\eta P_i e_s}{Z_a} + \frac{M_s}{Z_a} \leq f_{cw}$$

Multiplying the equation by loss factor β gives:

$$\beta\eta\left[\frac{P_i}{A} - \frac{P_i e_s}{Z_a}\right] + \frac{\beta M_s}{Z_a} \leq \beta f_{cw} \quad (9.5b)$$

By adding Eqs. (9.5a) and (9.5b), the required section modulus Z_a can be found out, i.e.,

$$\frac{(\beta M_s - \eta M_o)}{Z_a} \leq (\beta f_{cw} - \eta f_{tt})$$

$$\Rightarrow \qquad Z_a \geq \frac{(\beta M_s - \eta M_o)}{(\beta f_{cw} - \eta f_{tt})} \quad (9.6)$$

9.3.2 Minimum Section Modulus for the Bottom Fibre Z_b

The bottom fibre stress must meet the criteria given by Eqs. (9.4b) and (9.4d). From Eq. (9.4b),

$$\frac{\beta P_i}{A} + \frac{\beta P_i e_s}{Z_b} - \frac{M_o}{Z_b} \leq f_{ct}$$

$$\Rightarrow \qquad \beta\eta\left[\frac{P_i}{A} + \frac{P_i e_s}{Z_b}\right] - \frac{\eta M_o}{Z_b} \leq \eta f_{ct} \qquad (9.7a)$$

From Eq. (9.4d),

$$\frac{\eta P_i}{A} + \frac{\eta P_i e_s}{Z_b} - \frac{M_s}{Z_b} \geq f_{tw}$$

$$\Rightarrow \qquad \beta\eta\left[\frac{P_i}{A} + \frac{P_i e_s}{Z_b}\right] - \frac{\beta M_s}{Z_b} \geq \beta f_{tw}$$

$$\Rightarrow \qquad -\beta\eta\left[\frac{P_i}{A} + \frac{P_i e_s}{Z_b}\right] + \frac{\beta M_s}{Z_b} \leq -\beta f_{tw} \qquad (9.7b)$$

Adding Eqs. (9.7a) and (9.7b) gives the required section modulus Z_b. Thus,

$$Z_b \geq \frac{(\beta M_s - \eta M_o)}{(\eta f_{ct} - \beta f_{tw})} \qquad (9.8)$$

Comments

- In Eqs. (9.6) and (9.8), M_o and M_s are the bending moments at the critical section.
- From, Eqs. (9.6) and (9.8), it can be seen that the minimum values of Z_a and Z_b depends only on the difference between the applied moments (M_s and M_o) and on permissible stresses, and not on their absolute values.
- The minimum values of Z_a and Z_b depends on M_o and M_s. But M_o and M_s can be determined only if the self-weight of the member is known. The self-weight can be calculated only if the cross-section (hence Z_a and Z_b) is known. Hence the solution can be obtained only by a trial and error method or by using standard sections.

Example 9.1

A simply supported post-tensioned rectangular beam of span 15 m has to carry a live load of 10 kN/m. The permissible tensile stresses at transfer and in service condition are 2.5 N/mm^2 and 2 N/mm^2 respectively. The concrete is of grade M 60 with a compressive strength of 40 N/mm^2 at the transfer loading stage. Determine a appropriate section for the beam.

Solution:

Step 1: *Assuming a trial section*

For a span to depth ratio of 20, trial depth D is:

$$D = \frac{l}{20} = \frac{15 \times 10^3}{20} = 750 \text{ mm}$$

Assume a rectangular section of width b = 400 mm and depth D = 800 mm (trial $b/D = \frac{1}{2}$).

- Area of cross-section of the beam $A = b\,D = 400 \times 800 = 320000$ mm^2
- Section modulus $Z = \frac{1}{6}bD^2 = \frac{400 \times 800^2}{6} = 42.67 \times 10^6$ mm^3

Step 2: *Calculation of loads and bending moments at critical section*

- Self-weight of the beam w_{SW}

$$w_{SW} = \gamma_c A = 24 \times 320000 \times 10^{-6} = 7.68 \text{ kN/m}$$

- Dead load intensity $w_{DL} = w_{SW} = 7.68$ kN/m
- Live load intensity $w_{LL} = 10$ kN/m
- Load intensity at service condition $w_s = w_{DL} + w_{LL} = 17.68$ kN/m
- Moment at critical section

The critical section is taken as the mid-span of the beam.

Moment at transfer $M_o = \frac{w_{SW} l^2}{8} = \frac{7.68 \times 15^2}{8} = 216$ kN-m

Moment at service condition $M_s = \frac{w_s l^2}{8} = \frac{17.68 \times 15^2}{8} = 497.25$ kN-m

Step 3: *Calculation of allowable stresses*

- Allowable tensile stress
  Allowable tensile stress at transfer $f_{tt} = -2.5$ N/mm^2
  Allowable tensile stress at service condition $f_{tw} = -2$ N/mm^2

- Allowable compressive stress at transfer f_{ct}
  Compressive strength of concrete at transfer $f_{ci} = 40$ N/mm^2
  For M 60 grade concrete and for post-tensioned beam (Cl. 22.8.2.1 of the Code),
  $f_{ct} = 0.37 f_{ci} = 0.37 \times 40 = 14.8$ N/mm^2
- Allowable compressive stress at service condition f_{cw}
  From Cl. 22.8.1.1. of the Code (assuming Zone I),
  $f_{cw} = 0.35 f_{ck} = 0.35 \times 60 = 21$ N/mm^2

Step 4: *Calculation of minimum section modulus Z_{min}*
Assume loss ratios β and η as 0.9 and 0.75 respectively.

- Section modulus at the top fibre Z_a

$$(\beta M_s - \eta M_o) = (0.9 \times 497.2 \times 10^6) - (0.75 \times 216 \times 10^6)$$
$$= 285.48 \times 10^6 \text{ N-mm}$$
$$Z_a \geq \frac{(\beta M_s - \eta M_o)}{(\beta f_{cw} - \eta f_{tt})}$$
$$Z_a \geq \frac{285.48 \times 10^6}{[(0.9 \times 21) - (0.75 \times (-2.5))]}$$
$$Z_a \geq 13.74 \times 10^6 \text{mm}^3$$

- Section modulus at the bottom fibre Z_b

$$Z_b \geq \frac{(\beta M_s - \eta M_o)}{(\eta f_{ct} - \beta f_{tw})}$$
$$Z_b \geq \frac{285.48 \times 10^6}{[(0.75 \times 14.8) - (0.9 \times (-2))]}$$
$$Z_b \geq 22.13 \times 10^6 \text{ mm}^3$$

- *Minimum section modulus Z_{min}*
  For a rectangular section $Z_a = Z_b$. Hence in this example, the requirement for Z_b governs.
  $Z_{min} = 22.13 \times 10^6$ mm^2 (< 42.67×10^6 mm^3, OK)
  Since the section modulus of the trial section is greater than Z_{min}, the trial section is adequate.

9.4 Design of Prestressing Force

Once the section moduli Z_a and Z_b are determined, the next step is to find the required prestressing force *at the critical section* based on the maximum possible tendon eccentricity (which depends on the minimum cover requirements).

The stress inequalities Eqs. (9.4a–d) are rearranged to get inequalities for the required prestressing force for a given tendon eccentricity e_s. From Eq. (9.4a),

$$\frac{\beta P_i}{A} - \frac{\beta P_i e_s}{Z_a} + \frac{M_o}{Z_a} \geq f_{tt}$$

Multiplying the equation by Z_a gives:

$$\frac{Z_a \beta P_i}{A} - \beta P_i e_s + M_o \geq Z_a f_{tt}$$

$$\Rightarrow \qquad P_i\left(\frac{Z_a \beta}{A} - \beta e_s\right) \geq (Z_a f_{tt} - M_o)$$

Hence,

$$P_i \geq \frac{(Z_a f_{tt} - M_o)}{\beta\left(\frac{Z_a}{A} - e_s\right)} \tag{9.9a}$$

Similarly rearranging Eqs. (9.4b–d) gives:

$$P_i \leq \frac{(Z_b f_{ct} + M_o)}{\beta\left(\frac{Z_b}{A} + e_s\right)} \tag{9.9b}$$

$$P_i \leq \frac{(Z_a f_{cw} - M_s)}{\eta\left(\frac{Z_a}{A} - e_s\right)} \tag{9.9c}$$

$$P_i \geq \frac{(Z_b f_{tw} + M_s)}{\eta\left(\frac{Z_b}{A} + e_s\right)} \tag{9.9d}$$

Comments

- Eqs. (9.9a–d) are used to find the limits within which prestressing force should lie for a given value of e_s. Since the cost of prestressing steel has to be minimum, the lowest value of prestressing force is usually chosen.
- The sense of the inequality in Eqs. (9.9a) and (9.9c) should be reversed, if the denominator becomes a negative quantity.

Example 9.2

A type 1 post-tensioned simply supported beam 250 mm wide, 400 mm deep and of span 10 m supports a live load of intensity 3 kN/m. If the maximum eccentricity of the tendon at mid-span is 80 mm above the soffit, find the minimum prestressing force required. The concrete is of grade M 60 and the strength of concrete at transfer is 40 N/mm^2.

Solution:

Step 1: *Calculation of section properties*

Area of cross-section of the beam $A = b\,D = 250 \times 400 = 100000$ mm^2

Section modulus $Z_a = Z_b = \dfrac{1}{6} bD^2 = \dfrac{250 \times 400^2}{6} = 6666.67 \times 10^3$ mm^3

Kern distance $Z_a/A = Z_b/A = 66.67$ mm

Tendon eccentricity $e_s = (400/2) - 80 = 120$ mm

Step 2: *Calculation of loads and bending moment*

- Self-weight of the beam w_{SW}

$$w_{SW} = \gamma_c A = 24 \times 100000 \times 10^{-6} = 2.4 \text{ kN/m}$$

- Load intensity at service condition w_s

  Dead load $w_{DL} = w_{SW} = 2.4$ kN/m
  Live load $w_{LL} = 3$ kN/m
  $w_s = w_{DL} + w_{LL} = 2.4 + 3 = 5.4$ kN/m

- Moment at mid-span

  Moment at transfer $M_o = \dfrac{w_{SW} l^2}{8} = \dfrac{2.4 \times 10^2}{8} = 30$ kN-m

  Moment at service condition $M_s = \dfrac{w_s l^2}{8} = \dfrac{5.4 \times 10^2}{8} = 67.5$ kN-m

Step 3: *Calculation of allowable stresses*

- Allowable tensile stresses (Cl. 22.7.1 of the Code)

  At transfer $f_{tt} = 0$ N/mm^2
  At service condition $f_{tw} = 0$ N/mm^2

- Allowable compressive stress in flexure (Refer Example 9.1)

  At transfer $f_{ct} = 14.8$ N/mm^2
  At service condition $f_{cw} = 21$ N/mm^2

Step 4: *Calculation of required prestressing force*

Assume loss ratios β and η as 0.9 and 0.75 respectively.

- Prestressing force based on tensile stress at transfer

$$P_i \geq \frac{(Z_a f_{tt} - M_o)}{\beta(Z_a/A - e_s)}$$

$(Z_a f_{tt} - M_o) = 0 - 30 \times 10^6 = -30 \times 10^6$ N-mm
$\beta\,(Z_a/A - e_s) = 0.9\,(66.67 - 120) = -40$ mm
Since the denominator is negative, the sense of the inequality is reversed. Hence,

$$P_i \leq 625 \text{ kN}$$

- Prestressing force based on compressive stress at transfer

$$P_i \leq \frac{(Z_b f_{ct} + M_o)}{\beta(Z_b / A + e_s)}$$

$$P_i \leq \frac{[(6666.67 \times 10^3 \times 14.8) + (30 \times 10^6)]}{0.9(66.67 + 120)} \times \frac{1}{1000}$$

$$P_i \leq 765.8 \text{ kN}$$

- *Prestressing force based on compressive stress at service condition*

$$P_i \leq \frac{(Z_a f_{cw} - M_s)}{\eta(Z_a / A - e_s)}$$

$$(Z_a f_{cw} - M_s) = (6666.67 \times 10^3 \times 21) - 67.5 \times 10^6 = 725 \times 10^6 \text{ N-mm}$$

$$\eta\,(Z_a/A - e_s) = 0.75(66.67 - 120) = -40 \text{ mm}$$

The denominator is negative. Hence reversing the sense of inequality gives,

$$P_i \geq -1812.5 \text{ kN}$$

- *Prestressing force based on tensile stress at service condition*

$$P_i \geq \frac{(Z_b f_{tw} + M_s)}{\eta(Z_b / A + e_s)}$$

$$P_i \geq \frac{(0 + 67.5 \times 10^6)}{0.75(66.67 + 120) \times 1000}$$

$$P_i \geq 482.13 \text{ kN}$$

- *Limits for initial prestressing force P_i*

  P_i should lie between 482.13 kN and 625 kN. Therefore the required minimum initial prestressing force is 482.13 kN.

Step 5: *Design of prestressing steel*

- Properties of 8 mm diameter stress-relieved wire
  Characteristic strength f_p = 1800 N/mm^2 (assumed value).
  Initial prestress $f_{pi} = 0.8\ f_p = 0.8 \times 1800 = 1440$ N/mm^2 (as per Cl. 18.5.1 of the Code, $f_{pi} \leq 0.8\ f_p$)
- Number of 8 mm diameter wires (n)

$$n = \frac{482.13 \times 10^3}{1440 \times \dfrac{\pi}{4} \times 8^2} = 6.7$$

Provide 7 numbers of 8 mm diameter wire.

Initial prestressing force $P_i = \dfrac{7 \times 1440 \times \dfrac{\pi}{4} \times 8^2}{1000} = 506.68$ kN (P_i lies within the limits, OK)

9.5 Magnel Diagram

In Example 9.2, Eqs. (9.9a–d) yielded a possible range of P_i for e_s = 120 mm. This may not be the case for other values of e_s. For example, if the maximum eccentricity is increased to 150 mm in Example 9.2, use of Eqs. (9.9a–d) results in:

$$P_i \leq 400.02 \text{ kN} \qquad \text{(from Eq. (9.9a))}$$

$$P_i \leq 659.82 \text{ kN} \qquad \text{(from Eq. (9.9b))}$$

$$P_i \geq -1160 \text{ kN} \qquad \text{(from Eq. (9.9c))}$$

$$P_i \geq 415.38 \text{ kN} \qquad \text{(from Eq. (9.9d))}$$

Hence for e_s = 150 mm, it is not possible to find P_i which will satisfy all the stress inequalities. If such a situation is encountered, the general procedure is to try another value of e_s and check whether it is possible to find P_i. Hence a trial and error procedure is required to find a satisfactory combination of P_i and e_s.

As an alternative to the above method, a graphical procedure can be adopted to find suitable combinations of P_i and e_s. From Eqs. (9.9a–d), it can be seen that a plot of e_s versus P_i is a hyperbolic curve. But a plot of e_s versus $1/P_i$ results in a straight line. This diagram (i.e., plot of e_s versus $1/P_i$) can be conveniently used to find the possible combinations of P_i and e_s. This diagram was first developed by the Belgian Engineer, Magnel and hence it is known as *Magnel diagram.*

The Magnel diagram is constructed by rearranging Eqs. (9.4a–d). From Eq. (9.4a),

$$\frac{\beta P_i}{A} - \frac{\beta P_i e_s}{Z_a} + \frac{M_o}{Z_a} \geq f_{tt}$$

Multiplying the equation by Z_a gives:

$$\frac{Z_a \beta P_i}{A} - \beta P_i e_s + M_o \geq Z_a f_{tt}$$

$$\Rightarrow \qquad \beta P_i e_s \leq (M_o - Z_a f_{tt}) + \frac{Z_a \beta P_i}{A}$$

Hence,

$$e_s \leq \frac{(M_o - Z_a f_{tt})}{\beta} \frac{1}{P_i} + \frac{Z_a}{A} \tag{9.10a}$$

Similarly rearranging Eqs. (9.4b–d) results in:

$$e_s \leq \frac{(M_o + Z_b f_{ct})}{\beta} \frac{1}{P_i} - \frac{Z_b}{A} \tag{9.10b}$$

$$e_s \geq \frac{(M_s - Z_a f_{cw})}{\eta} \frac{1}{P_i} + \frac{Z_a}{A} \tag{9.10c}$$

$$e_s \geq \frac{(M_s + Z_b f_{tw})}{\eta} \frac{1}{P_i} - \frac{Z_b}{A} \tag{9.10d}$$

In the above equations, $Z_a/A = k_b$ and $Z_b/A = k_a$ i.e., the lower and upper kern distances.

The Eqs. (9.10a–d) are in the form $y = m\,x + c$, where m is the slope and c is the vertical axis intercept, and when plotted results in the Magnel diagram (Fig. 9.2(a)). In the diagram, the horizontal axis represents the inverse of initial prestressing force and vertical axis represents the tendon eccentricity (positive values are plotted in the downward direction).

The hatched area ABCD in the Magnel diagram represents the feasible domain. Any point inside the feasible domain is a acceptable combination of P_i and e_s that satisfies all the four stress inequalities. Point C in the feasible domain represents the minimum prestressing force and the

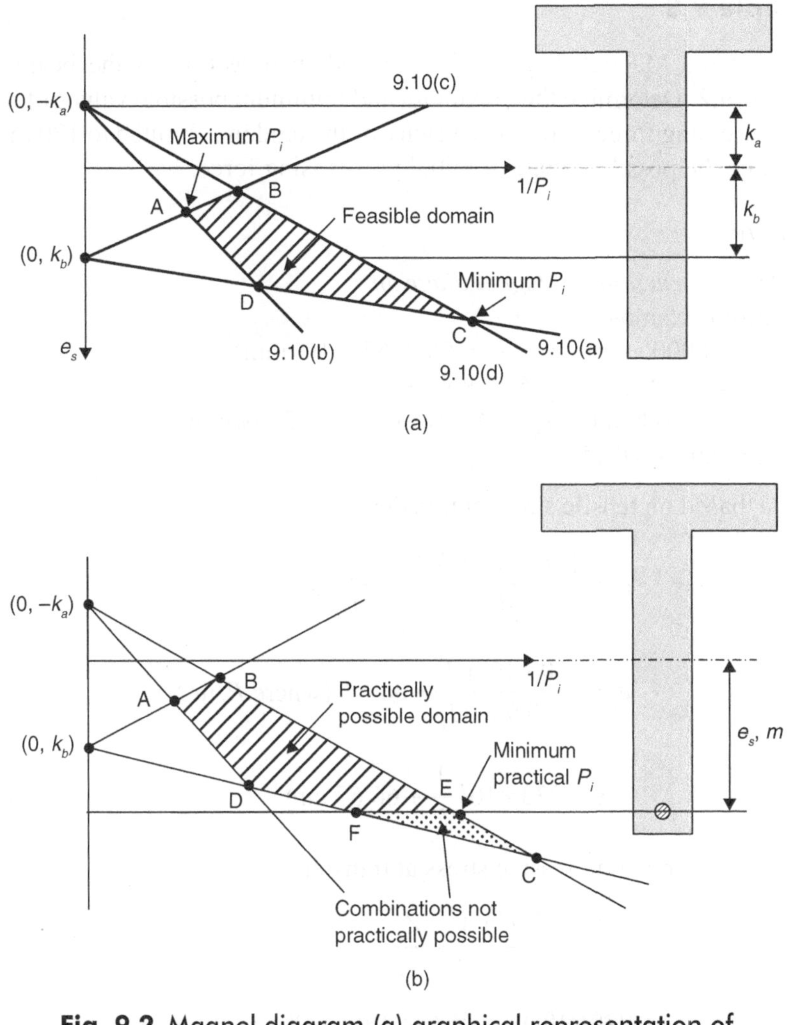

Fig. 9.2 Magnel diagram (a) graphical representation of governing inequalities and (b) reduced feasible domain

maximum eccentricity. Similarly, the maximum prestressing force and the minimum eccentricity that satisfies the stress inequalities is represented by point A.

The maximum practically possible tendon eccentricity ($e_{s,m}$) is governed by the minimum cover requirements. Some of the possible eccentricities can be greater than $e_{s,m}$. In such cases, these points should be excluded from the feasible domain as shown in Fig. 9.2(b). Now the feasible zone gets reduced to the hatched area ABEFD. The minimum prestressing force that is practically feasible is represented by the point E.

Example 9.3

Construct the Magnel diagram for the mid-span section of the beam in Example 9.2. Determine the maximum and minimum possible values of initial prestressing force. If the eccentricity of the tendon is limited to 120 mm, determine the possible range of initial prestressing force.

Solution:

Step 1: *Construction of Magnel diagram*

From Example 9.2,

$A = 100000$ mm^2, $Z_a = Z_b = 6666.67 \times 10^3$ mm^3

$M_o = 30$ kN-m, $M_s = 67.5$ kN-m

$f_{tt} = f_{tw} = 0$ N/mm^2, $f_{ct} = 14.8$ N/mm^2, $f_{cw} = 21$ N/mm^2

$\beta = 0.9$, $\eta = 0.75$

- e_s based on tensile stress at transfer

$$e_s \le \frac{(M_o - Z_a f_{tt})}{\beta}\frac{1}{P_i} + \frac{Z_a}{A}$$

$$e_s \le \frac{(30 \times 10^6)}{0.9}\frac{1}{P_i} + 66.67 \text{ (where } P_i \text{ in N)}$$

$$e_s \le 3.333 \times 10^7 \frac{1}{P_i} + 66.67 \text{ mm} \quad \text{(a)}$$

- e_s based on compressive stress at transfer

$$e_s \le \frac{(M_o + Z_a f_{ct})}{\beta}\frac{1}{P_i} - \frac{Z_b}{A}$$

$$e_s \le \frac{[(30 \times 10^6) + 6666.67 \times 10^3 \times 14.8)]}{0.9}\frac{1}{P_i} - 66.67 \text{ mm}$$

$$e_s \le 1.43 \times 10^8 \frac{1}{P_i} - 66.67 \text{ mm} \quad \text{(b)}$$

- e_s based on compressive stress at service condition

$$e_s \ge \frac{(M_s - Z_a f_{cw})}{\eta}\frac{1}{P_i} + \frac{Z_a}{A}$$

$$e_s \ge \frac{[(67.5 \times 10^6) - (6666.67 \times 10^3 \times 21)]}{0.75}\frac{1}{P_i} + 66.67 \text{ mm}$$

$$e_s \geq -9.7 \times 10^7 \frac{1}{P_i} + 66.67 \text{ mm} \qquad \text{(c)}$$

- e_s based on tensile stress at service condition

$$e_s \geq \frac{(M_s + Z_b f_{tw})}{\eta} \frac{1}{P_i} - \frac{Z_b}{A}$$

$$e_s \geq \frac{67.5 \times 10^6}{0.75} \frac{1}{P_i} - 66.67 \text{ mm}$$

$$e_s \geq 9 \times 10^7 \frac{1}{P_i} - 66.67 \text{ mm} \qquad \text{(d)}$$

The Mangel diagram is constructed by plotting Eqs. (a–d) and is shown in Fig. 9.3. In the figure, EFGH represents the feasible domain.

Step 2: *Finding the minimum and maximum values of P_i*

- Minimum prestressing force (coordinate of *G*)

  From the figure, maximum feasible $\frac{1}{P_i} = 2.38 \times 10^{-6} \frac{1}{N}$

  $\Rightarrow$ Minimum $P_i = 420$ kN ($e_s = 145$ mm)

- Maximum prestressing force (coordinate of *E*)

  Minimum feasible $\frac{1}{P_i} = 5.6 \times 10^{-7} \frac{1}{N}$

  $\Rightarrow$ Maximum $P_i = 1785.7$ kN ($e_s = 12.5$ mm)

Step 3: *Calculation of range of P_i for $e_s = 120$* mm

For $e_s = 120$ mm (line '*e*' in Fig. 9.3), the range of possible P_i is obtained from the coordinate of points p_1 and p_2 in Fig. 9.3.

- Lower limit of P_i (co-ordinate of p_2)

$$\frac{1}{P_i} = 2.8 \times 10^{-6} \frac{1}{N} \Rightarrow P_i = 480.8 \text{ kN}$$

- Upper limit of P_i (co-ordinate of p_1)

$$\frac{1}{P_i} = 1.68 \times 10^{-6} \frac{1}{N} \Rightarrow P_i = 625 \text{ kN}$$

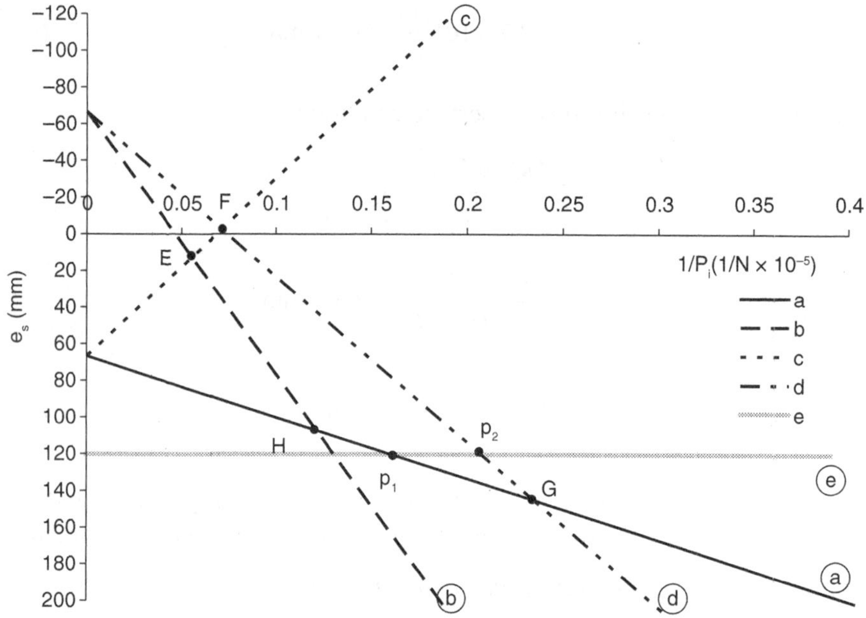

Fig. 9.3 Magnel diagram

Hence P_i should lie between 480.8 kN and 625 kN. The solution obtained from Magnel diagram matches with that obtained by using Eq. (9.9) (see Example 9.2).

9.6 Cable Zone

So far the prestressing force and tendon eccentricity was determined based on the conditions at critical section. The prestressing force (P_i) thus obtained will be a constant throughout the member. But the stress inequalities must be satisfied at all sections of the member. Since moments change along the length of the member, some aspect of the prestress should also be varied to meet the stress requirements. P_i being a constant, the only option is to change the tendon eccentricities over the length of the member.

It is possible to find the limits of e_s for a given P_i that satisfies the stress inequalities at all sections in the member. Thus a allowable cable zone can be determined within which the profile of the cable can take any shape. As long as the cable line lies within this zone, the stresses at all sections will be within the limits.

The term cable line refers to the resultant of all individual tendons (see Example 5.1). Hence even if some of the tendons physically lie outside the cable zone, only the resultant of all tendons, i.e., the cable line should lie within the cable zone.

The limits for the cable zone are obtained from Eqs. (9.10a–d). In this case, M_o and M_s are the bending moments at the section being considered. Out of the four inequalities, only two of them will be governing the allowable cable zone.

Example 9.4

Sketch the allowable cable zone for the beam in Example 9.2.

Solution:

Step 1: *Finding the bending moment at different loading stages*

Bending moment is calculated at a section x metres from the left end support.

- Bending moment at transfer

$$M_o = w_{SW}\frac{l}{2}x - w_{SW}\frac{x^2}{2}$$

$$= (2.4 \times (10/2)\,x) - (2.4x^2/2) = (12\,x - 1.2\,x^2)\text{ kN-m}$$

$$= (12x - 1.2x^2)10^6\text{ N-mm; } x \text{ in m}$$

- Bending moment at service condition

$$M_s = w_s\frac{l}{2}x - w_s\frac{x^2}{2}$$

$$= (5.4 \times (10/2)\,x) - (5.7x^2/2) = (27\,x - 2.7\,x^2)\text{ kN-m}$$

$$= (27x - 2.7x^2)10^6\text{ N-mm; } x \text{ in m}$$

Step 2: *Calculation of limits for cable zone*

$P_i = 506.68 \times 10^3$ N

$Z_a = Z_b = 6666.67 \times 10^3$ mm^3, $A = 100000$ mm^2, $Z_a/A = Z_b/A = 66.67$ mm

$f_{tt} = f_{tw} = 0$ N/mm^2, $f_{ct} = 14.8$ N/mm^2, $f_{cw} = 21$ N/mm^2

$\beta = 0.9$, $\eta = 0.75$

- e_s based on tensile stress at transfer

$$e_s \le \frac{(M_o - Z_a f_{tt})}{\beta P_i} + \frac{Z_a}{A};\quad e_s \le \frac{(12x - 1.2x^2)10^6}{(0.9 \times 506.68 \times 10^3)} + 66.67$$

$$e_s \le -2.632x^2 + 26.32x + 66.67\text{ mm} \tag{9.11a}$$

- e_s based on compressive stress at transfer

$$e_s \le \frac{(M_o + Z_b f_{ct})}{\beta P_i} - \frac{Z_b}{A}$$

$$e_s \leq \frac{[(12x - 1.2x^2)10^6 + 6666.67 \times 10^3 \times 14.8)]}{(0.9 \times 506.68 \times 10^3)} - 66.67 \text{ mm}$$

$$e_s \leq -2.632x^2 + 26.32x + 149.7 \text{ mm} \tag{9.11b}$$

- e_s based on compressive stress at service condition

$$e_s \geq \frac{(M_s - Z_a f_{cw})}{\eta P_i} + \frac{Z_a}{A}$$

$$e_s \geq \frac{[(27x - 2.7x^2)10^6 - (6666.67 \times 10^3 \times 21)]}{(0.75 \times 506.68 \times 10^3)} + 66.67 \text{ mm}$$

$$e_s \geq -7.1x^2 + 1x - 301.74 \text{ mm} \tag{9.11c}$$

- e_s based on tensile stress at service condition

$$e_s \geq \frac{(M_s + Z_b f_{tw})}{\eta P_i} - \frac{Z_b}{A}; \quad e_s \geq \frac{(27x - 2.7x^2) \times 10^6}{(0.75 \times 506.68 \times 10^3)} - 66.67 \text{ mm}$$

$$e_s \geq -7.1x^2 + 71.1x - 66.67 \text{ mm} \tag{9.11d}$$

Step 3: *Plotting the cable zone*

Eqs. (9.11a–d) are plotted to get the cable zone (Fig. 9.4(a)). In the figure, the distance of the tendon from the soffit d_p ($d_p = D/2 - e_s$) is plotted for the entire member.

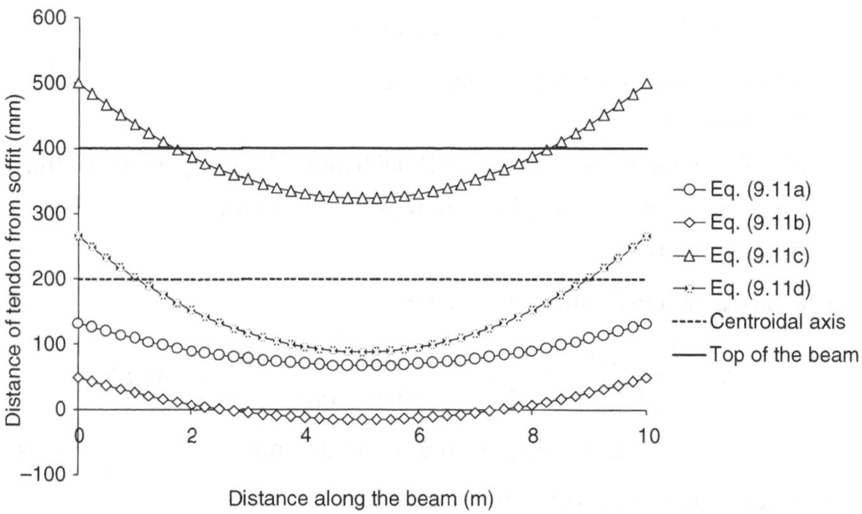

(a)

Fig. 9.4 (a) Limits for cable zone

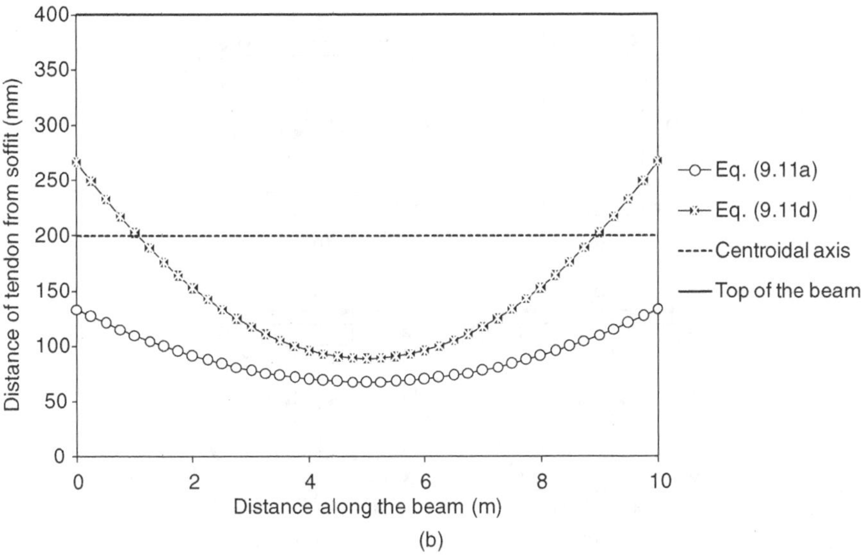

Fig. 9.4 (b) Allowable cable zone

From Fig. 9.4(a), it can be seen that only Eqs. (9.11a) and (9.11d) governs the size of the allowable cable zone. The cable zone for the given beam is shown in Fig. 9.4(b).

In this example, since the prestressing force is chosen close to the lower limit, the allowable upper and lower eccentricities are very close at the critical section (i.e., at mid-span). Choosing a higher prestressing force would lead to a larger width for cable zone at the critical section.

9.7 Selection of Cross-section

For beams with short spans, where dead load will be a small fraction of the total load, rectangular cross-sections can be chosen (Fig. 9.5(a)). For medium and long spans, flanged sections (Figs. 9.5(b)-(e)) are preferred.

In practice, first the depth of the beam is chosen. Having chosen the depth, the *flexural efficiency factor* of the section Q (discussed in Section 9.7.1) of the feasible sections are compared to arrive at the most optimum section.

9.7.1 Flexural Efficiency Factor Q

Consider a simply supported prestressed concrete beam in which tensile stresses are not permitted. Hence the stress distribution in the limiting condition at transfer and service conditions are shown in Fig. 9.6. For this to

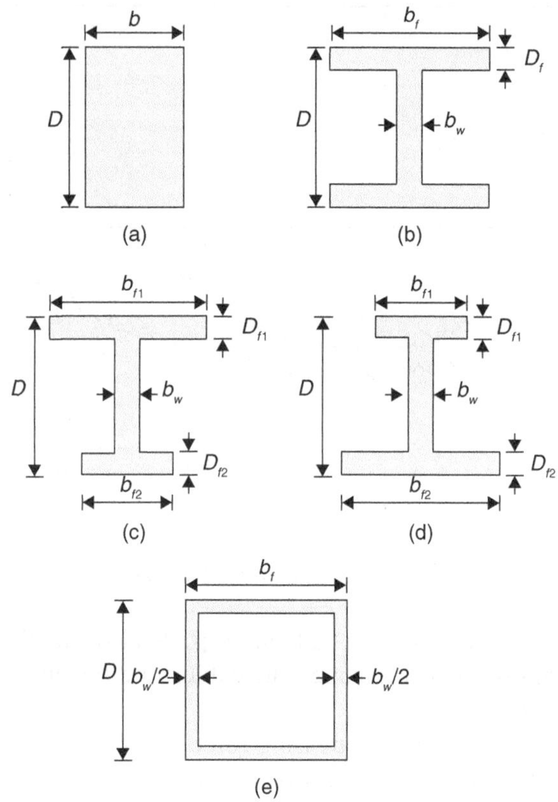

Fig. 9.5 Types of prestressed concrete cross-sections (a) rectangular section, (b) symmetrical I section, (c) T section, (d) inverted T section and (e) box section

happen, the resultant compressive force C must pass through the lower and upper kern points of the section (see Section 5.2.2.1) at the transfer loading stage (Fig. 9.6(a)) and at service condition (Fig. 9.6(b)) respectively. Equating the external moment to internal resting moment gives,

At Transfer: (Fig. 9.6(a))

$$M_o = Pz = P(e_s - k_b) \tag{9.12a}$$

At Service Condition: (Fig. 9.6(b))

$$M_s = Pz = P(e_s + k_a) \tag{9.12b}$$

The applied bending moment at transfer M_o is caused due to the self-weight of the beam and can be taken as the moment due to dead load M_{DL} (i.e., neglecting the super imposed dead load). Similarly, the bending

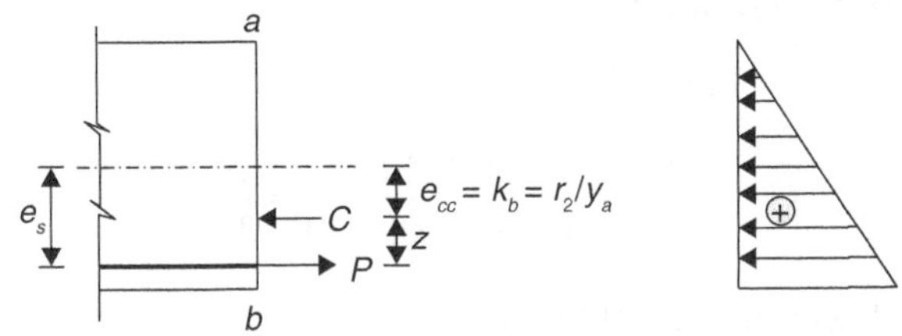

(i) *C* at lower kern point (ii) Longitudinal stress distribution

(a)

Fig. 9.6 (a) At transfer

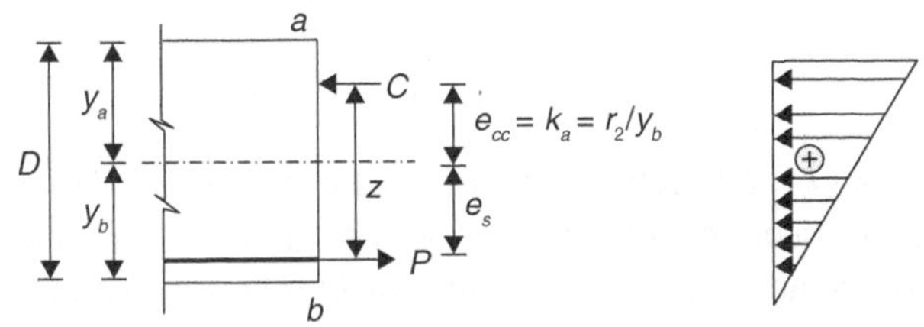

(i) *C* at upper kern point (ii) Longitudinal stress distribution

(b)

Fig. 9.6 (b) At service condition

moment at service condition M_s is obtained by adding moments due to live load M_{LL} and dead load M_{DL}. Hence,

$$M_s = M_{LL} + M_{DL} = M_{LL} + M_o = P(e_s + k_a) \tag{9.12c}$$

$$\Rightarrow \qquad M_{LL} = M_s - M_o = P(e_s + k_a) - P(e_s - k_b) = P(k_a + k_b) \tag{9.12d}$$

From Eq. (9.12d), it can be observed that greater the sum of the two kern distances $(k_a + k_b)$, greater will be capacity of the section to carry the external moment. If D is the depth of the beam, the flexural efficiency factor Q of the section is defined by the ratio:

$$Q = \frac{(k_a + k_b)}{D}$$

Substituting $K_a = r^2/y_b$ and $K_b = r^2/y_a$ gives,

$$Q = \frac{\left(\frac{r^2}{y_b} + \frac{r^2}{y_a}\right)}{D} = \frac{\left(\frac{r^2}{y_b} + \frac{r^2}{y_a}\right)}{(y_a + y_b)} \; ; (D = y_a + y_b)$$

Thus,

$$Q = \frac{r^2}{y_a y_b} = \frac{I}{A y_a y_b} \tag{9.13}$$

Q for Some Common Sections

- Rectangular section (Fig. 9.5(a))

$$Q = \frac{I}{Ay_a y_b} = \frac{bD^3/12}{(bD)(D/2)(D/2)} = \frac{1}{3}$$

- Symmetrical I section (Fig. 9.5(b))

$$Q = \frac{1}{3}\left[\frac{1-\left(1-\frac{b_w}{b_f}\right)\left(1-2\frac{D_f}{D}\right)^3}{1-\left(1-\frac{b_w}{b_f}\right)\left(1-2\frac{D_f}{D}\right)}\right]$$

Fig. 9.7 shows Q for I section for different b_w/b_f and D_f/D ratios. From the figure, it can be seen that Q for I sections are greater than for a rectangular section, and reaches a maximum value for D_f/D ratio close to 0.1. Also, smaller the ratio of b_w/b_f, greater is the flexural efficiency factor.

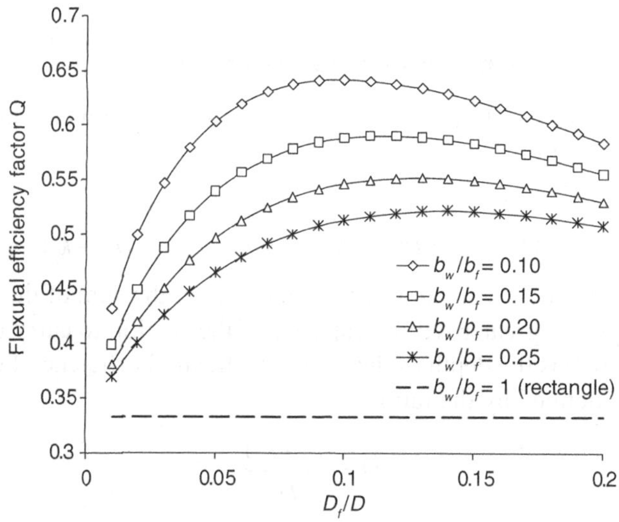

Fig. 9.7 Flexural efficiency factor for I section

9.7.2 General Guidelines for Beam Sections

The depth of the beam D is chosen between 1/24 and 1/16 of the span. For flanged sections, the width of the flange b_f is taken as 0.5-0.8 times the depth D. The flange thickness D_f is kept between $0.2D$ to $0.3D$. The web thickness b_w can be taken as 0.2-0.3 times the width of flange b_f.

9.8 Requirements for Flexural Reinforcement

9.8.1 Concrete Cover

The clear cover is defined as the distance measured from the concrete surface to the nearest surface of the tendon. A minimum cover is required to protect the steel reinforcement from corrosion and fire, and also to prevent slipping of steel relative to concrete (i.e., to improve bond between steel and concrete).

The Code (Cl. 11.1.6) recommends to provide a minimum clear cover of 20 mm for pre-tensioned and 30 mm for post-tensioned members respectively. The minimum cover should be increased by 10 mm for members located in aggressive environment.

9.8.2 Spacing of Tendons

The Code (Cl. 11.1.7) specifies minimum limits for spacing between the tendons so that during casting, concrete can be easily placed between the tendons. The spacing limits as per the Code are:

- For single wires used in pre-tensioned members, the minimum clear spacing should be the greater of 3 times wire diameter and $1^1/_3$ times the maximum size of aggregate.
- For cables or large bars, the minimum clear spacing is taken as the greater of 40 mm, maximum size of cable or bar and maximum size of aggregate plus 5 mm.
- In the case of grouped cables (not more than four tendons per group), the minimum horizontal spacing (Fig. 9.8) is taken as the greater of 40 mm and maximum size of aggregate plus 5 mm. The minimum vertical spacing should not be less than 50 mm.

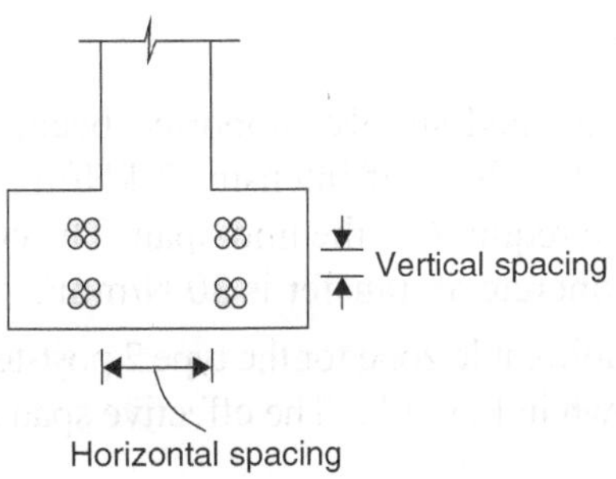

Fig. 9.8 Spacing of grouped cables

9.8.3 Minimum Longitudinal Steel

To avoid sudden failure due to fracture of steel, the Code (Cl.18.6.3.3(a)) specifies a minimum percentage of longitudinal reinforcement (including both tensioned and untensioned reinforcement) as 0.2 percent of concrete area. If high yield strength deformed steel is used, the reinforcement can be reduced to 0.15 percent.

9.8.4 Minimum Side Face Reinforcement

For relatively deep flexural members, it is necessary to provide side face reinforcement to control shrinkage and thermal cracks, and to improve the resistance against lateral buckling of the web. The Code (Cl. 18.6.3.3(b)) recommends to provide a minimum side face reinforcement of 0.05 percent of the web area on each face of the beam having depth of web greater than 500 mm. Further, the spacing of the individual bars of such reinforcement should not exceed 200 mm.

9.9 Design Procedure for Prestressed Concrete Members

After finding the required prestressing force and cable profile, the next step is to check the moment capacity at ultimate limit stage. Additional untensioned reinforcement can be provided in case of inadequate ultimate moment capacity. Then the member is designed for shear force and torsion. After the design of anchorage zones, the member is checked for deflection. The flow chart for the design procedure is shown in Fig. 9.9.

Problems

9.1 A type 2 post-tensioned simply supported beam shown in Fig. 9.10 has to support a live load of intensity 7 kN/m^2. Find the minimum prestressing force required at the mid-span. The compressive strength of M 50 grade concrete at transfer is 30 N/mm^2.

9.2 Sketch the allowable cable zone for the type 2 post-tensioned simply supported beam shown in Fig. 9.11. The effective span of the beam is 20 m.

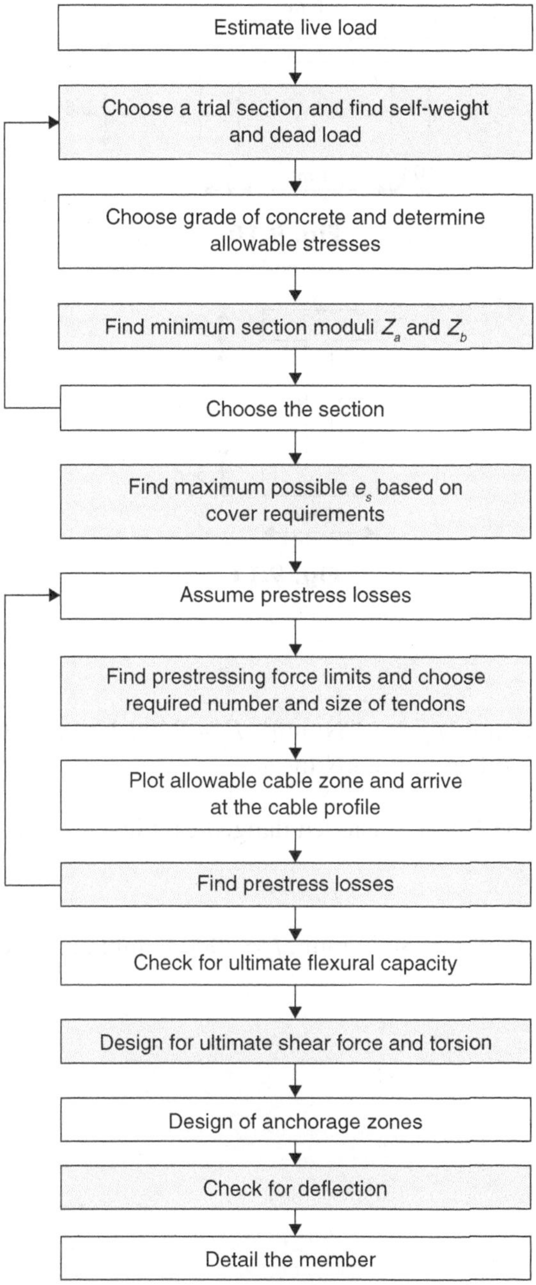

Fig. 9.9 Flow chart for the design of type 1 and type 2 prestressed concrete members

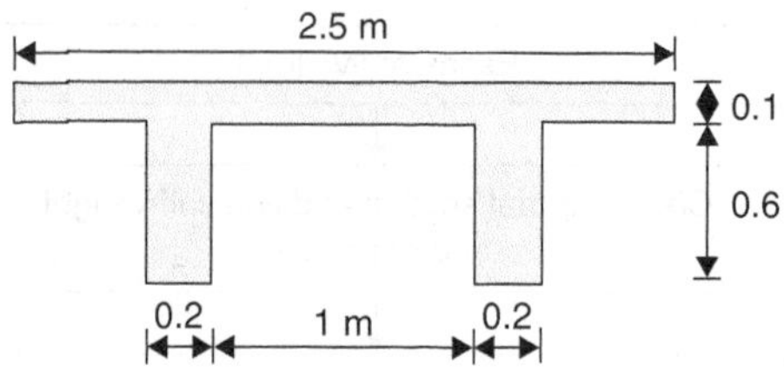

Fig. 9.10

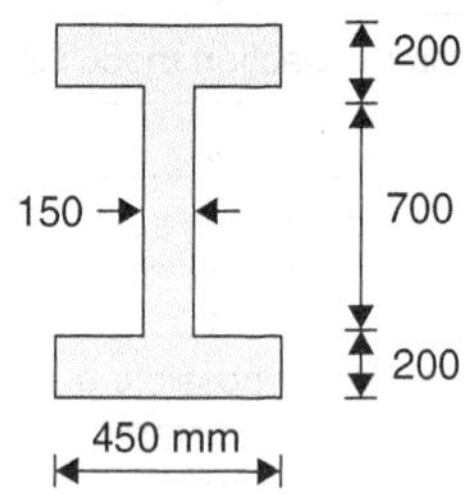

Fig. 9.11

Data:

f_{ck} = 40 N/mm^2, f_{ci} = 30 N/mm^2, P_i = 2400 kN, β = 0.9, η = 0.8, w_{LL} = 16 kN/m, w_{SIDL} = 2 kN/m

9.3 Design a type 1 post-tensioned flanged cantilever beam of span 8 m.

Data:

f_{ck} = 50 N/mm^2, f_{ci} = 30 N/mm^2, f_p = 2000 N/mm^2, w_{LL} = 8 kN/m

chapter 10

Composite Members

10.1 Introduction

The term composite members in the context of prestressed concrete refers to the members made by combining precast (PC) prestressed concrete beams (or slabs) with cast-in-situ (CI) reinforced concrete slabs. The important advantage of using composite construction lies in the use of standard precast prestressed members which are manufactured under good quality control. The precast members act as a permanent formwork for the CI slab, and hence there will be savings in formwork and scaffolding. Further the construction can be carried out at a faster pace with little interference to traffic or work below the structure. Some typical forms of composite members are shown in Fig. 10.1.

In the composite member, the precast prestressed concrete section is designed to carry its self-weight and the weight of the wet concrete of the in-situ slab. Once the concrete in the slab has hardened, the cast-in-situ slab and the precast section act as a single unit (composite section) and they are designed to resist the super imposed dead load and live load.

10.2 Analysis at Serviceability Limit State

While calculating the elastic flexural stresses, it is necessary to identify the loads acting on the precast prestressed beam and on the composite section. Those loads which are applied before the hardening of the cast-in-situ slab

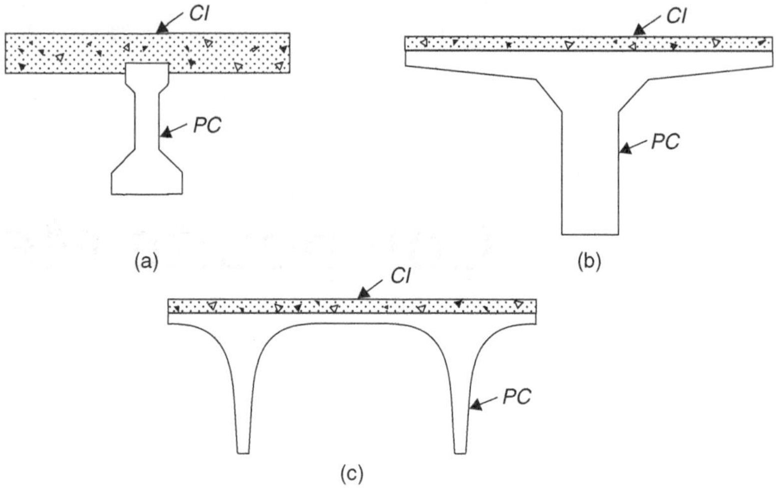

Fig. 10.1 Typical composite sections

acts on the precast section and causes bending about the centroidal axis of the precast section. But the loads applied after hardening of the slab cause bending about the centroid of the composite section. The stresses developed in the composite section due to bending are added with the stresses already present in the precast beam section.

The procedure to find the longitudinal stresses developed in a simply supported composite member at different loading stages is discussed below.

10.2.1 Stresses in Precast Web at Transfer

At this loading stage, the loads acting on the precast web are the prestressing force P_o (prestressing force after immediate losses) and the self-weight of the precast beam.

The stresses at the top (σ_a) and bottom (σ_b) fibres of the precast beam section are (Fig. 10.2(a)):

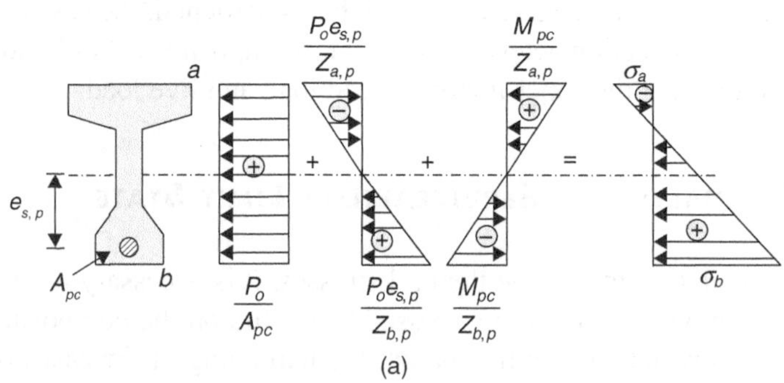

Fig. 10.2 (a) Stresses in precast web at transfer

$$\sigma_a = \frac{P_o}{A_{pc}} - \frac{P_o e_{s,p}}{Z_{a,p}} + \frac{M_{pc}}{Z_{a,p}} \quad (10.1a)$$

$$\sigma_b = \frac{P_o}{A_{pc}} + \frac{P_o e_{s,p}}{Z_{b,p}} - \frac{M_{pc}}{Z_{b,p}} \quad (10.1b)$$

where M_{pc} is the bending moment due to self-weight of precast beam. $e_{s,p}$ is the tendon eccentricity measured from the centroid of the precast beam section of area A_{pc}. $Z_{a,p}$ and $Z_{b,p}$ are the section moduli of the top and bottom fibres of precast beam section.

10.2.2 Stresses in Precast Web After Time Dependent Losses

The stresses at the top and bottom fibres of the precast beam section at this loading stage are (Fig. 10.2(b)):

$$\sigma_a' = \frac{P_e}{A_{pc}} - \frac{P_e e_{s,p}}{Z_{a,p}} + \frac{M_{pc}}{Z_{a,p}} \quad (10.2a)$$

$$\sigma_b = \frac{P_e}{A_{pc}} + \frac{P_e e_{s,p}}{Z_{b,p}} - \frac{M_{pc}}{Z_{b,p}} \quad (10.2b)$$

where P_e is the effective prestressing force.

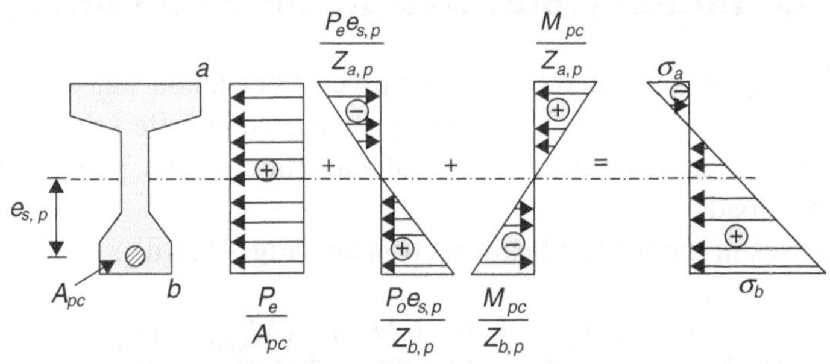

Fig. 10.2 (b) Stresses in precast web after time dependent losses

10.2.3 Stresses in Precast Web After Casting of In-situ Slab

It can be assumed that casting of slab is done after the occurrence of all time dependent losses in precast beam. The loads acting on the prestressed beam at this loading stage are its self-weight, effective prestressing force P_e

and the weight of the wet concrete slab. The stresses due to these loads are (Fig. 10.2(c)):

$$\sigma_a = \frac{P_e}{A_{pc}} - \frac{P_e e_{s,p}}{Z_{a,p}} + \frac{(M_{pc} + M_{CI})}{Z_{a,p}} \quad (10.3a)$$

$$\sigma_b = \frac{P_e}{A_{pc}} + \frac{P_e e_{s,p}}{Z_{b,p}} - \frac{(M_{pc} + M_{CI})}{Z_{b,p}} \quad (10.3b)$$

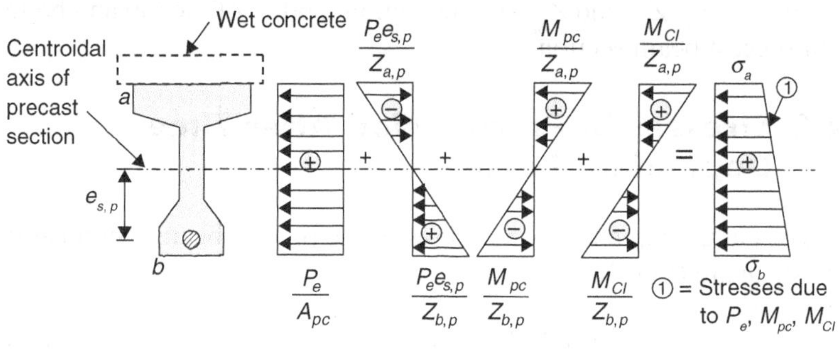

Fig. 10.2 (c) Stresses in precast web after casting of concrete slab

where M_{CI} is the bending moment due to the weight of wet concrete slab.

10.2.4 Stresses in Composite Section at Service Condition (Total Design Load Condition)

After hardening of the concrete slab, the precast beam and slab act together as a composite section. The loads acting on the composite section are the superimposed dead load (self-weight of pavements, foot paths, finishes, etc.) and live load.

The stresses in the precast beam section are (Fig. 10.2(d)):

$$\sigma_a = \frac{P_e}{A_{pc}} - \frac{P_e e_{s,p}}{Z_{a,p}} + \frac{(M_{pc} + M_{CI})}{Z_{a,p}} + \frac{(M_{SIDL} + M_{LL})}{Z_{a,c}} \quad (10.4a)$$

$$\sigma_b = \frac{P_e}{A_{pc}} + \frac{P_e e_{s,p}}{Z_{b,p}} - \frac{(M_{pc} + M_{CI})}{Z_{b,p}} - \frac{(M_{SIDL} + M_{LL})}{Z_{b,c}} \quad (10.4b)$$

where M_{SIDL} and M_{LL} are the bending moments due to superimposed dead load and live load. $Z_{a,c}$ and $Z_{b,c}$ are the section moduli of the top and bottom fibre of the web (precast beam) of the composite section.

Similarly, the stresses at the top (σ_c) and the bottom (σ_d) fibres of the flange (in-situ concrete slab) are:

$$\sigma_c = \frac{(M_{SIDL} + M_{LL})}{Z_{c,c}} \quad (10.5a)$$

$$\sigma_d = \frac{(M_{SIDL} + M_{LL})}{Z_{d,c}} \quad (10.5b)$$

where $Z_{c,c}$ and $Z_{d,c}$ are the section moduli of the top and bottom fibres of the flange in the composite section.

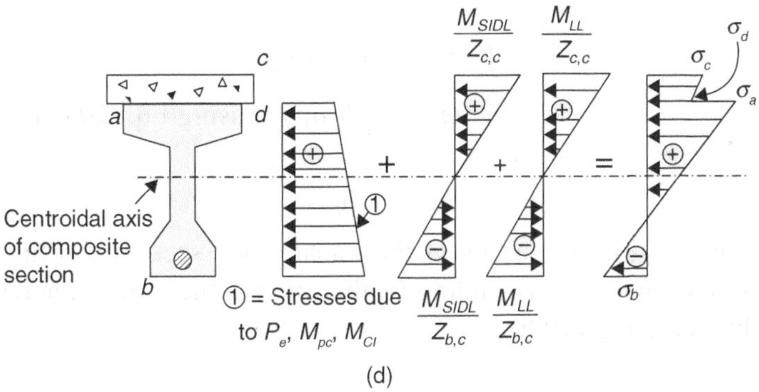

Fig. 10.2 (d) Stresses in the composite section

From Fig. 10.2(d), it can be seen that there is a stress discontinuity at the junction between the precast beam and slab. This is due to the fact that stresses are already present in the precast beam before the development of composite action. But in the case of slab, stresses occur only after it becomes a part of the composite section.

Transformed Section

Since the slab is cast in field conditions, its compressive strength and modulus of elasticity will be less than the corresponding values in precast beam. This difference is accounted in the analysis by using the concept of transformed section in which the cast-in-situ slab is replaced by a smaller but equivalent section having the properties of precast concrete.

Fig. 10.3 shows the actual composite section and the corresponding transformed section. In the figure, b_f is the actual width of flange and b_{ft} is the width of flange in transformed section.

At any horizontal section distance y from the centroid, the longitudinal strain ε is same in both the actual and transformed section. Hence,

$$\varepsilon = \frac{f_{c1}}{E_{c1}} = \frac{f_{c2}}{E_{c2}}$$

$$\Rightarrow \qquad f_{c1} = \frac{E_{c1}}{E_{c2}} f_{c2} = m_c\, f_{c2} \text{ (Since } m_c = E_{c1}/E_{c2}) \qquad (10.6)$$

f_{c1} and f_{c2} are the stresses in the actual and transformed section. E_{c1} and E_{c2} are the modulus of elasticity of the in-situ concrete and precast concrete. m_c is the modular ratio, which is normally less than one.

If dC is the compressive force in the infinitesimal element of thickness dy located at a distance y from the centroid of composite section, then:

$$dC = b_f\, dy\, f_{cl} = dy\, f_{c2}$$

$$\Rightarrow \qquad b_f\, dy\, (m_c f_{c2}) = b_f\, dy\, f_{c2} \text{ (using Eq. (10.6))}$$

Hence,

$$b_{ft} = m_c\, b_f \qquad (10.7)$$

Using b_{ft}, the section properties of the transformed section and hence the stress f_{c2} is determined. After finding f_{c2}, the stress in the actual concrete f_{c1} is obtained by using Eq. (10.6).

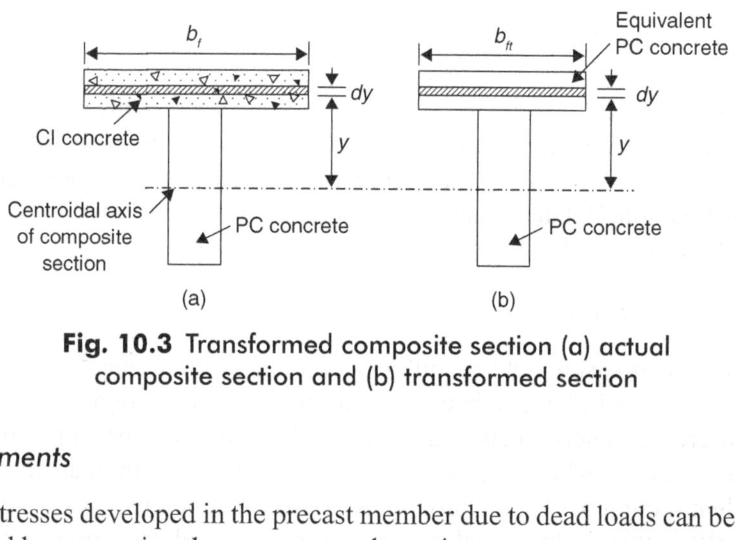

Fig. 10.3 Transformed composite section (a) actual composite section and (b) transformed section

Comments

The stresses developed in the precast member due to dead loads can be minimized by supporting the precast member using temporary supports (*props*) at the time of casting of in-situ concrete slab. The props are removed only after hardening of the in-situ concrete. Hence the weight of slab is resisted by the composite section. This type of construction is known as *propped composite construction* (see Example 10.2).

Example 10.1

A precast post-tensioned beam of rectangular section 300 mm wide, 1200 mm deep and span 20 m is prestressed using parabolic cable having eccentricity of 200 mm from the soffit at the mid-span. The area of prestressing steel is 1400 mm^2 and the initial prestress is 1500 N/mm^2. The grade of concrete for the precast beam is M 60. After the precast beam is erected in place, a top slab of width 1200 mm and thickness 180 mm is cast over it. The resulting composite beam carries superimposed dead load and live load of intensities 1.5 kN/m and 6 kN/m respectively. If unpropped composite construction is used, determine the stresses in concrete at the mid-span at different loading stages. The compressive strength of concrete in slab is 40 N/mm^2.

Solution:

Step 1: *Calculation of section properties (Fig. 10.4)*

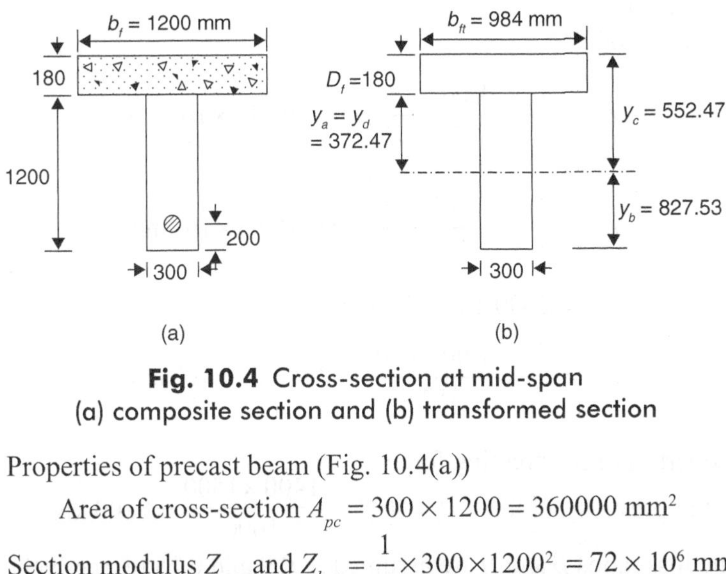

Fig. 10.4 Cross-section at mid-span (a) composite section and (b) transformed section

- Properties of precast beam (Fig. 10.4(a))

$$\text{Area of cross-section } A_{pc} = 300 \times 1200 = 360000 \text{ mm}^2$$

$$\text{Section modulus } Z_{a,p} \text{ and } Z_{b,p} = \frac{1}{6} \times 300 \times 1200^2 = 72 \times 10^6 \text{ mm}^3$$

$$\text{Tendon eccentricity } e_{s,p} = (1200/2) - 200 = 400 \text{ mm}$$

$$\text{Self-weight } w_{SW,p} = \gamma_c A_{pc} = 24 \times 360000 \times 10^{-6} = 9 \text{ kN/m}$$

$$\text{Modulus of elasticity } E_{c2} = 5000\sqrt{60} = 38729.83 \text{ N/mm}^2$$

- Properties of cast-in-situ slab (Fig. 10.4(a))

$$\text{Area of cross-section } A_{CI} = 1200 \times 180 = 216000 \text{ mm}^2$$

$$\text{Self-weight } w_{SW,c} = \gamma_c A_{CI} = 24 \times 216000 \times 10^{-6} = 5.18 \text{ kN/m}$$

$$\text{Modulus of elasticity } E_{c1} = 5000\sqrt{40} = 31622.78 \text{ N/mm}^2$$

- Properties of transformed section (Fig. 10.4(b))

Modular ratio $m_c = \frac{E_{c1}}{E_{c2}} = \frac{31622.78}{38729.83} = 0.82$

Width of flange $b_{ft} = m_c\, b_f = 0.82 \times 1200 = 984$ mm

Area of cross-section $A_{tr} = (984 \times 180) + (1200 \times 300) = 537120$ mm^2

Distance of centroid from bottom fibre y_b:

$$y_b = \frac{(984 \times 180 \times 1290) + (1200 \times 300 \times 600)}{537120} = 827.53 \text{ mm}$$

Distance of centroid from top fibre $y_c = 1380 - 827.53 = 552.47$ mm

Moment of inertia of transformed section I_{tr}:

$$I_{tr} = \left[\frac{984 \times 180^3}{12}\right] + [984 \times 180 \times (552.47 - 90)^2] + \left[\frac{300 \times 1200^3}{12}\right]$$
$$+ [1200 \times 300 \times (827.53 - 600)^2]$$
$$= 1.002 \times 10^{11} \text{ mm}^4$$

Section modulus:

$$Z_{a,t} = \frac{I_{tr}}{y_a} = \frac{1.002 \times 10^{11}}{372.47} = 2690.15 \times 10^5 \text{ mm}^3$$

$$Z_{b,t} = \frac{I_{tr}}{y_b} = \frac{1.002 \times 10^{11}}{827.53} = 1210.83 \times 10^5 \text{ mm}^3$$

$$Z_{d,t} = Z_{a,t} = 2690.15 \times 105 \text{ mm}^3$$

$$Z_{c,t} = \frac{I_{tr}}{y_c} = \frac{1.002 \times 10^{11}}{552.47} = 1813.67 \times 10^5 \text{ mm}^3$$

- Properties of prestressing force

Initial prestressing force $P_i = A_p f_{pi} = \frac{1400 \times 1500}{1000} = 2100$ kN

The loss ratios β and η are assumed as 0.9 and 0.75 respectively.

Prestressing force immediately after transfer $P_o = \beta P_i = 1890$ kN

Effective prestressing force $P_e = \eta P_i = 1575$ kN

Step 2: *Calculation of stresses in precast beam at transfer (Stage 1)*

- Bending moment at mid-span due to self-weight of precast beam M_{pc}

$$M_{pc} = \frac{w_{SW,p}\, l^2}{8} = \frac{9 \times 20^2}{8} = 450 \text{ kN-m}$$

- Stress in top fibre of the precast beam σ_a

$$\sigma_a = \frac{P_o}{A_{pc}} - \frac{P_o e_{s,p}}{Z_{a,p}} + \frac{M_{pc}}{Z_{a,p}}$$

$$= \left(\frac{1890 \times 10^3}{360000}\right) - \left(\frac{1890 \times 10^3 \times 400}{72 \times 10^6}\right) + \left(\frac{450 \times 10^6}{72 \times 10^6}\right)$$

$$= 5.25 - 10.5 + 6.25 = 1 \text{ N/mm}^2$$

- Stress in bottom fibre of the precast beam σ_b

$$\sigma_b = \frac{P_o}{A_{pc}} + \frac{P_o e_{s,p}}{Z_{b,p}} - \frac{M_{pc}}{Z_{b,p}}$$

$$= 5 + 10.5 - 6.25 = 9.5 \text{ N/mm}^2$$

Step 3: *Calculation of stresses in precast beam after time dependent losses (Stage 2)*

- Stress in top fibre of the precast beam σ_a

$$\sigma_a = \frac{P_e}{A_{pc}} - \frac{P_e e_{s,p}}{Z_{a,p}} + \frac{M_{pc}}{Z_{a,p}}$$

$$= \left(\frac{1575 \times 10^3}{360000}\right) - \left(\frac{1575 \times 10^3 \times 400}{72 \times 10^6}\right) + \left(\frac{450 \times 10^6}{72 \times 10^6}\right)$$

$$= 4.38 - 8.75 + 6.25 = 1.88 \text{ N/mm}^2$$

- Stress in bottom fibre of the precast beam σ_b

$$\sigma_b = \frac{P_e}{A_{pc}} + \frac{P_e e_{s,p}}{Z_{b,p}} - \frac{M_{pc}}{Z_{b,p}}$$

$$= 4.38 + 8.75 - 6.25 = 6.88 \text{ N/mm}^2$$

Step 4: *Calculation of stresses in precast beam after casting of in-situ slab (Stage 3)*

Bending moment at mid-span due to self-weight of in-situ slab M_{CI}

$$M_{CI} = \frac{w_{SW,c}\, l^2}{8} = \frac{5.18 \times 20^2}{8} = 259 \text{ kN-m}$$

- Stress in top fibre of the precast beam σ_a

$$\sigma_a = \frac{P_e}{A_{pc}} - \frac{P_e e_{s,p}}{Z_{a,p}} + \frac{(M_{pc} + M_{CI})}{Z_{a,p}}$$

$$= \left(\frac{1575\times10^3}{360000}\right) - \left(\frac{1575\times10^3\times400}{72\times10^6}\right) + \left(\frac{[450+259]\times10^6}{72\times10^6}\right)$$

$$= 4.38 - 8.75 + 9.85 = 5.48 \text{ N/mm}^2$$

- Stress in bottom fibre of the precast beam σ_b

$$\sigma_b = \frac{P_e}{A_{pc}} + \frac{P_e e_{s,p}}{Z_{b,p}} - \frac{(M_{pc} + M_{CI})}{Z_{b,p}}$$

$$= 4.38 + 8.75 - 9.85 = 3.28 \text{ N/mm}^2$$

Step 5: *Calculation of stresses in composite section (Stage 4)*

- Bending moment at mid-span due to superimposed dead load M_{SIDL}

$$M_{SIDL} = \frac{w_{SIDL} l^2}{8} = \frac{1.5\times20^2}{8} = 75 \text{ kN-m}$$

- Bending moment at mid-span due to live load M_{LL}

$$M_{LL} = \frac{w_{LL} l^2}{8} = \frac{6\times20^2}{8} = 300 \text{ kN-m}$$

- Stresses at top and bottom fibres of the web in composite section (σ_a and σ_b)

$$\sigma_a = \frac{P_e}{A_{pc}} - \frac{P_e e_{s,p}}{Z_{a,p}} + \frac{(M_{pc} + M_{CI})}{Z_{a,p}} + \frac{(M_{SIDL} + M_{LL})}{Z_{a,t}}$$

$$= 5.48 + \frac{[(75+300)\times10^6]}{2690.15\times10^5} = 6.88 \text{ N/mm}^2$$

$$\sigma_b = \frac{P_e}{A_{pc}} + \frac{P_e e_{s,p}}{Z_{b,p}} - \frac{(M_{pc} + M_{CI})}{Z_{b,p}} - \frac{(M_{SIDL} + M_{LL})}{Z_{b,t}}$$

$$= 3.28 - \frac{[(75+300)\times10^6]}{1210.83\times10^5} = 0.18 \text{ N/mm}^2$$

- Stresses in top and bottom fibres of the flange in transformed section ($\sigma_{c,t}$ and $\sigma_{d,t}$)

$$\sigma_{c,t} = \frac{(M_{SIDL} + M_{LL})}{Z_{c,t}} = \frac{(75+300)\times10^6}{1813.67\times10^5} = 2.07 \text{ N/mm}^2$$

$$\sigma_{d,t} = \frac{(M_{SIDL} + M_{LL})}{Z_{d,t}} = \frac{(75+300)\times 10^6}{2690.15\times 10^5} = 1.39 \text{ N/mm}^2$$

- Stresses in top and bottom fibres of the flange in the composite section (σ_c and σ_d)

$$\sigma_c = m_c\, \sigma_{c,t} = 0.82 \times 1.43 = 1.17 \text{ N/mm}^2$$
$$\sigma_d = m_c\, \sigma_{d,t} = 0.82 \times 0.96 = 0.79 \text{ N/mm}^2$$

The stresses at the various loading stages are shown in Fig. 10.5

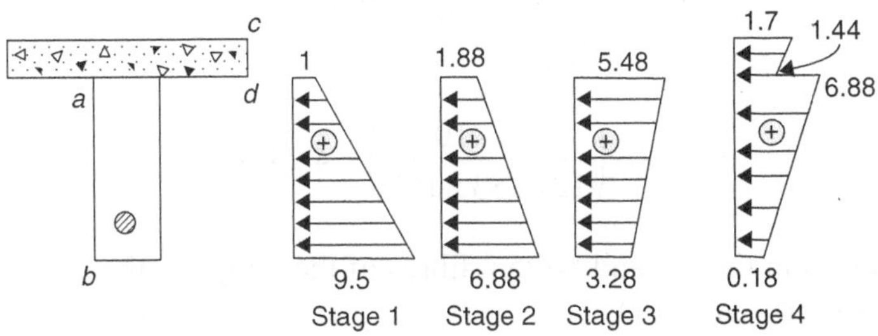

Fig. 10.5 Longitudinal stress distribution in N/mm²

Example 10.2

Determine the stresses in the composite beam in Example 10.1 at different loading stages, if the precast beam was propped during the casting of slab.

Solution:

Step 1: *Calculation of stresses in precast beam at transfer (Stage 1)*

The stresses in the top and bottom fibres of the beam are same as in Example 10.1.

$$\sigma_a = 1 \text{ N/mm}^2$$
$$\sigma_b = 9.5 \text{ N/mm}^2$$

Step 2: *Calculation of stresses in precast beam after time dependent losses (Stage 2)*

The stresses in the precast beam are same as in Example 10.1.

$$\sigma_a = 1.88 \text{ N/mm}^2$$
$$\sigma_b = 6.88 \text{ N/mm}^2$$

Step 3: *Calculation of stresses in the composite beam after the casting of in-situ slab (Stage 3)*

Since the precast beam was propped during casting of slab, the self-weight of the slab is resisted by the composite section.

- Stresses in the top and bottom fibres of the web in composite section (σ_a and σ_b)

$$\sigma_a = \frac{P_e}{A_{pc}} - \frac{P_e e_{s,p}}{Z_{a,p}} + \frac{M_{pc}}{Z_{a,p}} + \frac{M_{CI}}{Z_{a,t}}$$

$$= 1.88 + \left(\frac{259 \times 10^6}{2690.15 \times 10^5} \right) = 2.84 \text{ N/mm}^2$$

$$\sigma_b = \frac{P_e}{A_{pc}} + \frac{P_e e_{s,p}}{Z_{b,p}} - \frac{M_{pc}}{Z_{b,p}} - \frac{M_{CI}}{Z_{b,t}}$$

$$= 6.88 - \left(\frac{259 \times 10^6}{1210.83 \times 10^5} \right) = 4.74 \text{ N/mm}^2$$

- Stresses in the top and bottom fibres of the flange in the transformed section ($\sigma_{c,t}$ and $\sigma_{d,t}$)

$$\sigma_{c,t} = \frac{M_{CI}}{Z_{c,t}} = \frac{259 \times 10^6}{1813.67 \times 10^5} = 1.43 \text{ N/mm}^2$$

$$\sigma_{d,t} = \frac{M_{CI}}{Z_{d,t}} = \frac{259 \times 10^6}{2690.15 \times 10^5} = 0.96 \text{ N/mm}^2$$

- Stresses in the top and bottom fibres of the flange in the composite section (σ_c and σ_d)

$$\sigma_c = m_c\, \sigma_{c,t} = 0.82 \times 1.43 = 1.17 \text{ N/mm}^2$$
$$\sigma_d = m_c\, \sigma_{d,t} = 0.82 \times 0.96 = 0.79 \text{ N/mm}^2$$

Step 4: *Calculation of stresses in the composite section after the application of superimposed dead and live loads (Stage 4)*

- Stresses in the top and bottom fibres of the web in composite section (σ_a and σ_b)

$$\sigma_a = \frac{P_e}{A_{pc}} - \frac{P_e e_{s,p}}{Z_{a,p}} + \frac{M_{pc}}{Z_{a,p}} + \frac{M_{CI}}{Z_{a,t}} + \frac{(M_{SIDL} + M_{LL})}{Z_{a,t}}$$

$$= 2.84 + \left(\frac{375 \times 10^6}{2690.15 \times 10^5} \right) = 4.23 \text{ N/mm}^2$$

$$\sigma_b = \frac{P_e}{A_{pc}} + \frac{P_e e_{s,p}}{Z_{b,p}} - \frac{M_{pc}}{Z_{b,p}} - \frac{M_{CI}}{Z_{b,t}} - \frac{(M_{SIDL} + M_{LL})}{Z_{b,t}}$$

$$= 4.74 - \left(\frac{375 \times 10^6}{1210.83 \times 10^5}\right) = 1.64 \text{ N/mm}^2$$

- Stresses in the top and bottom fibres of the flange in the transformed section ($\sigma_{c,t}$ and $\sigma_{d,t}$)

$$\sigma_{c,t} = \frac{M_{CI}}{Z_{c,t}} + \frac{(M_{SIDL} + M_{LL})}{Z_{c,t}}$$

$$= 1.43 + 2.07 = 3.5 \text{ N/mm}^2$$

$$\sigma_{d,t} = \frac{M_{CI}}{Z_{d,t}} + \frac{(M_{SIDL} + M_{LL})}{Z_{d,t}}$$

$$= 0.96 + 1.39 = 2.35 \text{ N/mm}^2$$

- Stresses in the top and bottom fibres of the flange in the composite section (σ_c and σ_d)

$$\sigma_c = m_c \, \sigma_{c,t} = 0.82 \times 3.5 = 2.87 \text{ N/mm}^2$$
$$\sigma_d = m_d \, \sigma_{d,t} = 0.82 \times 2.35 = 1.93 \text{ N/mm}^2$$

The stresses in the composite beam at different loading stages are shown in Fig. 10.6.

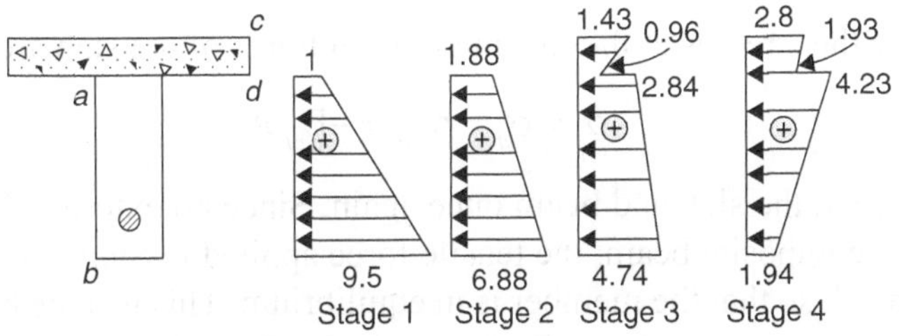

Fig. 10.6 Longitudinal stress distribution in N/mm²

10.3 Stresses Due to Differential Shrinkage

Since the in-situ slab is cast at a much later stage than the precast beam, most of the shrinkage in the precast beam would have taken place before the concrete in the slab gets hardened. Further, due to less compressive strength of concrete in slab, the water content in the in-situ concrete will be higher than in precast concrete leading to considerable shrinkage in the slab region. Thus differential shrinkage occurs between the in-situ slab and the precast beam.

Due to bond between the slab and precast beam, as the slab tries to shrink, the beam prevents the movement. This results in the development of stresses in both slab and beam, which should be considered in the design process.

The stresses due to shrinkage can be determined by using the following procedure:

(i) *De-bond the slab and the precast beam and allow the slab to shrink* freely (Fig. 10.7(a)). In the figure, ε_{shd} is the differential shrinkage strain, and in the absence of experimental data, it can be taken as 100×10^{-6}.

(ii) Apply a external tensile force F_T at the centroid of the slab to nullify the effect of shrinkage and to bring the slab to its original length (Fig. 10.7(b)). The tensile force F_T is:

$$F_T = -A_{CI} E_{cl\ shd} \tag{10.8a}$$

where A_{CI} is the cross-sectional area of the in-situ slab and E_{c1} is the modulus of elasticity of in-situ concrete. As per the sign convention, the tensile force F_T is treated as a negative quantity. This force causes a uniform tensile stress σ_{slab} in the slab.

$$\sigma_{slab} = \frac{F_T}{A_{CI}} = -E_{c1}\,\varepsilon_{shd} \tag{10.8b}$$

Hence the stresses at the top and bottom fibres of the slab are:

$$\sigma_c = \sigma_d = \sigma_{slab} = -E_{cl}\,\varepsilon_{shd} \tag{10.8c}$$

(iii) Now bond the slab and beam once again. Since no external force acts on the composite beam, the tensile force applied in Step (ii) has to be balanced, so that the member is in equilibrium. This is done by applying a compressive force F of magnitude F_T at the centroid of slab region of the composite member (Fig. 10.7(c)).

This force causes stresses in the composite section and transformed section properties should be used to find the stresses. After finding the stresses in the transformed section, the stresses in the actual material are obtained using Eq. (10.6).

The stresses in the top and bottom fibres of the flange in the composite section due to compressive force F are:

$$\sigma_c = \left(\frac{F}{A_{tr}} - \frac{Fe_f}{Z_{c,t}} \right) m_c \qquad (10.9a)$$

$$\sigma_d = \left(\frac{F}{A_{tr}} - \frac{Fe_f}{Z_{d,t}} \right) m_c \qquad (10.9b)$$

where A_{tr} is the area of cross-section of the transformed section. e_f is the eccentricity of F from the centroid of the transformed section.

The stresses in the top and bottom fibres of the web are:

$$\sigma_a = \frac{F}{A_{tr}} - \frac{Fe_f}{Z_{a,t}} \qquad (10.9c)$$

$$\sigma_b = \frac{F}{A_{tr}} + \frac{Fe_f}{Z_{b,t}} \qquad (10.9d)$$

(iv) The stresses obtained in steps (ii) and (iii) are added to get the stress caused due to differential shrinkage.

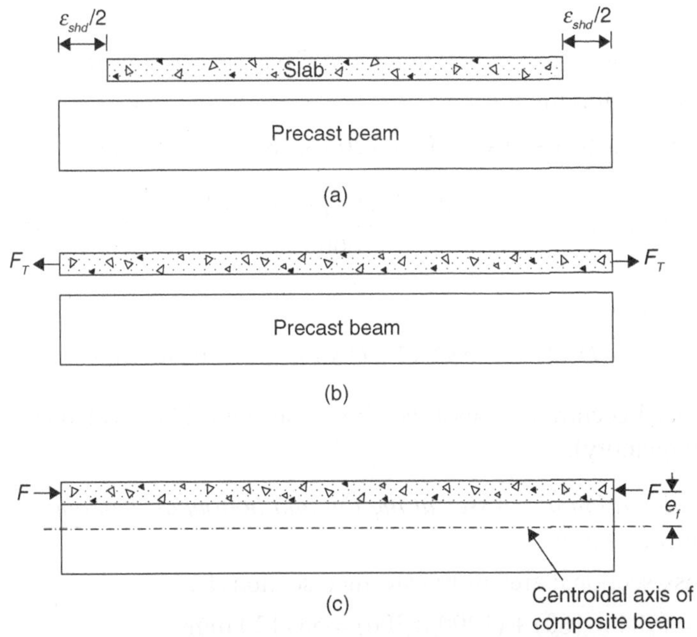

Fig. 10.7 Calculation of stresses due to differential shrinkage (a) free shrinkage of slab, (b) free shrinkage of slab compensated by tensile force F_t and (c) applying compressive force F in the composite beam

Comments

The compressive stresses in the web of the composite beam will be higher than in the flange region. Hence the creep deformation of web is higher than in the flange. Further creep in web would have taken place even before casting of slab. This leads to differential creep deformation between the web and flange resulting in the development of stresses. These stresses will cause reduction in the stresses caused due to differential shrinkage.

Also, stresses will be developed due to variation of temperature across the composite section. The hotter upper surface will expand more than the bottom surface causing stresses in the composite section.

The procedure used to find the stresses due to differential shrinkage can be used to find the stresses caused by differential creep and temperature movements.

Example 10.3

For the composite beam in Example 10.1, determine the stress distribution across the section, if the slab undergoes a differential shrinkage strain of 100×10^{-6}.

Solution:

Step 1: *Calculation of compression force F and its eccentricity e_f*

- Compression force F

  Area of cast-in-situ slab $A_{CI} = 1200 \times 180 = 216000$ mm^2

$$F = A_{CI}\, E_{c1}\, \varepsilon_{shd} = \frac{216000 \times 31622.78 \times 100 \times 10^{-6}}{1000} = 683.05 \text{ kN}$$

- Eccentricity e_f

  $e_f = -\,[y_c - D_f/2] = -\,[552.47 - (180/2)] = -\,462.47$ mm

  (*Note:* Eccentricity measured above the centroid is treated as a negative quantity).

Step 2: *Calculation of stresses at the top and bottom of the flange in composite section*

Cross-sectional area of transformed section A_{tr}:

$A_{tr} = (984 \times 180) + (1200 \times 300) = 537120$ mm^2

$$\sigma_c = -E_{c1}\varepsilon_{shd} + m_c\left[\frac{F}{A_{tr}} - \frac{F e_f}{Z_{c,t}}\right]$$

$$= -(31622.78 \times 100 \times 10^{-6})$$

$$+0.82\left[\left(\frac{683.05\times10^{3}}{537120}\right)-\left(\frac{683.05\times10^{3}\times(-462.47)}{1813.67\times10^{5}}\right)\right]$$

$$= -3.16 + 0.82\,[1.27 + 1.74] = -0.69\ \text{N/mm}^2$$

$$\sigma_d = -E_{c1}\varepsilon_{shd} + m_c\left[\frac{F}{A_{tr}} - \frac{Fe_f}{Z_{d,t}}\right]$$

$$= -3.16 + 0.82\left[1.27 - \left(\frac{683.05\times10^{3}\times(-462.47)}{2690.15\times10^{5}}\right)\right]$$

$$= -3.16 + 0.82[1.27 + 1.17] = -1.16\ \text{N/mm}^2$$

Step 3: *Calculation of stresses at the top and bottom fibres of the web in composite section*

$$\sigma_a = \left(\frac{F}{A_{tr}} - \frac{Fe_f}{Z_{a,t}}\right)$$

$$= \left(\frac{683.05\times10^{3}}{537120}\right)-\left(\frac{683.05\times10^{3}\times(-462.47)}{2690.15\times10^{5}}\right)$$

$$= 1.27 + 1.17 = 2.44\ \text{N/mm}^2$$

$$\sigma_b = \frac{F}{A_{tr}} + \frac{Fe_f}{Z_{b,t}}$$

$$= 1.27 + \left(\frac{683.05\times10^{3}\times(-462.47)}{1210.83\times10^{5}}\right)$$

$$= 1.27 - 2.61 = -1.34\ \text{N/mm}^2$$

The stresses developed due to differential shrinkage are shown in Fig. 10.8. These stresses must be added to the stresses caused by prestress and applied loads.

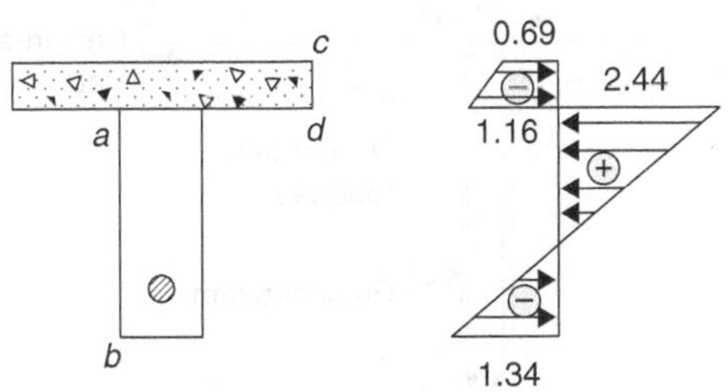

Fig. 10.8 Longitudinal stress distribution in N/mm²

10.4 Horizontal Shear Transfer

Adequate bond between the precast section and cast-in-situ concrete is required for the composite action of the member. Due to bending, there will be a tendency of the in-situ slab to slip horizontally. This causes the precast beam and the slab to behave as two independent beams kept one over the other (Fig. 10.9(a)).

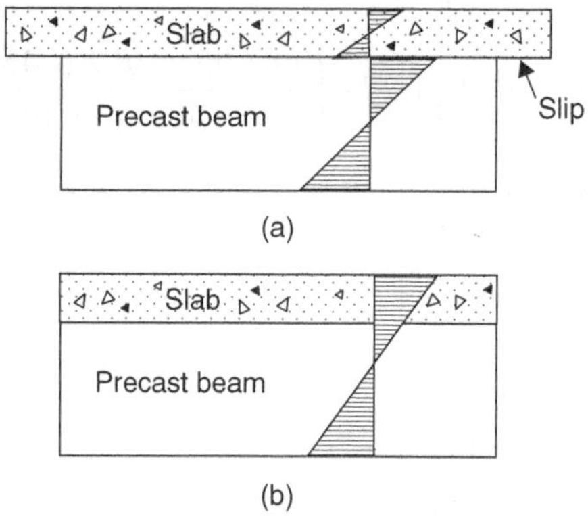

Fig. 10.9 Strain distribution across the cross-section (a) non-composite behaviour and (b) composite behaviour

The slip can be prevented by providing resistance to the horizontal shear stress developed at the interface between the slab and the precast beam. The resistance to shear at interface is provided by the adhesion and friction between the slab and precast beam. To enhance the resistance developed by friction, the top surface of the precast beam is deliberately roughened. This method is sufficient for composite sections having large contact areas (Figs 10.1(b) and (c)). For members having smaller contact area (Fig. 10.1(a)), the shear reinforcement in the web can be extended and anchored in the in-situ slab (Fig. 10.10) to act as shear friction reinforcement.

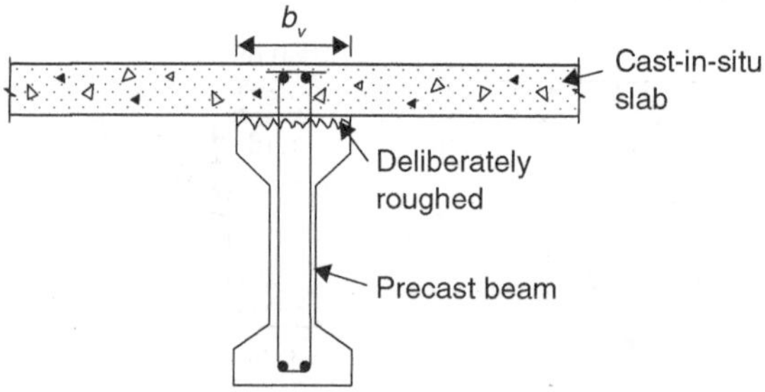

Fig. 10.10 Shear reinforcement for horizontal shear transfer

The required shear friction reinforcement A_h (mm²/m) is calculated as:

$$A_h = \frac{1000 b_v \tau_h}{0.87 f_v} \tag{10.10}$$

where b_v is the width of interface between the precast beam and slab (Fig. 10.10). f_v is the yield strength of shear reinforcement. τ_h is the horizontal shear stress at the interface approximated as:

$$\tau_h = \frac{V_u}{b_v d} \tag{10.11}$$

In the above equation, V_u is the factored shear force and d is the effective depth measured from the extreme compression fibre of the composite section to the centroid of the prestressing steel.

10.5 Ultimate Moment of Resistance

The differences in the elastic modulus of precast concrete and in-situ concrete will not affect the ultimate moment of resistance M_{uR}. Hence M_{uR} is calculated using the actual dimensions of the composite section (i.e., transformed section is not needed). Also the strain discontinuity at the junction between the slab and precast beam caused due to prior bending of precast beam can be neglected while calculating M_{uR}. But the differences in the concrete stress-strain relations for the in-situ concrete and precast concrete causes stress discontinuity in concrete at the interface between slab and beam (Fig. 10.11).

However, this discontinuity in the concrete stress distribution can be neglected because in most of the situation the neutral axis lies within the flange region. Further M_{uR} for an under-reinforced section depends on the strength of steel and not on the strength of concrete. Hence for finding M_{uR}, the stress-strain curve for in-situ concrete is used, and M_{uR} is calculated using the procedures discussed in Section 5.8.3.

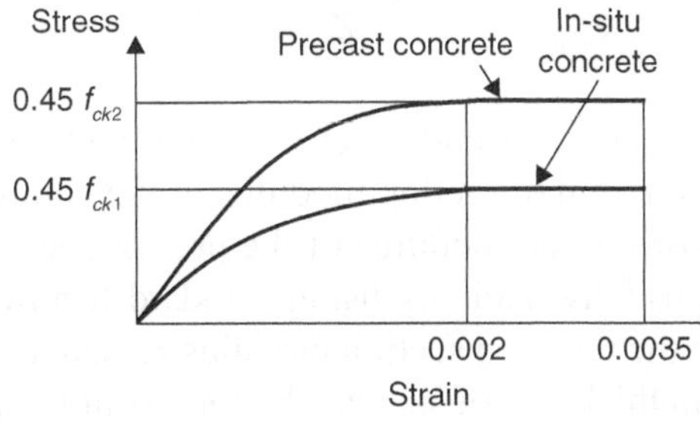

(a)

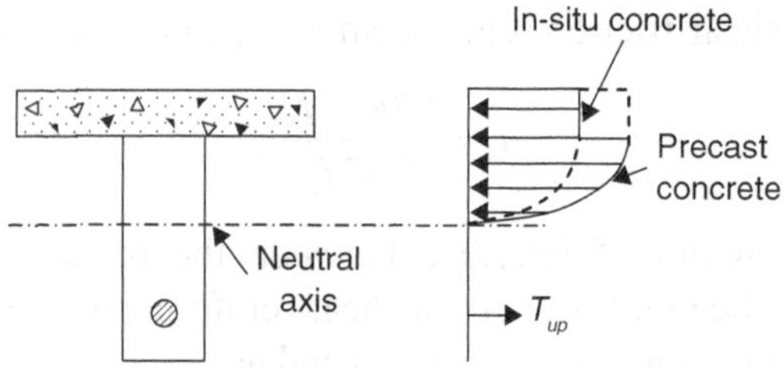

(b)

Fig. 10.11 Stress distribution in concrete for a composite section (a) design stress-strain curves for concrete at ultimate limit state and (b) stress discontinuity in concrete at the interface

10.6 Design of Composite Members

The procedures used for finding the required section modulus, prestressing force and allowable cable zone in a prestressed concrete member (Chapter 9) are used for the design of composite member. But in this case, since four loading stages are considered, there will be eight stress inequalities. Out of these, only a few will be governing the design of composite member.

10.6.1 Minimum Section Modulus

The minimum section modulus of the composite section can be determined by considering the stress conditions at the bottom fibre of the section during transfer loading stage and at total design load condition, i.e.,

$$\frac{\beta P_i}{A_{pc}} + \frac{\beta P_i e_{s,p}}{Z_{b,p}} - \frac{M_{pc}}{Z_{b,p}} \le f_{ct} \tag{10.12a}$$

$$\frac{\eta P_i}{A_{pc}} + \frac{\eta P_i e_{s,p}}{Z_{b,p}} - \frac{(M_{pc} + M_{CI})}{Z_{b,p}} - \frac{(M_{SIDL} + M_{LL})}{Z_{b,c}} \ge f_{tw} \tag{10.12b}$$

Using the above two equations and by eliminating the terms containing P_i and $e_{s,p}$, the minimum required section modulus of the composite section $Z_{b,c}$ is written in terms of section modulus of the precast section $Z_{b,p}$. Since the precast member is usually made by using standard formwork, $Z_{b,p}$ is a known quantity. Hence the required section modulus $Z_{b,c}$ can be determined. Using $Z_{b,c}$, the minimum thickness of slab can be found out for a given value of flange width.

10.6.2 Required Prestressing Force and Allowable Cable Zone

The prestressing force P_i and its eccentricity $e_{s,p}$ are governed by the stress conditions at the top fibre of the precast web at transfer and at the bottom fibre of the composite section during the total design load condition.

$$\frac{\beta P_i}{A_{pc}} - \frac{\beta P_i e_{s,p}}{Z_{a,p}} + \frac{M_{pc}}{Z_{a,p}} \geq f_{tt} \tag{10.13a}$$

$$\frac{\eta P_i}{A_{pc}} + \frac{\eta P_i e_{s,p}}{Z_{b,p}} - \frac{(M_{pc} + M_{CI})}{Z_{b,p}} - \frac{(M_{SIDL} + M_{LL})}{Z_{b,c}} \geq f_{tw} \tag{10.13b}$$

The limits for the prestressing force are determined using Eqs. (10.13a) and (b) for a given value of $e_{s,p}$. After choosing P_i, the allowable cable zone is determined using these stress inequalities.

Problems

10.1 A type 1 composite beam formed by propped construction method has a T-section consisting of the in-situ flange 1.2 m × 0.25 m and precast prestressed web 0.2 m × 0.8 m. The beam is simply supported and the effective span is 15 m. The compressive strength of concrete in the flange and web are 40 N/mm^2 and 60 N/mm^2 respectively. An initial prestressing force of 800 kN is applied at 200 mm from the soffit. Determine the maximum uniformly distributed load that the beam can carry.

10.2 Determine the ultimate moment of reistance of the composite T-section in Problem 10.1 for the following data:

$$f_p = 2000 \text{ N/mm}^2, f_{pi} = 0.7 f_p$$

10.3 Determine the intensity of shear stress at the web-flange junction of the beam section in Problem 10.1 when the section carries a factored shear force of 250 kN. Also design the shear friction reinforcement.

chapter 11

Indeterminate Structures

11.1 Introduction

A statically indeterminate structure has the following benefits when compared with a determinate structure:

- The continuity of the indeterminate structure greatly reduces the bending moments and deflections leading to economic design.
- Indeterminate structure offers better resistance to horizontal forces developed by wind, blast or earthquake.
- Tensioning of cables over several supports in a continuous member greatly reduce the number of expensive anchorages and the labour cost involved during the prestressing operation.
- Due to the presence of multiple load paths, failure of one member (or section) does not result in the collapse of the entire structure.
- Due to redistribution of moments, the load carrying capacity of indeterminate structure is higher than a determinate structure.

But the indeterminate structure has some limitations which are given below:

- The sections at the continuous supports are subjected to both high bending moment and shear force. The shear force causes reduction in the moment carrying capacity of the section, and hence additional flexural reinforcement has to be provided.

- Due to larger length of member and presence of number of bends in the cable profile, the prestress loss due to friction will be higher in continuous members.
- Stresses will be developed due to shrinkage, creep, temperature changes and support settlements.
- Moment reversal can occur in the structure due to alternate loading of spans.
- In a indeterminate structure, secondary moments are developed due to prestressing which complicates the design process.

In spite of these limitations, indeterminate structure is an attractive solution when compared with determinate structure, especially in the case of long span structures like bridges. Most of the limitations of indeterminate prestressed concrete structures can be eliminated by properly designing the structure.

11.2 Effects of Prestress in Indeterminate Structure

Prestressing a indeterminate structure usually induces support reactions which in turn causes bending moments and shear forces. The effects of prestress in a indeterminate structure are explained by considering a two-span continuous beam.

11.2.1 Primary and Secondary Moments and Shears

Fig. 11.1(a) shows a two-span continuous rectangular beam prestressed using a straight tendon at an eccentricity of 0.1 m and tensioned to 1000 kN. The beam is 250 mm wide and 600 mm deep. The modulus of elasticity of concrete is 30000 N/mm^2.

Assume that the support B is initially removed and the beam is not subjected to any external loads including its self-weight. The prestressing force exerts a constant compressive force P and a constant negative (hogging) moment M_1 ($M_1 = -Pe_s$) throughout the length of the beam. This moment M_1 causes the beam to bend upwards and the deflection at B due to M_1 is 37.04 mm (Fig. 11.1(b)).

But the continuous beam should have zero deflection at B. Hence a downward force of 30 kN is to be applied at B so that the beam has a downward deflection of 37.04 mm (Fig. 11.1(c)). Application of 30 kN force causes support reactions at A and D. These reactions induces bending moments M_2

which increases linearly from zero at A and D to a maximum value of 150 kN-m at B (Fig. 11.1(e)).

The moment M_1 acting *on concrete at a section* is known as the *primary bending moment* (Fig. 11.1(d)) and is given by:

$$M_1 = -P\, e_s \tag{11.1}$$

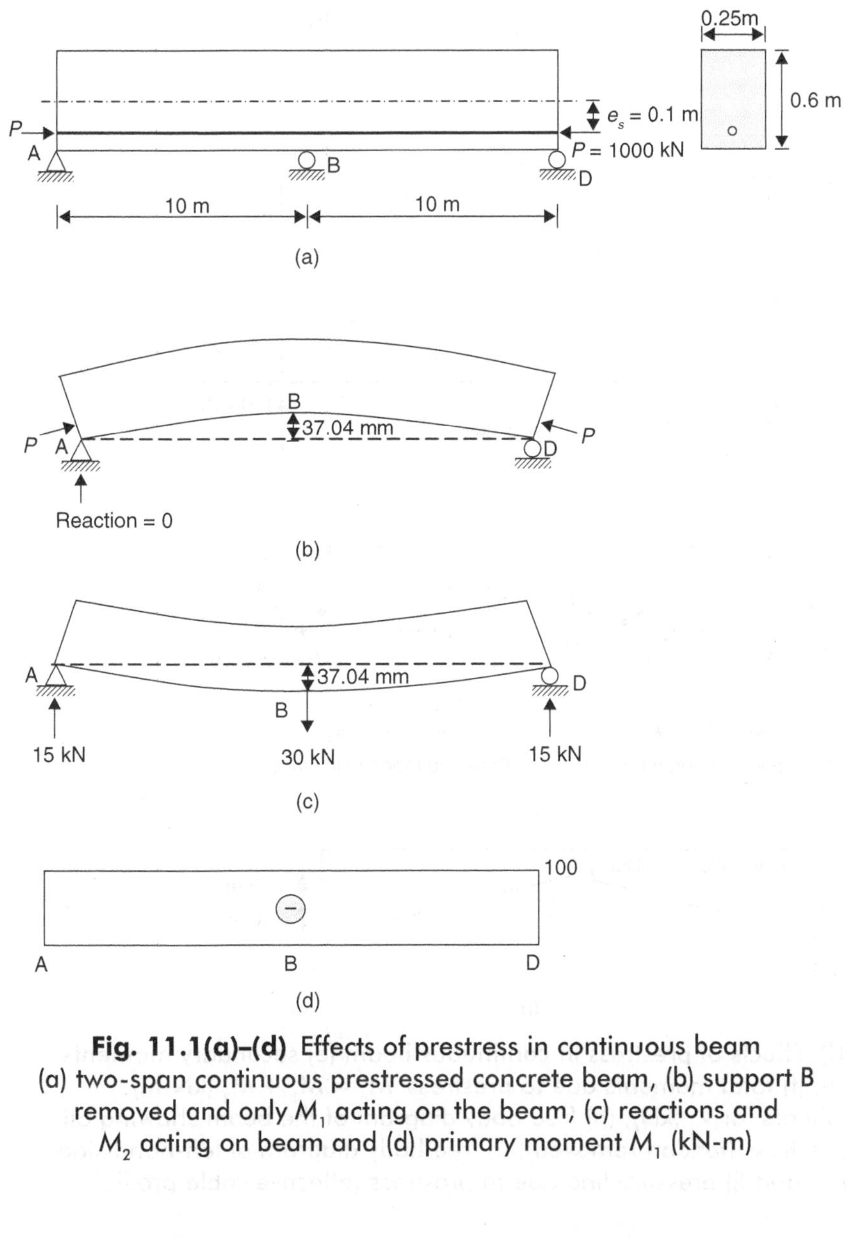

Fig. 11.1(a)–(d) Effects of prestress in continuous beam (a) two-span continuous prestressed concrete beam, (b) support B removed and only M_1 acting on the beam, (c) reactions and M_2 acting on beam and (d) primary moment M_1 (kN-m)

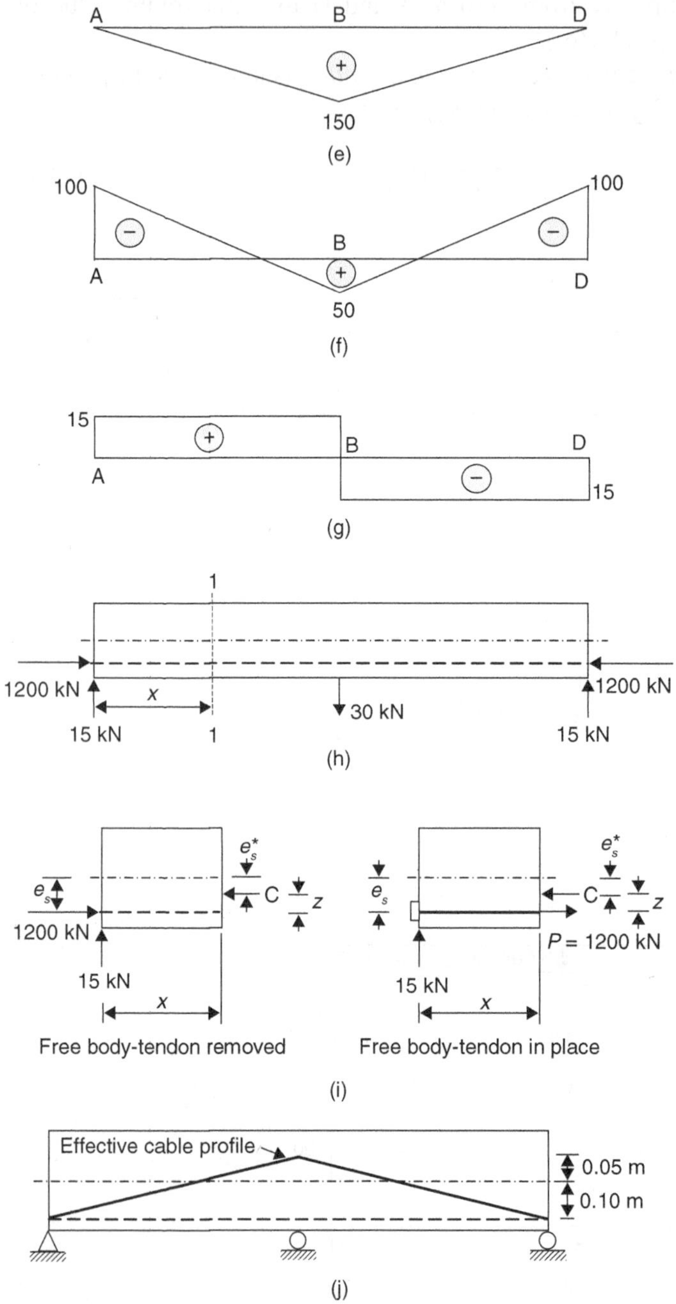

Fig. 11.1(e)–(j) Effects of prestress in continuous beam (e) secondary moments M_2 (kN-m), (f) total moments due to prestress $M_p = M_1 + M_2$ (kN-m), (g) secondary shears for V_2 (kN), (h) free body diagram of the beam showing all forces due to prestress (tendon removed), (i) free body diagram of left-hand side of section 1–1 and (j) pressure line due to prestress (effective cable profile)

The negative sign is used because M_1 will be a hogging moment for a positive value of e_s.

The support reactions developed due to prestressing force are known as *hyperstatic reactions*, and the moment M_2 developed *at a section* due to hyperstatic reaction is known as *secondary moment* due to prestress.

The *total bending moment* acting *on concrete at any section* due to prestress M_p is obtained by adding M_1 and M_2 (Fig. 11.1(f)). Hence,

$$M_p = M_1 + M_2 \tag{11.2}$$

The prestress also develops shear force in a beam. At any section, if the tendon is inclined at an angle θ to the horizontal, then the prestress causes in concrete a *primary shear force* $V_1 = P\theta$. The shear force due to hyperstatic reactions are known as the *secondary shear force* V_2. Then the *total shear force* at a cross-section due to prestress (V_p) is:

$$V_p = V_1 + V_2 \tag{11.3}$$

In the case of beam in Fig. 11.1(a), since the tendon is horizontal, $V_1 = 0$. The shear force due to hyperstatic reaction V_2 is shown Fig. 11.1(g). Then the total shear force due to prestress V_p is equal to V_2.

Since the secondary moments M_2 and shears V_2 are developed due to support reactions, M_2 can have only linear variation between the supports and V_2 will be a constant between the supports.

The term secondary does not mean that the effects are small in magnitude. Usually the secondary moments M_2 are large in magnitude and hence cannot be neglected.

11.2.2 Section Moments and Concrete Moments

While calculating the bending moment (or shear force) due to prestress at a cross-section, it is important to understand the difference between the moment acting on section and the moment acting on concrete of the cross-section. For this purpose first consider a simply supported prestressed concrete beam shown in Fig. 11.2(a).

Figs. 11.2(b) and (c) shows two possible ways of representing the free body diagram of region AB. In these figures, *C* denotes the resultant of longitudinal stress distribution acting on concrete. Using the free body diagram, the calculated moment is:

The moment acting on Section 1–1 = 0 kN-m

Also from the free body of the entire beam shown in Fig. 11.2(a), it can be seen that, the support reaction is equal to zero. Hence the moment at

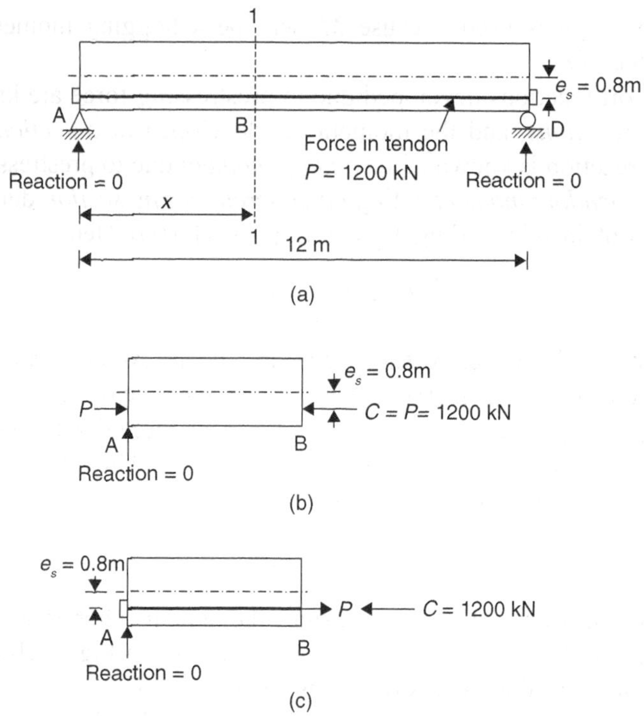

Fig. 11.2 Free body diagram of simply supported beam (a) simply supported prestressed concrete beam, (b) free body diagram of AB (tendon removed) and (c) free body diagram of AB (tendon in place)

section 1–1 is zero. The moment acting on concrete at section 1–1 (due to C) = $-C\,e_s = -P\,e_s = -960$ kN-m.

From this example, it can be concluded that for a determinate beam, there is no moment caused due to prestress on the total cross-section. The moment acting on concrete at the section is equal to the product of prestressing force and tendon eccentricity ($-P\,e_s$).

Now consider the force body diagram of the continuous beam (Fig. 11.1(h)). From this free body diagram, it can be seen that the moment at any section is caused by the hyperstatic reaction. Hence for indeterminate structures, moment acting on a section due to prestress is equal to the secondary moment M_2. The total moment acting on concrete at the section 1–1 (due to $C = M_p$) is determined from the free body of section (Fig. 11.1(i)) and is given by:

$$M_p = M_1 + M_2$$

11.2.3 Pressure Line Due to Prestressing Force

Pressure line represents the locus of location of C, the resultant of longitudinal stress acting on concrete (see Section 5.2.2.2). The distance z between the pressure line and the cable line at any section (Eq. (5.8b)) is:

$$z = \frac{M}{P}$$

where M is the bending moment acting on the section.

In the case of indeterminate structures, it is important to consider the pressure line due to prestressing force alone which is also known as the *effective cable profile*. For a determinate structure, since the prestressing force does not cause any moment at a section, the effective cable profile coincides with the actual cable profile (i.e., $M = 0 \Rightarrow z = 0$). While in the case of indeterminate structure, the moment at a section due to prestress is equal to M_2. Hence the distance between the effective cable profile and the actual cable is:

$$z = \frac{M_2}{P} \tag{11.4}$$

The eccentricity* of the effective cable profile e_s^* (Fig. 11.1(i)) is:

$$e_s^* = e_s - z = \left(\frac{-M_1}{P}\right) - \left(\frac{M_2}{P}\right) = \frac{-(M_1 + M_2)}{P}, \ (\because M_1 = -Pe_s)$$

$$\Rightarrow \qquad e_s^* = \frac{-M_p}{P} \tag{11.5}$$

The effective cable profile of the continuous beam in Fig. 11.1(a) can be plotted using the above equation.

At Supports A and D

$$M_p = -\,100 \text{ kN-m}$$

Hence,

$$e_s^* = \frac{-M_p}{P} = -\left(\frac{-100}{1000}\right) = 0.1 \text{ m}$$

A positive value of e_s^* indicate that C is below the centroidal axis.

* The eccentricity of pressure line due to combined effect of prestress and applied moment is denoted as e_{cc} and the eccentricity of pressure line due to prestress alone is represented by e_s^*.

At Support B

$$M_p = 50 \text{ kN-m. Hence,}$$

$$e_s^* = \frac{-M_p}{P} = \left(\frac{-50}{1000}\right) = -0.05 \text{ m}$$

The effective cable profile thus obtained is shown in Fig. 11.1(j).

Comments

Since the moment M_2 is caused due to support reactions, it can have only linear variation between the supports. Hence the distance z between the actual and effective cable profile also vary linearly between the supports ($\because z = M_2/P$ and P is constant). This means that the effective cable profile can be obtained by raising or lowering the actual cable profile by appropriate deviation at internal supports while maintaining the intrinsic shape of the cable profile.

11.2.4 Calculating the Effects of Prestress Using the Concept of Equivalent Loads

In Section 11.2, the effects of prestress in the continuous beam (Fig. 11.1) were calculated using the superposition method. Though this method can be used for any type of indeterminate structures, the practical use of this method is limited for simple structures. For a general indeterminate structure, the effects of prestress are determined using the concept of equivalent loads discussed in Section 5.2.3.

In this method, the equivalent loads due to prestress are applied on the structure and the structure is analyzed using any convenient method. The total moment due to prestress M_p is obtained from analysis and the secondary moment M_2 at any section is obtained by deducting the primary moment ($M_1 = -P\,e_s$) from M_p.

The procedure for finding the effects of prestress using this method is illustrated by considering the two-span continuous beam shown in Fig. 11.1.

The equivalent loads due to prestress acting on the continuous beam are indicated in Fig. 11.3(a). The horizontal and vertical equivalent loads acting on the supports do not cause any bending moment. Hence the equivalent loads required for finding bending moment can be simplified by removing the concentrated forces acting on the supports (Fig. 11.3(b)), and the structure is analyzed for these simplified loads to find M_p (Fig. 11.3(c)).

The secondary moment M_2 is determined by deducting the primary moment M_1 (Fig. 11.3(d)) from M_p. The secondary shear forces and hyperstatic reactions are found out by using the free body diagrams of individual spans (Figs. 11.3(f) and (g)).

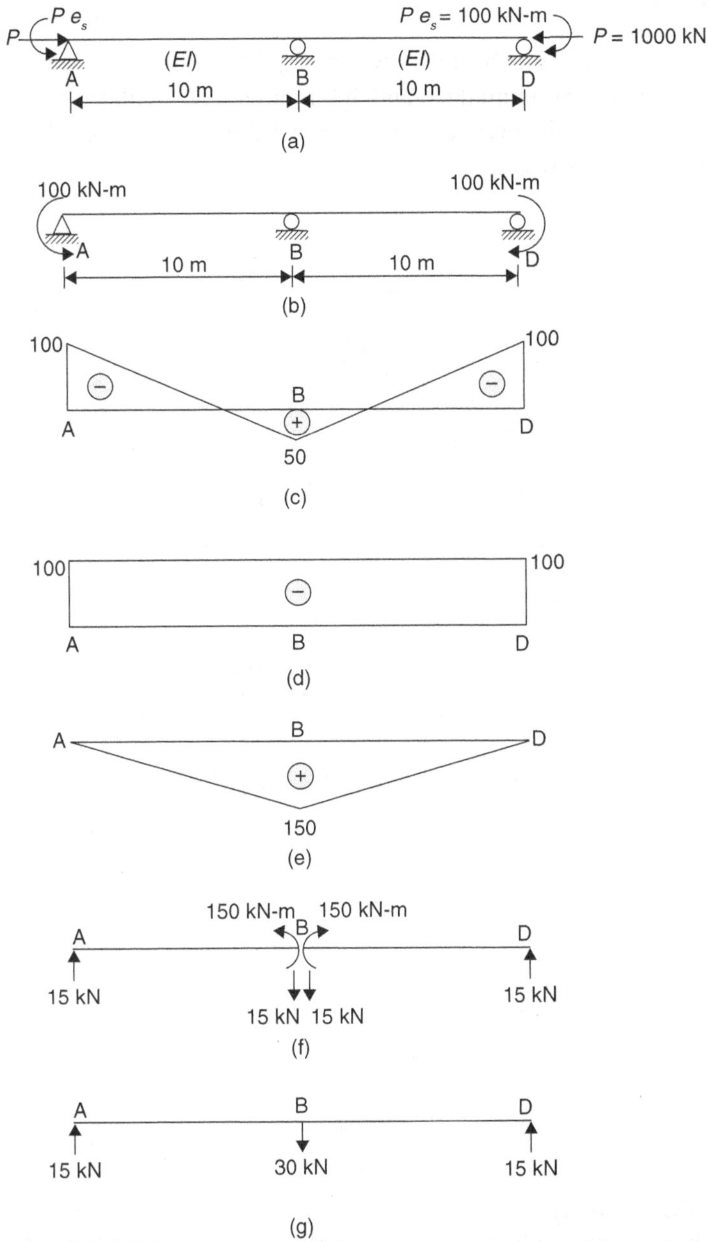

Fig. 11.3(a)–(g) Application of equivalent load approach (a) equivalent loads due to prestress, (b) simplified loading for analysis, (c) bending moment diagram obtained from analysis in kN-m (M_p), (d) primary moments in kN-m ($M_1 = -P\ e_s$), (e) secondary moments in kN-m ($M_2 = M_p - M_1$), (f) secondary moments and shears and (g) application of equivalent load approach

Example 11.1

The two-span continuous beam shown in Fig. 11.4(a) has a uniform rectangular cross-section 800 mm deep and 300 mm wide. Calculate the moments and reactions due to the prestressing force of 1200 kN. Also sketch the pressure line due to prestressing force.

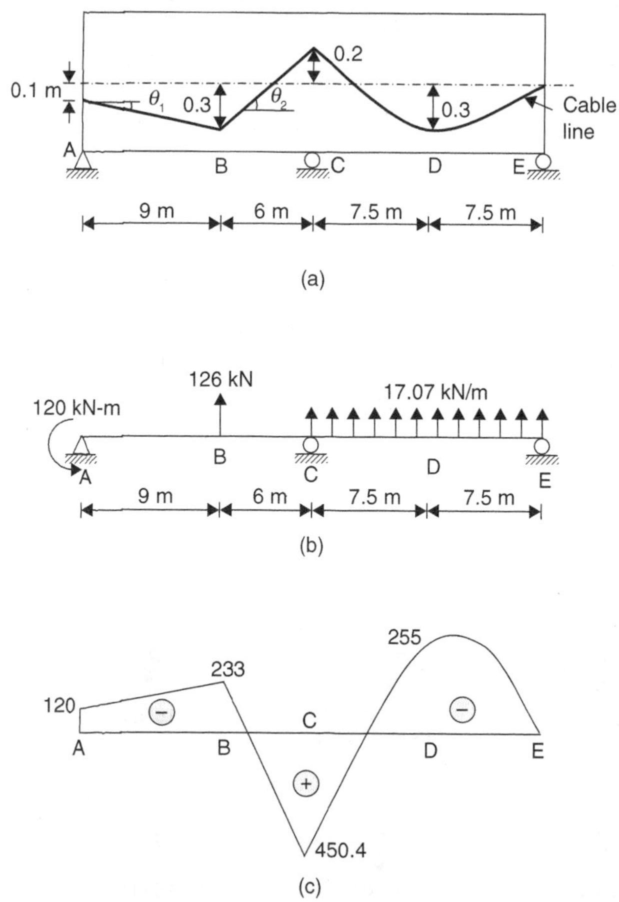

Fig. 11.4(a)–(c) Effects of prestress on two-span beam (a) beam and cable profile, (b) simplified equivaleant loads to find M_p and (c) total moments due to prestress M_p (kN-m)

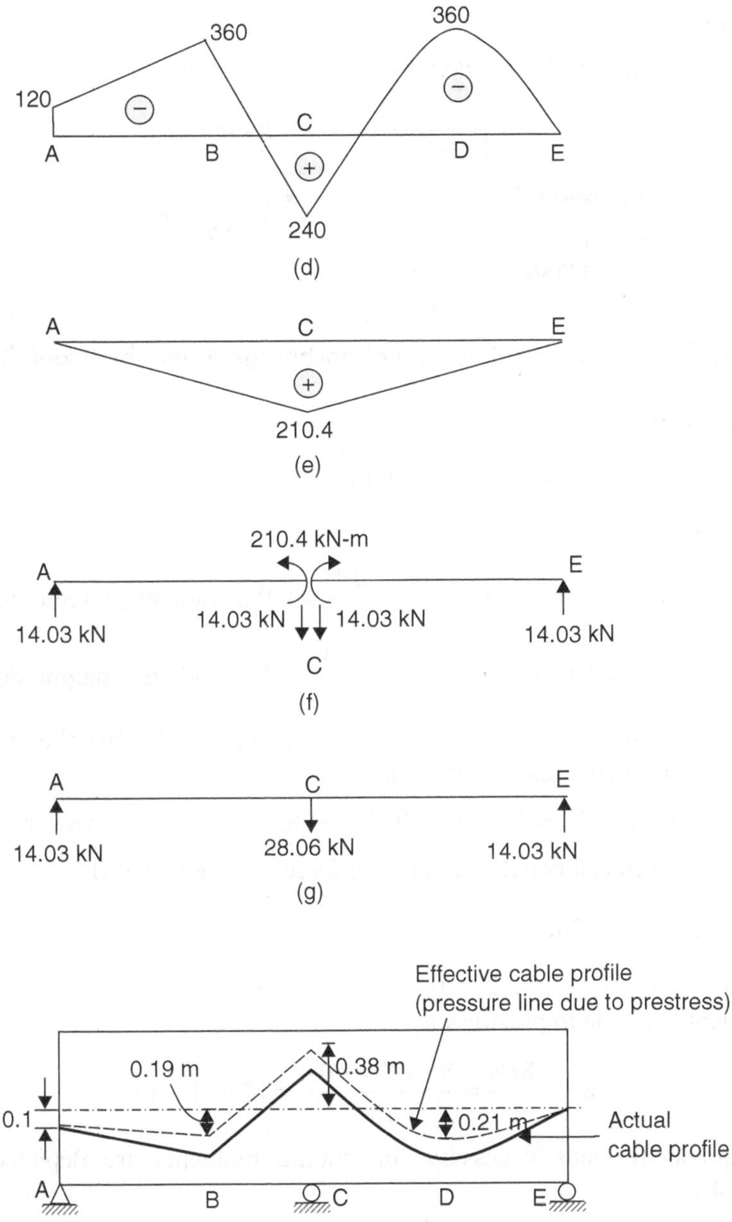

Fig. 11.4(d)–(h) Effects of prestress on two-span beam (d) primary moments M_1 (kN-m), (e) secondary moments M_2 (kN-m), (f) secondary moments M_2 and shears V_2, (g) hyperstatic reactions and (h) effective cable profile

Solution:

Step 1: *Calculation of equivalent loads due to prestress*

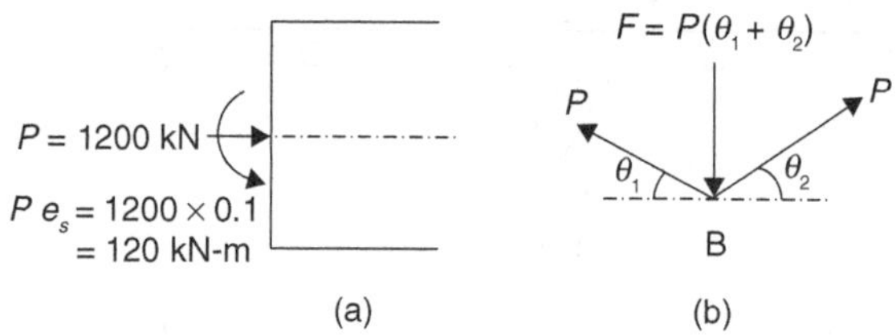

Fig. 11.5 Equivalent loads (a) at anchorage A and (b) at kink B

At Anchorage A

The equivalent loads at A is shown in Fig. 11.5(a).

At Kink B

$$\text{Slope of tendon AB, } \theta_1 = \frac{(0.3-0.1)}{9} = 0.022 \text{ radians (magnitude)}$$

$$\text{Slope of tendon BC, } \theta_2 = \frac{(0.3+0.1)}{6} = 0.083 \text{ radians (magnitude)}$$

Using the free body diagram of tendon at kink B (Fig. 11.5(b)), the force F transferred from concrete to tendon is:

$$F = P(\theta_1 + \theta_2) = 1200 \times (0.022 + 0.083) = 126 \text{ kN (downward)}$$

The tendon exerts an equal and opposite force on concrete at B.

Due to Parabolic Cable CE

Sag of cable $h = 0.3 + (0.2/2) = 0.4$ m
Equivalent load due to prestress w_p:

$$w_p = \frac{8Ph}{l^2} = \frac{8 \times 1200 \times 0.4}{15^2} = 17.07 \text{ kN/m}$$

The equivalent loads necessary for finding moments are depicted in Fig. 11.4(b).

Step 2: *Calculation of moments due to prestress M_p*

- The continuous beam subjected to equivalent loads is analyzed and the bending moment M_p is indicated in Fig. 11.4(c).

Step 3: *Calculation of primary moments M_1*

- The primary moment is calculated using Eq. (11.1) and is given in Table 11.1. The moments thus obtained are shown in Fig. 11.4(d).

Table 11.1 Calculation of primary moment M_1

Location	A	B	C	D	E
e_s (m)	0.1	0.3	–0.2	0.3	0
$M_1 = -P\,e_s$ (kN-m) where $P = 1200$ kN	–120	–360	240	–360	0

Step 4: *Calculation of secondary moments M_2*

- The secondary moments are obtained by subtracting M_1 from M_p (Fig. 11.4(e)).

Step 5: *Calculation of secondary shear force V_2 and hyperstatic reactions*

- The secondary shear force and reactions are calculated using the free body diagrams of spans AC and CE (Fig. 11.4(f) and (g)).

Step 6: *Plotting the pressure line due to prestress*

- The eccentricity of pressure line due to prestress or the effective cable profile e_s^* is calculated using Eq. (11.5). e_s^* at salient points are given in Table 11.2 and the effective cable profile is plotted (Fig. 11.4(h)).

Table 11.2 Eccentricity of effective cable profile e_s^*

Location	A	B	C	D	E
M_p (kN-m)	–120	–233	450.4	–255	0
$e_s^* = -M_p/P$ (m)	0.1	0.19	–0.38	0.21	0

The distance between the actual and effective cable profile (z) is given in Table 11.3.

Table 11.3 Calculation of z

Location	A	B	C	D	E
e_s (m)	0.1	0.3	–0.2	0.3	0
e_s^* (m)	0.1	0.19	–0.38	0.21	0
$z = \lvert e_s - e_s^* \rvert$ (m)	0	0.11	0.18	0.09	0

Alternatively once z at the supports are known, z at intermediate points can be obtained by linearly interpolating the values between the supports (refer the comments given at the end of Section 11.2.3).

At the end supports, $M_2 = 0$. Hence the effective cable profile coincides with the actual cable line at these points (i.e., $z = 0$).

z at Intermediate Points

- For span AC of length (l_3) 15 m
  z at C, $z_c = a = 0.18$ m
  z at A, $z_a = 0$ m
  z at B located at a distance $l_1 = 9$ m from A (z_b):

$$z_B = \frac{|z_c - z_a|}{l_3} \times l_1 = \frac{al_1}{l_3} = \frac{0.18 \times 9}{15} = 0.11 \text{ m}$$

- For span CE of length (l_4) 15 m
  z at E, $z_E = 0$ m
  z at D, located at a distance $l_2 = 7.5$ m from E (z_D):

$$z_D = \frac{|z_c - z_E|}{l_4} \times l_2 = \frac{al_2}{l_4} = \frac{0.18 \times 7.5}{15} = 0.09 \text{ m}$$

The effective cable profile obtained using this approach is shown in Fig. 11.6.

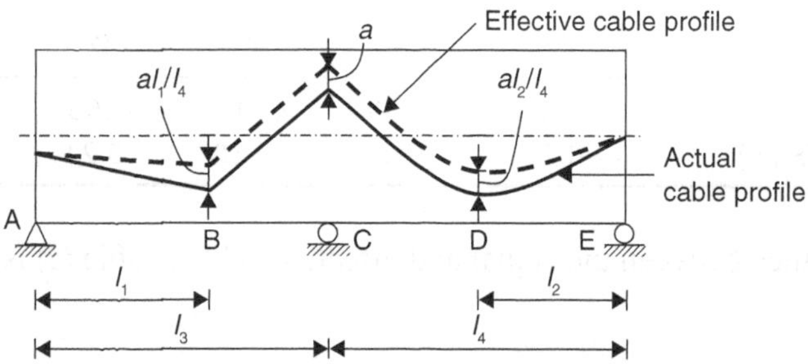

Fig. 11.6 Effective cable profile

11.3 Linear Transformation of Cable Profile

Linear transformation of cable profile indicates the process of raising or lowering the tendon at internal supports without changing the intrinsic shape of the profile between the supports. Hence the tendon eccentricities at the free ends, curvature of tendon and location of bends does not change by linear transformation.

The effects of linear transformation of the cable can be appreciated by considering a two-span continuous beam shown in Fig. 11.7(a). This beam is similar to the beam in Fig. 11.1(a) except that the cable eccentricity varies linearly. But the end eccentricity and the shape of the cable is same as in Fig. 11.1(a). In other words, the cable profile in Fig. 11.7(a) is obtained by lifting the tendon in Fig. 11.1(a) by 0.1 m at the interior support B (i.e. by linear transformation of cable profile).

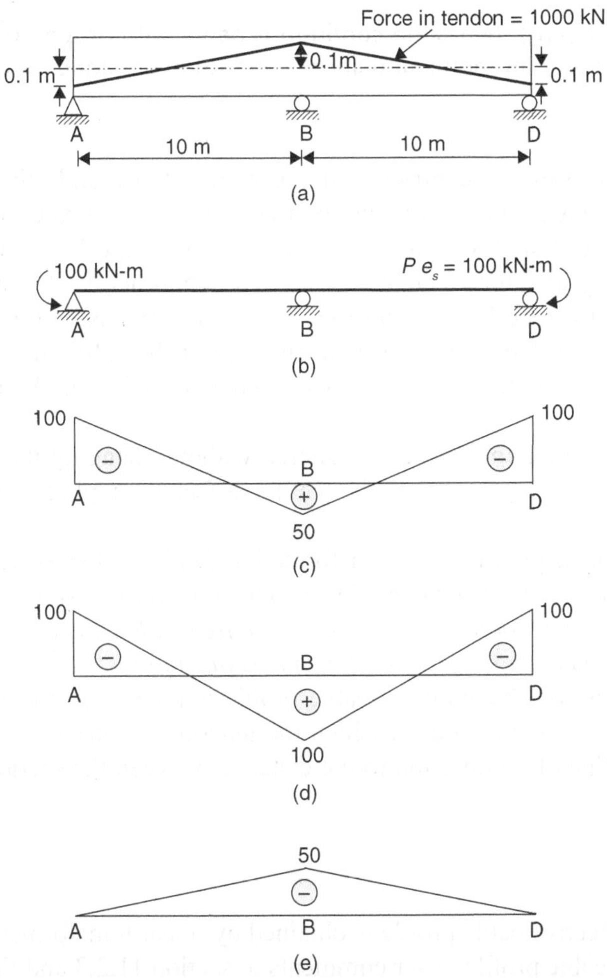

Fig. 11.7(a)–(e) Two-span continuous beam with draped tendon (a) two-span continuous beam, (b) equivalent loads for finding moments, (c) total moments due to prestress M_p (KN-m), (d) primary moments M_1 (KN-m) and (e) secondary moments M_2 (KN-m)

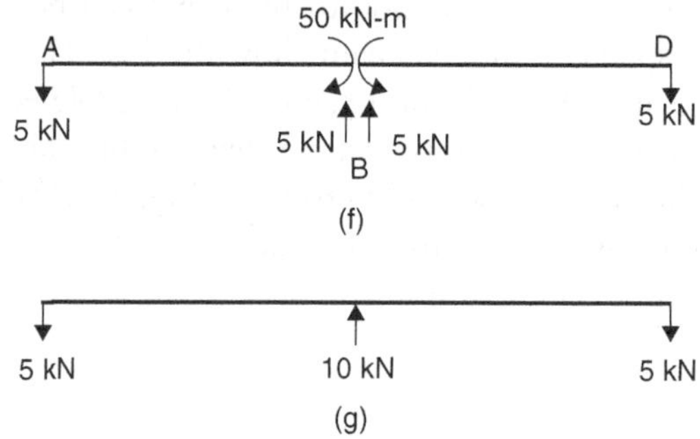

Fig. 11.7(f)–(g) Two-span continuous beam with draped tendon, (f) secondary moments and shears and (g) hyperstatic reactions

Due to the same eccentricity of the cable at the end supports, the equivalent loads causing moment and the total moments due to prestress M_p are same for both the beams shown in Figs. 11.1 and 11.7. The change in eccentricity at internal support causes a difference in M_1, M_2, V_1, V_2 and hyperstatic reactions in the two beams. From this example, it can be observed that due to linear transformation of the cable, through there is a change in primary and secondary moments, their sum M_p remains a constant.

Similarly if a parabolic cable is altered without changing its curvature, there will be no change in the equivalent loads (see Example 5.8) and hence no change in M_p.

Thus the principle of linear transformation can be stated as: *varying the tendon eccentricities over internal supports in a continuous member without changing the tendon eccentricities at the free ends and curvature of the cable causes no change in the total moments due to prestress.*

This process of changing the cable profile is known as linear transformation because the amount by which the tendon eccentricity at a section is altered is linearly proportion to the distance between the section and the support.

Comments

- The effective cable profile is obtained by linear transformation of the actual cable profile (refer comments in section 11.2.3 and Fig. 11.6).
- Linear transformation of the cable does not affect the position of effective cable profile. This is because, by linear transformation, M_p does not change. Hence the eccentricity of the effective cable profile ($e_s^* = -M_p/P$) also remains unaltered by linear transformation.

11.4 Concordant Cable

A tendon in a continuous member is known as *concordant* if it produces a effective cable profile coincident with the actual cable profile. Therefore by this definition, for a concordant cable profile,

$$e_s^* = e_s$$

$$\Rightarrow \qquad -Pe_s^* = Pe_s$$

Hence,

$$M_p = M_1$$

The above equation indicates that the secondary moments will be zero at all sections. In other words, the cable profile which does not induce hyperstatic reactions and hence no secondary moments is known as concordant, otherwise the cable profile is *non-concordant.*

Comments

- The effective cable profile obtained for any cable line itself is a concordant cable profile.
- Any bending moment diagram for a continuous member with non-yielding supports caused by any set of transverse forces or moments provides a basis of concordant profile. The tendon eccentricity is taken as a constant times the moment ordinate at that section.

  This fact can be proved very easily. The bending moment diagram due to the external loads is drawn based on the condition that deflection at the supports is zero. If this moment diagram coincides with the primary moment diagram for a cable profile, then the primary moments will not cause any deflection at the supports. Due to this, no reactions will be developed at the supports due to prestress. Hence there will be no secondary moments and the cable profile is concordant.
- The superposition of any number of concordant profiles will also result in a concordant profile.

11.5 Analysis of Real Cables

Practical Cable Profile

In the analysis of continuous members discussed so far, it was assumed that the direction of the cable changes abruptly over the supports (at support C in Fig. 11.8(a)). But in reality, the required change in the curvature occurs over a finite length. Frequently, a practical cable profile (Fig. 11.8(b)) consists of a combination of parabolic cables, concave in the span and convex over the interior supports. The convex segment is required to avoid sudden change in curvature of cable at the support.

For the span AC in Fig. 11.8(b), the slope of the parabolas 1 and 2 is zero at B, the point having maximum eccentricity. Similarly the slope of parabola 3 is zero at C. The point of inflection F can be positioned at one-tenth of the span ($l_3 = l_1/10$). At F, the slope of parabolas 2 and 3 are same. Using this condition, it can be shown that F lies on a straight line joining the points of maximum eccentricity B and C.

The equivalent loads for this practical cable profile are indicated in Fig.11.8(c) and the structure is analyzed for this load to find M_p.

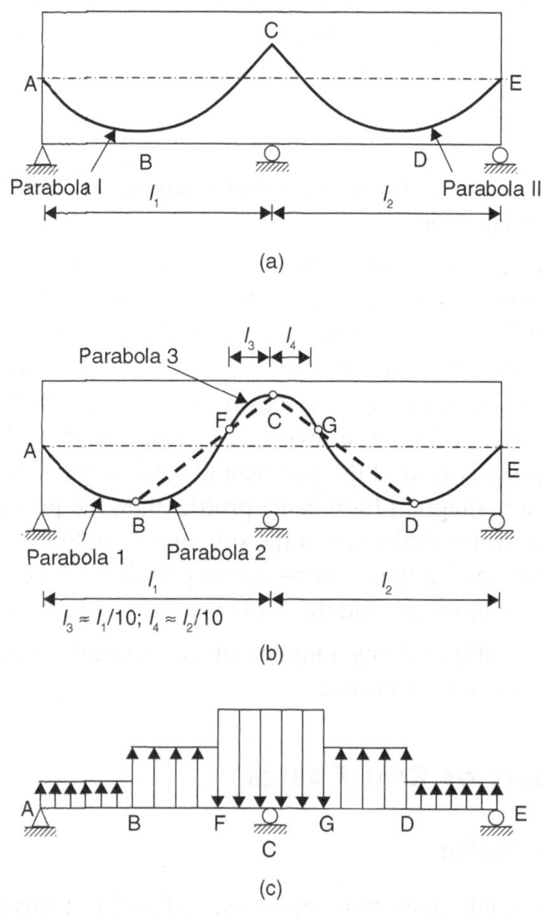

Fig. 11.8 Two-span continuous beam (a) idealized cable profile, (b) practical cable profile and (c) equivalent loads to find M_p

Accounting for Prestress Losses Due to Friction

Due to friction losses, the force in tendon decreases with distance from the jacking end. For small losses due to friction, the analysis can be done by using the average value of prestressing force in each span.

In order to get more accurate results, each span is divided into number of segments or zones, and in each zone, the average value of prestressing force in that zone is used to find the equivalent load. But increasing the number of zones in a span, the accuracy of the results can be improved.

11.6 Calculation of Elastic Stresses in Concrete

The elastic stress distribution in a determinate structure is calculated based on the actual cable profile. But in the case of indeterminate structure, the stress is calculated using the effective cable profile. Hence the stress in concrete σ at a distance y below the centroid in an indeterminate structure is:

$$\sigma = \frac{P}{A} + \frac{(Pe_s^*)y}{I} - \frac{M\,y}{I} \tag{11.6}$$

Eq. (11.6) is obtained by replacing the tendon eccentricity e_s in Eq. (5.4a) by e_s^* the eccentricity of effective cable profile.

Example 11.2

Calculate the extreme fibre stresses at the interior support C of the beam in Example 11.1, when the beam is acted upon by a total uniformly distributed load of 15 kN/m (including the self-weight).

Solution:

Step 1: *Calculation of section properties*
Cross-sectional area A = 300 × 800 = 240000 mm^2

Section modulus $Z_a = Z_b = 1/6 \times 300 \times 800^2 = 32 \times 10^6$ mm^3

Step 2: *Calculation of fibre stresses at interior support C*
Force in tendon P = 1200 kN
Eccentricity of effective cable profile $e_s^* = -380$ mm

- Moment due to applied loads at C (M)

$$M = -\frac{wl^2}{8} = \frac{-15\times 15^2}{8} = -421.88 \text{ kN-m (hogging moment)}$$

- Stress at the top fibre of the section σ_a

$$\sigma_a = \frac{P}{A} - \frac{Pe_s^*}{Z_a} + \frac{M}{Z_a}$$

$$= \left[\frac{1200\times10^3}{240000}\right] - \left[\frac{1200\times10^3\times(-380)}{32\times10^6}\right] + \left[\frac{-421.88\times10^6}{32\times10^6}\right]$$

$$= 5 + 14.25 - 13.18 = 6.07 \text{ N/mm}^2$$

- Stress at the bottom fibre of the section σ_b

$$\sigma_b = \frac{P}{A} + \frac{Pe_s^*}{Z_b} - \frac{M}{Z_b}$$

$$= 5 - 14.25 + 13.18 = 3.93 \text{ N/mm}^2$$

11.7 Behaviour at Ultimate Load

The analysis of prestressed concrete members at ultimate limit state discussed in Section 5.8 is applicable for statically determinate structures. In these structures, a failure mechanism develops when the ultimate moment of resistance (M_{uR}) has reached at any section of the member. But in the case of indeterminate structures, even if a section has reached M_{uR}, a failure mechanism will not occur. Instead a plastic hinge will be developed at that section. If this plastic hinge has adequate rotation capacity, the structure can carry additional load, and this load will be redistributed to less loaded regions till sufficient plastic hinges are developed to form a mechanism[1].

Consider a two-span continuous beam (Fig. 11.9) subjected to a factored load of 15 kN/m. Fig. 11.10(a) shows the elastic bending moment diagram for this load. If M_{uR} of the beam at section B is only 150 kN-m, then the bending moment distribution shown in Fig. 11.10(a) can never be achieved by the structure. At a load of 12 kN/m, the bending moment at B reaches 150 kN-m (Fig. 11.10(b)). If the load is further increased, a plastic hinge develops at B (Fig. 11.10(c)). The bending moments at the load of 15 kN-m can be found out using the free body diagram of the individual spans (Fig. 11.10(d)). Comparing this bending moment diagram (Fig. 11.10(e)) with Fig. 11.10(a) shows that a redistribution of moment by 20 percent has occurred at B to bring the moment from 187.5 kN-m to 150 kN-m. Due to this, the moment at mid-span has increased to 112.5 kN-m.

[1] For more details, the reader may refer the topic 'Plastic analysis' in books on Structural Analysis.

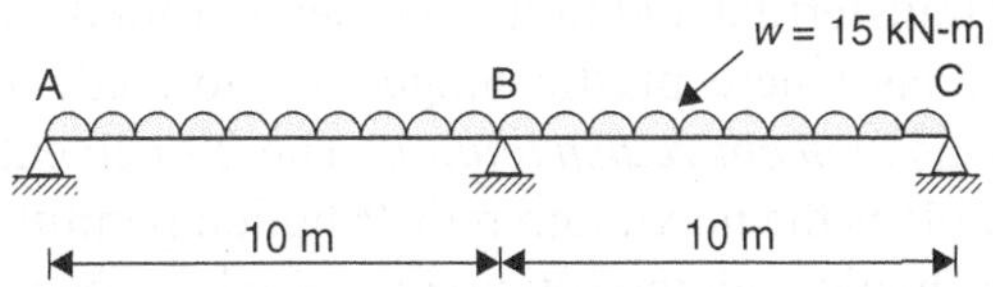

Fig. 11.9 Two-span continuous beam

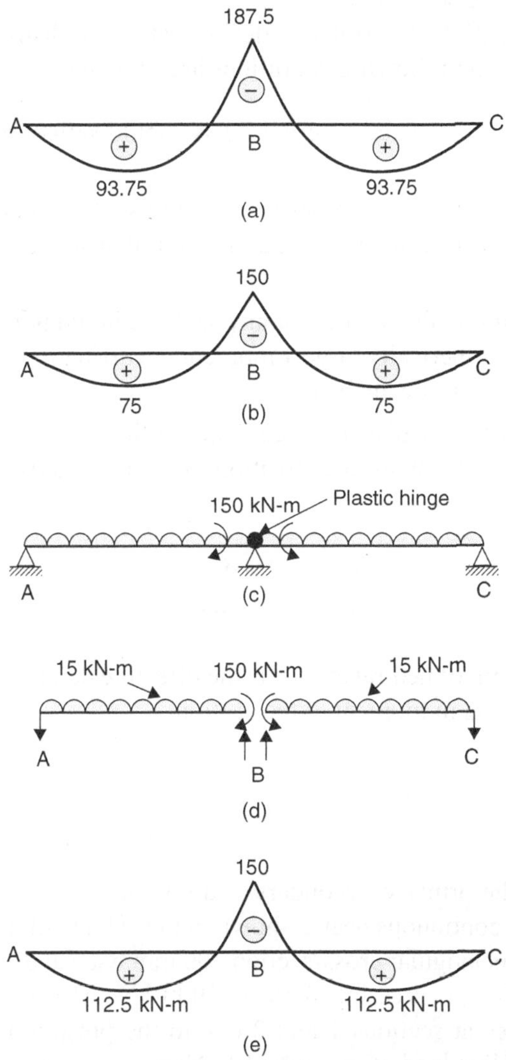

Fig. 11.10 Illustration of moment redistribution (a) elastic bending moment diagram at $w = 15$ kN/m, (b) elastic bending moment diagram at $w = 12$ kN/m, (c) development of plastic hinge at B, (d) free body diagram of individual spans at $w = 15$ kN/m and (e) bending moment diagram after redistribution of moments

The process of transferring additional moments from the highly stressed sections (i.e., sections where plastic hinges are formed) to less stressed sections is known as *moment redistribution*. The 20 percent redistribution shown in the example is the maximum redistribution permitted in the Code. The plastic hinges should undergo rotation required for the redistribution of moments. The rotation ability of the hinges in fact depends on the position of neutral axis in the section.

The Code (Cl. 21.1.1) permits to design using a redistributed moment diagram provided the following conditions are satisfied:

- The internal forces and external loads should satisfy the equilibrium conditions.
- The ultimate moment of resistance at any section should not be less than 80 percent of the bending moment at that section obtained from elastic analysis.
- The reduction in the moment at any section should not exceed 20 percent of the numerically largest moment occurring in the member calculated using elastic analysis.
- At the location of plastic hinges, the section should be sufficiently under-reinforced with the position of neutral axis satisfying the condition:

$$\frac{x_u}{d} \le 0.5 - \frac{|\delta M|}{100} \tag{11.7}$$

where x_u is the depth of neutral axis, d is the effective depth and $|\delta M|$ is the percentage reduction in moment at the section.

PROBLEM

11.1 Calculate the primary, secondary and total moments due to prestress in the three continuous beams shown in Fig. 11.11. All the beams have a uniform rectangular cross-section 300 mm wide and 1000 mm deep. The effective prestressing force is 2000 kN. Also find the extreme fibre stresses at sections 1 and 2 due to the prestressing force, self-weight and live load of intensity 10 kN/m.

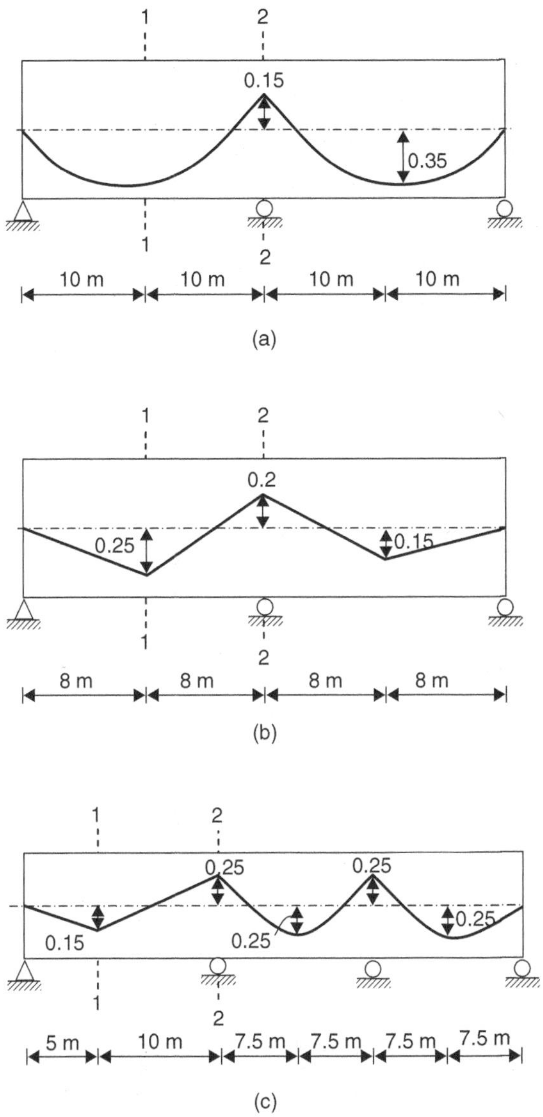

Fig. 11.11 (a) Continuous beam–1, (b) continuous beam–2 and (c) continuous beam–3

chapter 12

Slabs

12.1 Introduction

Prestressed concrete slabs are widely used as floors in structures like warehouses, shopping centres, offices, apartments etc. The deflection of the slab which governs the thickness is better controlled in prestressed concrete slabs. By properly choosing the amount of prestress, the designer can in fact eliminate deflection. Thus more slender and crack-free slab systems are possible by using prestressed concrete.

For design purposes, it is necessary to distinguish slabs as one-way slabs and two-way slabs. A one-way slab acts like a wide beam with predominant bending in one direction. Hence in these slabs, the main reinforcement is placed in this direction and secondary reinforcement is provided in the other direction to control cracking due to shrinkage and temperature changes. In the case of two-way slabs, bending moments of comparable magnitude occurs in both directions and reinforcements are placed in both directions to resist these bending moments.

A rectangular slab can be classified into one-way or two-way slab based on its support conditions and the long span (l_y) to short span (l_x) ratio. If a rectangular slab is supported on two opposite edges, then it is a one-way slab. In the case of slabs supported on all four edges by either walls or beams (*edge-supported slabs*), if $l_y/l_x > 2$, then such slabs are treated as one-way slabs. Otherwise, they are designed as two-way slabs. The design of one-way and edge-supported slabs (Fig. 12.1) is discussed in this chapter.

Most of the post-tensioned slabs are made by using tendons having flat ducts containing a maximum of five strands with fan-shaped anchorage.

The strands are stressed one-by-one by using hydraulic jacks. A typical detail of the flat-ducted tendon with four strands is indicated in Fig. 12.2.

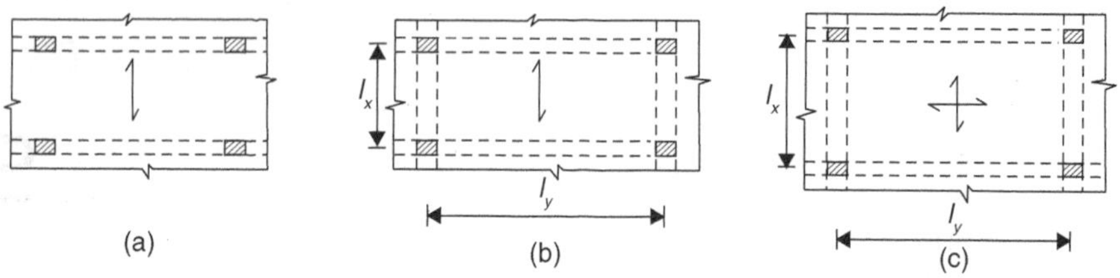

Fig. 12.1 Types of slab systems (a) one-way slab, (b) edge-supported one-way slab ($l_y/l_x>2$) and (c) edge-supported two-way slab ($l_y/l_x\leq 2$)

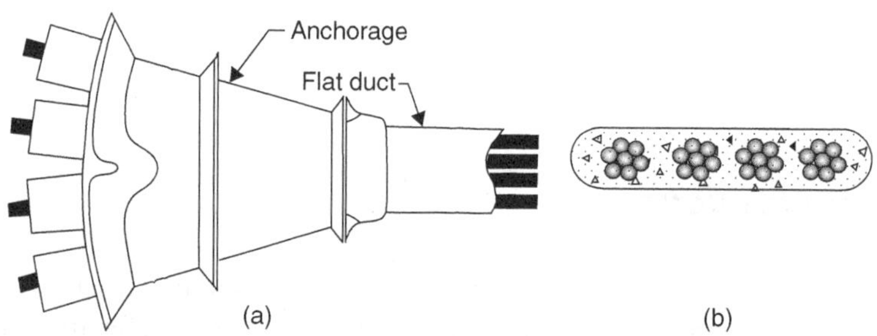

Fig. 12.2 Details of flat-ducted tendon (a) fan-shaped anchorage and (b) cross-section of duct

12.2 General Design Procedure

The first step in the design of prestressed concrete slab is to arrive at the initial slab thickness. After deciding the slab thickness, the next step is to select the load that has to be balanced by prestress. To have good serviceability behaviour, a major portion of the sustained load should be balanced by prestress. Due to balanced load (w_b), the slab will not undergo any deflection and will be subjected to only axial compressive stress (refer Section 5.3). The remaining unbalanced loads (w_n) should be used to check the behaviour of slab at serviceability limit state, i.e. to find deflections and bending stresses in concrete.

Due to non linear behaviour of slab at ultimate conditions, the principle of superposition is not valid at this stage. Hence instead of factored unbalanced loads, the total factored loads should be used to find the design moments and shears at critical sections. The factored moments and shears are compared with the ultimate moment of resistance and shear capacity of the section.

At the collapse condition, full redistribution of moments occur due to the ductile behaviour of slab. Under this situation, the secondary moments and shear forces need not be considered in the analysis.

12.3 One-way Slabs

A one-way slab is designed as a beam of unit width. For a simply supported slab, the span to depth ratio (l/D) is selected between 35 and 40. l/D of 45 would be usually adequate for continuous one-way slabs.

The equivalent loads due to prestress acting in a continuous one-way slab are shown in Fig. 12.3. If the spacing of the cable is uniform across the width of the slab and P is the prestressing force per unit width of the slab, then the equivalent load due to prestress w_p is:

$$w_p = w_b = \frac{8Ph}{l^2} \quad (12.1)$$

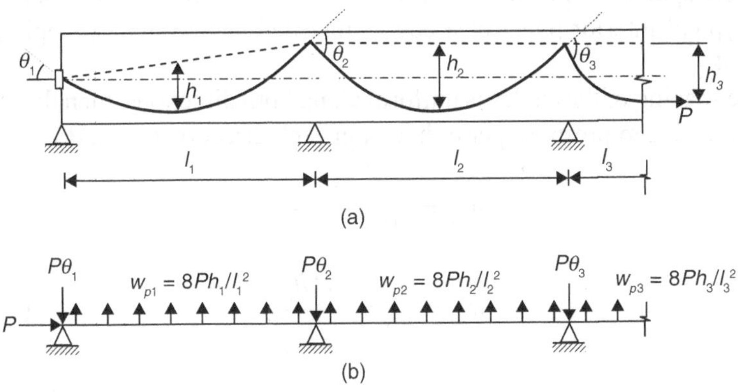

Fig. 12.3 Continuous one-way slab (a) cable profile and (b) equivalent loads due to prestress

The sag of the cable h depends on the clear cover requirements. By knowing the balanced load w_b, the required prestressing force P is calculated using Eq. (12.1).

In the transverse direction, minimum untensioned reinforcement is provided to control cracks due to temperature changes and shrinkage. As per Cl. 26.5.2.1 of IS 456:2000, the minimum flexural reinforcement for slab ($A_{st, min}$) is:

$A_{st, min} = 0.0015\, A_g$, for mid steel reinforcement

$= 0.0012\, A_g$, for high strength deformed bars or welded wire fibre

where A_g is the gross area of the section.

The moments and shears in a one-way continuous slabs subjected to uniformly distributed loads can be determined using the moment and shear coefficients given in Tables 12 and 13 of IS 456:2000. The use of these tables are limited to one-way continuous slabs having at least three spans which do not differ by more than 15 percent of the longest.

For checking the shear capacity of slabs, the critical section is taken at a distance effective depth from the face of the support. The factored shear force at the critical section V_u should be less than or equal to 0.5 V_{uc}, so that there is no need for shear reinforcement (Cl. 22.4.3.1 of the Code). Where V_{uc} is the ultimate shear capacity, and for slabs, V_{uc} can be taken as V_{co}. If V_u is greater than 0.5 V_{uc}, the depth of the slab is increased to avoid shear reinforcement.

12.4 Edge-supported Two-way Slabs

The short span to depth ratio (l_x/D) of a two-way slab lies between 45 to 50. A typical interior panel of a edge-supported slab system is depicted in Fig. 12.4.

If the spacing of the cables is uniform along both directions, then the equivalent loads due to prestress per unit area in each direction (w_{px} and w_{py}) are:

$$w_{px} = w_{bx} = \frac{8P_x h_x}{l_x^2} \tag{12.2a}$$

$$w_{py} = w_{by} = \frac{8P_y h_y}{l_y^2} \tag{12.2b}$$

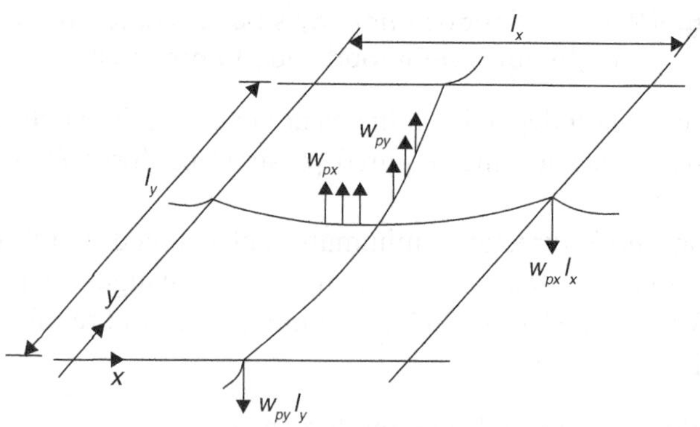

Fig. 12.4 Interior panel (two-way slab)

where P_x and P_y are the prestressing force per unit width in x and y direction. h_x and h_y are the sag of the cables. If w_{bx} and w_{by} are the loads balanced by prestress in the short and long span directions, then the total load balanced by prestressed w_b is:

$$w_b = w_{bx} + w_{by} \tag{12.3}$$

Any combination of w_{bx} and w_{by} which satisfies Eq. (12.3) can be used for the design of two-way slabs. Generally it is found to be economical to carry the major portion of the balanced load in the shorter direction. After deciding w_{bx} and w_{by}, the required prestressing forces are calculated using Eq. (12.2).

The bending moment coefficients given in Table 26 of IS 456:2000 are used to find the slab moments caused by the unbalanced service loads and factored design loads. In order to perform the check for shear strength, it is necessary to find the proportion of the total load transferred to different edges of the slab. The distribution of shear forces at the various edges of a rectangular two-way slab subjected to uniformly distributed load is assumed to occur as shown in Fig. 12.5 (Cl. 24.5 of IS 456:2000). The critical section for shear is taken at a distance of effective depth (d) from the face of support, and the factored shear force at this section V_u is compared with the shear capacity of the section.

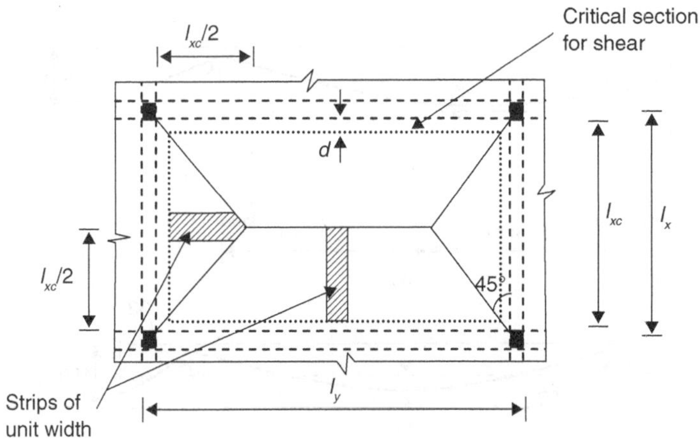

w_u = Factored load, l_{xc} = Clear span in shorter direction
Minimum shear force per unit width $V_u = w_u \times$ Shaded area
$= w_u\,(0.5\,l_{xc} - d)$

Fig. 12.5 Distribution of shear forces in a edge-supported slab

Example 12.1

Design a two-way slab supported on four edges by stiff beams and discontinuous on one long edge (Fig. 12.6(a)) to carry a live load of 5 kN/m². Use M 60 grade concrete.

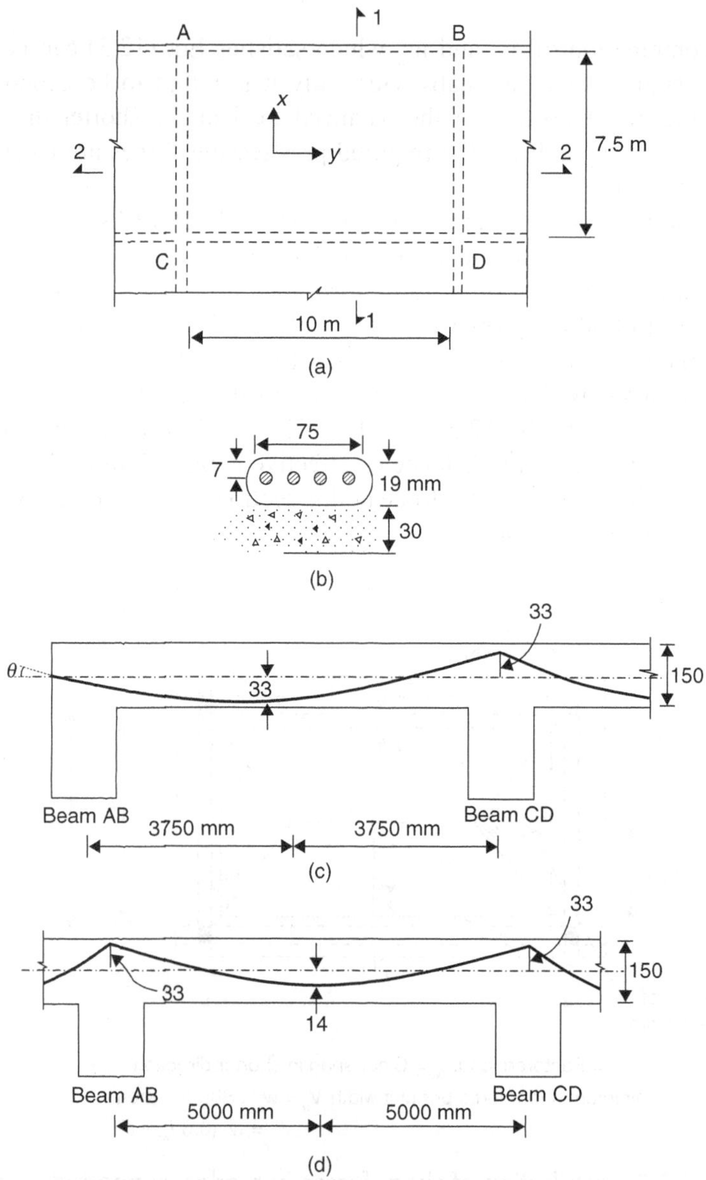

Fig. 12.6 Edge-supported slab (a) plan, (b) section through duct, (c) section 1–1 and (d) section 2–2

The properties of a 7 wire strand are:
Nominal diameter = 12.8 mm, nominal area = 99.3 mm^2,
Characteristic strength f_p = 1800 N/mm^2

Solution:

Step 1: *Selection of thickness of slab and calculation of effective depth*

- Thickness of slab D

  Effective short span l_x = 7500 mm

  Assuming l_x/D = 50, thickness of slab $D = \frac{l_x}{50} = \frac{7500}{50} = 150$ mm

- Effective depth and properties of cable line in x direction

  Flat ducted tendons containing four 12.8 mm strands are used. The size of the duct is 75 mm × 19 mm (Fig. 12.6(b)).
  Maximum effective depth of slab in x direction (providing 30 mm clear cover to duct):

  $d_x = 150 - 30 - (19 - 7) = 108$ mm
  Tendon eccentricity at the mid-span $e_s = 108 - (150/2) = 33$ mm
  Tendon eccentricity above the supporting beam CD, e_s = 33 mm (above the centroid of slab)
  Sag of cable h_x (Fig. 12.6(c)) = 33 + (33/2) = 49.5 mm

- Effective depth and properties of cable line in y direction

  The effective depth along y direction is less than d_x by the thickness of duct.
  Maximum effective depth of slab in y direction $d_y = 108 - 19 = 89$ mm
  Tendon eccentricity at mid-span $e_s = 89 - (150/2) = 14$ mm
  Tendon eccentricity at supporting beams e_s = 33 mm (above the centroid of slab)
  Sag of cable h_y (Fig. 12.6(d)) = 14 + 33 = 47 mm

Step 2: *Calculation of loads acting on slab*

- Dead load w_{DL}

  Self-weight $w_{SW} = 24 \times 0.15 = 3.6$ kN/m^2

  Weight of finishes = 1 kN/m^2 (assumed value)

  Total dead load $w_{DL} = 3.6 + 1 = 4.6$ kN/m^2

- Total load acting on slab

  Dead load w_{DL} = 4.6 kN/m^2

Live load $w_{LL} = 5$ kN/m^2
Total load $w = w_{DL} + w_{LL} = 9.6$ kN/m^2
Factored design load $w_u = 1.5\ w = 14.4$ kN/m^2

Step 3: *Calculation of balanced load and the required prestressing force*

- Load balanced by prestress w_b

  Let all the dead load and one-fifth of the live load is balanced by prestress.

$$w_b = w_{DL} + \frac{1}{5} w_{LL} = 4.6 + (5/5) = 5.6 \text{ kN/m}^2$$

- Required prestressing force and load balanced in y direction

  It is economical to carry most of the balanced load along the shorter direction (i.e., along x direction). Hence it is assumed that prestressing force which induces a uniform compression of 2 N/mm^2 is sufficient along the y direction.

$$\sigma_{py} = \frac{P_{ey}}{A} = 2 \text{ N/mm}^2 = 2 \times 10^6 \text{ N/m}^2$$

Effective prestressing force in y direction P_{ey}:

$$P_{ey} = \sigma_{py} A = \frac{2 \times 10^6 \times 1 \times 0.15}{1000} = 300 \text{ kN/m}$$

Load balanced in y direction w_{by}:

$$w_{by} = \frac{8 P_{ey} h_y}{l_y^2} = \frac{8 \times 300 \times 0.047}{10^2} = 1.13 \text{ kN/m}^2$$

Initial prestressing force in y direction P_{iy}:

$$P_{iy} = \frac{P_{ey}}{\eta} = \frac{300}{0.75} = 400 \text{ kN/m (assuming } \eta = 0.75)$$

- Required prestressing force and load balanced in x direction

  Load balanced in x direction w_{bx}:

$$w_{bx} = w_b - w_{by} = 5.6 - 1.13 = 4.47 \text{ kN/m}^2$$

Effective prestressing force in x direction P_{ex}:

$$P_{ex} = \frac{w_{bx} l_x^2}{8 h_x} = \frac{4.47 \times 7.5^2}{8 \times 0.0495} = 634.94 \text{ kN/m}$$

Initial prestressing force in x direction P_{ix}:

$$P_{ix} = \frac{P_{ex}}{\eta} = \frac{634.94}{0.75} = 846.6 \text{ kN/m}$$

- Spacing of tendons (s)

  Cross-sectional area of a tendon A_p:

  $$A_p = 4 \times 99.3 = 397.2 \text{ mm}^2$$

  Initial prestress $f_{pi} = 0.7 f_p = 0.7 \times 1800 = 1260$ N/mm^2 (assuming $f_{pi}/f_p = 0.7$)

  Initial prestressing force in a tendon P_i:

  $$P_i = f_{pi} A_p = \frac{1260 \times 397.2}{1000} = 500.47 \text{ kN}$$

  Number of tendons required for 1m width in x direction (n_x):

  $$n_x = \frac{P_{ix}}{P_i} = \frac{846.6}{500.47} = 1.69$$

  Spacing of tendons in x direction (s_x):

  $$s_x = \frac{1000}{n_x} = \frac{1000}{1.69} = 591.7 \text{ mm}$$

  Spacing of tendons in y direction (s_y):

  $$s_y = \frac{1000}{n_y} = \frac{1000}{(P_{iy}/P_i)} = \frac{1000}{(400/500.47)} = 1251.18 \text{ mm}$$

  Choose $s_x = 600$ mm and $s_y = 1000$ mm.

- Actual prestressing force and load balanced in x direction

  Initial prestressing force $P_{ix} = \dfrac{1000P_i}{n_x} = \dfrac{1000 \times 500.47}{600} = 834.12$ kN/m

  Effective prestressing force $P_{ex} = \eta P_{ix} = 0.75 \times 834.12 = 625.6$ kN/m

  Load balanced in x direction w_{bx}:

  $$w_{bx} = \frac{8P_{ex}h_x}{l_x^2} = \frac{8 \times 625.6 \times 0.0495}{7.5^2} = 4.4 \text{ kN/m}^2$$

- Actual prestressing force and load balanced in y direction

Initial prestressing force $P_{iy} = \dfrac{1000P_i}{s_y} = \dfrac{1000 \times 500.47}{1000} = 500.47$ kN/m

Effective prestressing force $P_{ey} = \eta P_{iy} = 0.75 \times 500.47 = 375.35$ kN/m
Load balanced in y direction w_{by}:

$$w_{by} = \frac{8P_{ey}h_y}{l_y^2} = \frac{8 \times 375.35 \times 0.047}{10^2} = 1.41 \text{ kN/m}^2$$

- Load balanced by prestress w_b

$$w_b = w_{bx} + w_{by} = 4.4 + 1.41 = 5.81 \text{ kN/m}^2$$

Step 4: *Check for stresses in concrete*

- Permissible stresses in concrete (Cl. 22.7 and 22.8 of the Code)

Maximum permissible compressive stress at service condition f_{cw} (assuming zone I):

$$f_{cw} = 0.35 f_{ck} = 0.35 \times 60 = 21 \text{ N/mm}^2$$

Maximum permissible tensile stress at service condition f_{tw}:

$$f_{tw} = 0 \text{ N/mm}^2$$

- Axial compressive stress due to prestressing force σ_p

At the balanced load, the slab is subjected to axial compression.

Axial stress in x direction, $\sigma_{px} = \dfrac{P_{ex}}{A} = \dfrac{625.6 \times 10^3}{1000 \times 150} = 4.17$ N/mm²

Axial stress in y direction, $\sigma_{py} = \dfrac{P_{ey}}{A} = \dfrac{375.35 \times 10^3}{1000 \times 150} = 2.5$ N/mm²

- Net unbalanced load w_n

$$w_n = w - w_b = 9.6 - 5.81 = 3.79 \text{ kN/m}^2$$

It is the net unbalanced load that causes bending in slab.

- Calculation of moments in slab at service condition due to w_n

The maximum moment M_{CD} occurs over the supporting beam CD (i.e., over the longer continuous edge) and is given by:

$$M_{CD} = \alpha_x^- w_n l_x^2$$

where is α_x^- the negative moment coefficient for the short span.

From Table 26 of IS 456:2000,

for $$\frac{l_y}{l_x} = \frac{10}{7.5} = 1.33; \alpha_x^- = 0.059$$

Hence,

$M_{CD} = 0.059 \times 3.79 \times 7.5^2 = 12.58$ kN-m/m (hogging moment)

- Stress in the extreme fibres of slab due to M_{CD} (σ_m)

$$\sigma_m = \pm \frac{M_{CD}}{Z} = \pm \frac{12.58 \times 10^6}{\left(\frac{1}{6} \times 1000 \times 180^2\right)} = \pm 3.35 \text{ N/mm}^2$$

The hogging moment causes tensile stress at the top fibre and compressive stress at bottom fibre.

- Total stress in the extreme fibre of slab

  Maximum stress at the top fibre in x direction (σ_a):

  $\sigma_a = 4.17 - 3.35 = 0.82$ N/mm² ($> f_{tw}$, OK)

  Maximum stress at the bottom fibre in x direction (σ_b):

  $\sigma_b = 4.17 + 3.35 = 7.82$ N/mm² ($< f_{cw}$, OK)

  Similarly, check for stresses should be also done along y direction.

Step 5: *Check for flexural capacity of slab*

- Maximum moment in x direction due to factored load M_u

  $M_u = \alpha_x^- w_u l_x^2 = 0.059 \times 14.4 \times 7.5^2 = 47.79$ kN-m/m (hogging moment)

- Ultimate moment of resistance of slab M_{uR} in x direction

  Area of prestressing steel per metre in x direction (A_{px}):

$$A_{px} = \frac{A_p}{s_x} = \frac{397.2}{0.6} = 622 \text{ mm}^2/\text{m}$$

Prestressed reinforcement index ω_p:

$$\omega_p = \frac{A_{px} f_p}{b d_x f_{ck}} = \frac{622 \times 1800}{1000 \times 108 \times 60} = 0.173$$

From Table 11 of the Code,

$$\text{For } \omega_p = 0.173; \ \frac{f_{pu}}{0.87 \times f_p} = 0.977, \ \frac{x_u}{d_x} = 0.361$$

$$f_{pu} = 0.977 \times 0.87 f_p = 0.977 \times 0.87 \times 1800 = 1530 \text{ N/mm}^2$$

$$x_u = 0.361 \times d_x = 0.361 \times 108 = 39 \text{ mm}$$

$$M_{uR} = f_{pu} \times A_{px} [d_x - 0.42\, x_u]$$

$$= 1530 \times 622 \times [108 - (0.42 \times 39)]/10^6$$

$$= 87.2 \text{ kN-m/m } (>M_u, \text{ OK})$$

In a similar way, the moment of resistance of the slab should be checked at the mid-span and at supports along y direction.

Step 6: *Check for shear capacity*

- Factored shear force at critical section V_u

$$V_u \approx w_u \times 1 \times \frac{l_x}{2} = 14.4 \times 1 \times \frac{7.5}{2} = 54 \text{ kN/m}$$

- Calculation of V_{uc} ($V_{uc} = V_{co}$)

Stress at centroid due to prestress $f_{cp} = \sigma_{px} = 4.17$ N/mm²

Tensile strength of concrete $f_t = 0.25\sqrt{60} = 1.93$ N/mm²

Slope of cable at support θ (Fig. 12.6(c)):

$$\theta = \frac{4h_x + e_a - e_b}{l} \qquad \text{(see Example 5.8)}$$

$$\theta = \frac{(4 \times 0.495) + 0 - 0.033}{7.5} = 0.022 \text{ radians}$$

Vertical component of prestressing force P_v:

$$P_v = P_{ex}\,\theta = 625.6 \times 0.022 = 13.76 \text{ kN/m}$$

Ultimate shear resistance of section uncracked in flexure V_{co}:

$$V_{co} = 0.67bD\sqrt{f_t^2 + 0.8 f_{cp} f_t} + P_v$$

$$= \left[\frac{0.67 \times 1000 \times 150\sqrt{1.93^2 + (0.8 \times 4.17 \times 1.93)}}{1000}\right] + 13.76$$

$$= 334.\ 15 \text{ kN/m}$$

$$V_{uc} = V_{co} = 334.15 \text{ kN/m}$$

$$0.5\, V_{uc} = 0.5 \times 334.15 = 167.08 \text{ kN/m } (>V_u)$$

Since V_u is less than $0.5V_{uc}$, there is no need for shear reinforcement

Step 7: *Check for deflection*
Since the slab is uncracked at service load and all dead load and a part of the live load is balanced by prestress, it can be concluded that deflection criteria will be satisfied in this case. The deflection can be determined using linear elastic analysis and checked with the deflection limits given in the Code.

PROBLEMS

12.1. Design a simply supported isolated rectangular slab 6m × 8m with unrestrained corners to carry a live load of intensity 5 kN/m^2. The grade of concrete is M 50 and use the properties of the 7 wire strand given in Example 12.1.

12.2. Re-do the above problem, if the slab has restrained corners.

12.3. Design a simply supported one-way slab of span 10 m to carry a live load of intensity 6 kN/m^2. Use M 60 grade concrete.

chapter 13

Circular Prestressing

13.1 Introduction

Circular prestressing refers to the process of prestressing circular structures like pipes and tanks in which tendons are wound in circles. But linear prestressing indicates all other types of prestressing where the tendons are straight or curved, but not wound in a circular shape. Most often in the case of prestressed concrete circular structures, prestressing is done in both circumferential (circular prestressing) and longitudinal (linear prestressing) direction. The design of some typical prestressed circular structures like pipes, water tanks and domes are briefly covered in this chapter.

13.2 Circumferential Prestressing

The prestress in the circumferential direction is designed to resist the hoop tension developed due to internal pressure. The tendons for this purpose are placed outside the concrete core. But the stress developed in concrete due to circumferential prestress will be always axial in nature. This is due to the fact that the purpose of circular prestress is to induce hoop compression in concrete, which will be always axial in nature irrespective of the point of application of prestressing force.

13.2.1 Analysis at Transfer (Fig. 13.1(a))

The stresses in concrete at different loading stages are determined by considering the free body diagram of one half of the slice of unit length of the circular member taken along the longitudinal direction (Fig. 13.1).

At the transfer loading stage, the compression force C in the concrete core is equal to the prestressing force at transfer P_o, i.e., $C = P_o$. Hence the stress in concrete σ_c is:

$$\sigma_c = \frac{C}{A_c} = \frac{P_o}{A_c} \tag{13.1}$$

where A_c is the area of concrete core that resists P_o.

13.2.2 Analysis at Service Condition (Fig. 13.1(b))

At service condition, the prestressing force is P_e. Hence the stress in concrete due to prestressing force is P_e/A_c.

The internal pressure p is resisted by both tendon and concrete together. If T is the hoop tension caused by p, then the stress in concrete due to internal pressure is:

$$\sigma_c = \frac{T}{A_t}$$

where

A_t = area of transformed section = $A_c + mA_p$
m = modular ratio = E_p/E_c

Note: Stress in concrete due to T is tensile in nature and as per sign convention is a negative stress.

Thus the stress in concrete at service condition due to effective prestressing force P_e and internal pressure p is:

$$\sigma_c = \frac{P_e}{A_c} + \frac{T}{A_t} \tag{13.2}$$

13.2.3 Design

The area of tendon is generally determined using the assumption that the hoop tension due to p is entirely resisted by the effective prestressing force P_e. Hence

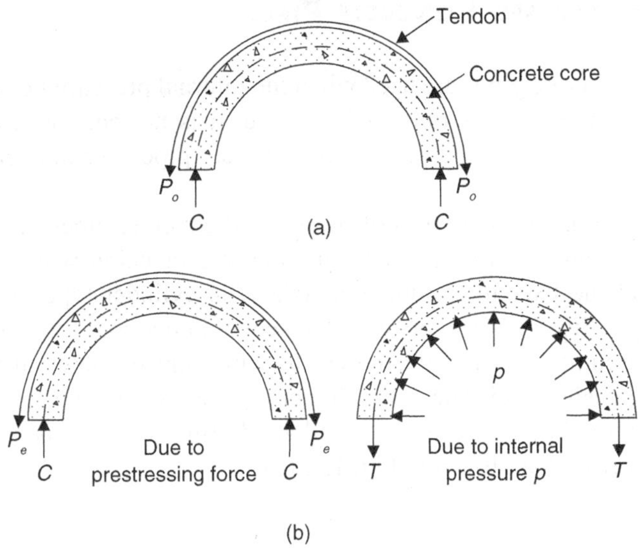

Fig. 13.1 Forces at different loading stages (a) at transfer and(b) at service condition

$$P_e = A_p f_{pe} = T$$

$$\Rightarrow \qquad A_p = \frac{T}{f_{pe}} \qquad (13.3)$$

Once the area of prestressing steel A_p is determined, the prestressing force at transfer P_o is found out using the relation given below.

$$P_o = A_p f_{po}$$

After finding P_o, the required area of the concrete core is calculated from the equation:

$$A_c = \frac{C}{f_{ct}} = \frac{P_o}{f_{ct}} \qquad (13.4)$$

where f_{ct} is the allowable compressive stress in concrete at transfer. Using A_c, the thickness of the concrete core t is determined.

For the selected values of P_e and A_c, the stress in concrete at service condition is calculated using Eq. (13.2) and it should be less than the allowable stress at service condition.

13.3 Prestressed Concrete Pipes

Prestressed concrete pipes are used when the internal pressure lies between 0.5 to 2.0 N/mm^2. At this range of pressure, it is not economical to use steel or cast iron pipes, and reinforced concrete pipes are not practically possible.

There are basically two types of prestressed concrete pipes, namely the *cylinder type* and the *non-cylinder type*, the former being widely used in America and the later in Europe and Asia. The cylinder type prestressed concrete pipe (Fig. 13.2(a)) consists of a steel cylinder core lined with concrete and circumferentially prestressed to withstand the internal pressure and external design load. The non-cylinder pipe has a concrete core embedded with longitudinally prestressed tendons. Further the core is circumferentially prestressed as shown in Fig. 13.2(b).

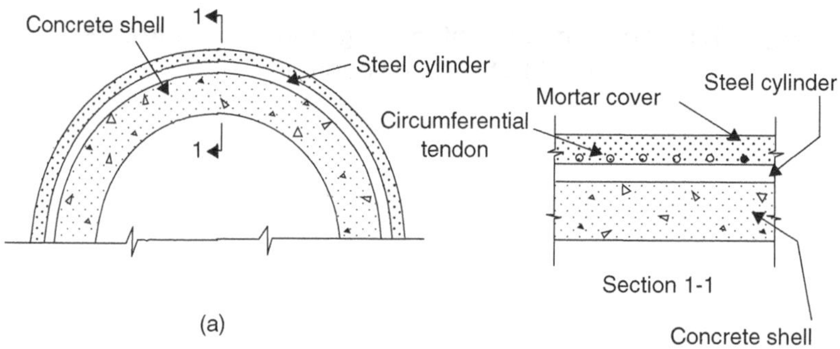

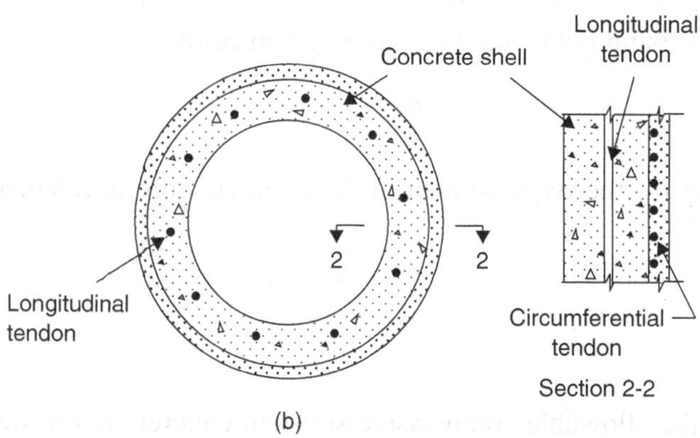

Fig. 13.2 Different types of prestressed concrete pipes (a) prestressed concrete cylinder pipe and (b) prestressed concrete non-cylinder pipe

The guidelines for design of prestressed concrete pipes are given in IS 784:2001[1]. The design of a thin prestressed concrete pipe subjected to internal fluid pressure is discussed in Example 13.1. For more information on the design and manufacture of different types of prestressed concrete pipes, the reader may refer the work of Joshi[2].

Example 13.1

Design a prestressed concrete pipe of diameter 600 mm to withstand a fluid pressure of 1 N/mm^2. The thickness of the pipe is 50 mm. The permissible compressive stress in concrete at transfer is 14 N/mm^2. The pipe should have a residual compression of 0.9 N/mm^2 at service loads. High tensile wires of 2 mm diameter initially stressed to 1800 N/mm^2 are available for use.

Solution:

Step 1: *Diameter of maximum hoop tension T*

Fluid pressure $p = 1$ N/mm^2 = 10 kN/m^2

Diameter of pipe $D = 600$ mm = 0.6 m

- Hoop tension due to fluid pressure T

$$T = -\frac{pD}{2} = -\frac{10^3 \times 0.6}{2} = -300 \text{ kN/m} = -300 \text{ N/mm}$$

Step 2: *Determination of effective pressure*

Thickness of pipe $t = 50$ mm

Residual compressive stress at service condition $f_{res} = 0.9$ N/mm^2

- Stress in concrete due to T

$$\sigma_c = \frac{T}{A} = -\frac{300 \times 1000}{1000 \times 50} = -6 \text{ N/mm}^2$$

In order to obtain a residual compressive stress of 0.9 N/mm^2 at service condition, the effective prestressing force should produce a hoop compressive stress in concrete equal to 6 + 0.9 = 6.9 N/mm^2

Step 3: *Check for stress in concrete*

Assuming the loss ratios $\beta = 0.9$ and $\eta = 0.75$

Permissible stress in concrete at transfer $f_{ct} = 14$ N/mm^2

---

[1] IS:784-2001, *Prestressed Concrete Pipes (including specials) – Specification*, Bureau of Indian Standards, Manak Bhavan, New Delhi.

[2] Joshi N. G., *Prestressed Concrete Pipes and Pipelines*, Narosa Publishing House, 2012.

- Stress in concrete due to prestressing force at service condition

$$\sigma_c = 6.9 \text{ N/mm}^2$$

- Stress in concrete due to prestressing force at transfer

$$\sigma_c = \frac{\beta}{\eta} 6.9 = \frac{0.9}{0.75} \times 6.9 = 8.28 \text{ N/mm}^2 < 14 \text{ N/mm}^2, \text{ OK}$$

Step 4: *Determination of pitch of wire winding*

- Required effective prestressing force P_e

$$P_e = \frac{6.9 \times 50 \times 1000}{1000} = 345 \text{ kN/m}$$

- Required initial prestressing force P_i

$$P_i = \frac{P_e}{\eta} = \frac{345}{0.75} = 460 \text{ kN/m}$$

- Initial prestressing force in a wire F_i

$$F_i = \frac{\pi}{4} \times 2^2 \times 1800 \times \frac{1}{1000} = 5.65 \text{ kN}$$

- Number of wires n

$$n = \frac{P_i}{F_i} = \frac{460}{5.65} = 81.42$$

- Spacing or pitch of wire winding s

$$s = \frac{1000}{n} = \frac{1000}{81.42} = 12.3 \text{ mm}$$

Comments

In the above example, for sake of simplicity, the stresses in concrete are determined using the gross area of concrete A instead of using A_c and A_t.

13.4 Prestressed Concrete Circular Tanks

Prestressed concrete tanks are extensively used for storing liquids like water, petroleum products, sewage, etc. They are also used as bunkers and silos for storing cement, grains, chemicals, fertilizers etc. and as pressure vessels for nuclear power plants.

13.4.1 Behaviour

The horizontal elements of the tank (i.e. closed circular rings) are subjected to hoop stress and the vertical elements are subjected to bending. Hence the internal pressure p is resisted partly by the hoop action of the circular ring and by the beam action of the vertical element.

The magnitude of hoop stress and bending moment in the vertical element depends on the support condition at the bottom of the wall. If the wall is free to slide relative to the base of the tank, then only hoop stress is developed and the vertical elements are not subjected to any bending. If the base of the wall is either hinged or fixed, both hoop stress and vertical moments are developed. The different types of support conditions for the base of wall are shown in Fig. 13.3. A typical variation of hoop tension and vertical moments along the height H of a circular water tank for the different bases are depicted in Fig. 13.4.

The other factors that affect the hoop stress and vertical moment are:

- Support condition at the top of the wall, i.e., whether it is free or hinged or fixed or partially restrained.

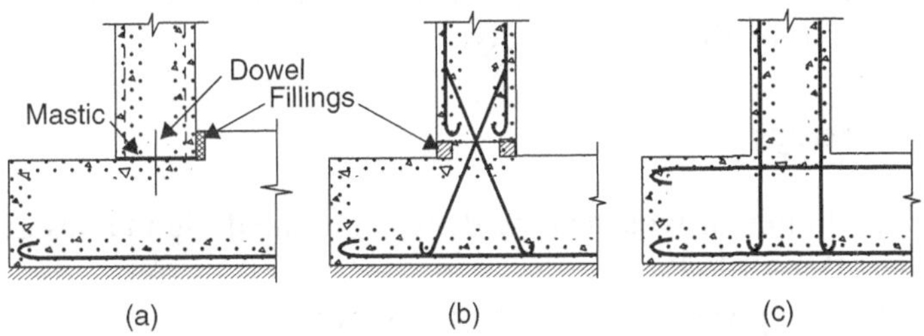

Fig. 13.3 Different types of bases for tanks (a) sliding base, (b) hinged base and (c) fixed base

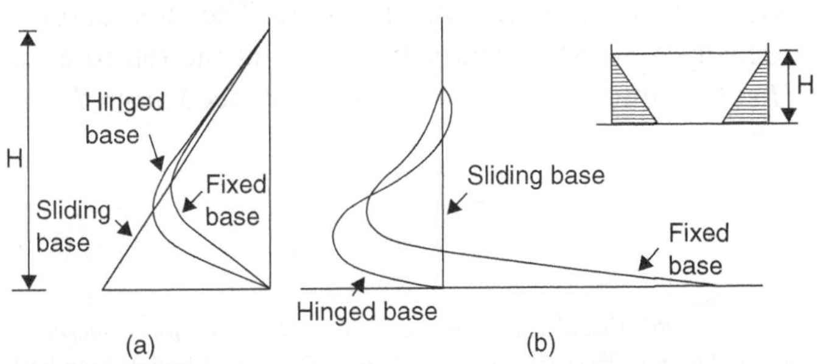

Fig. 13.4 Hoop tension and vertical bending moment in a tank (a) hoop tension and (b) vertical bending moment

- Distribution of internal pressure along the depth, i.e., whether uniform or triangular or trapezoidal.
- Thickness of wall, i.e., constant thickness or varying.
- Ratio of height of tank to its diameter.

In prestressed concrete tanks, the horizontal pressure developed by circumferential prestressing induces vertical moments in walls. The vertical prestress is provided to resist these moments. If the vertical tendons are placed at an eccentricity, then it will develop force components which will affect the circumferential prestress. Hence the analysis becomes complicated. To avoid such difficulties, generally the vertical tendons are placed concentric with respect to the wall so that only axial stress is developed.

The vertical or longitudinal prestress is designed to resist the stresses caused by the following forces:

- Partial winding of circumferential prestress.
- Vertical moments developed by internal fluid pressure.
- Vertical moments due to circumferential prestress.

The stress developed by circumferential prestress is counterbalanced by the stress caused by internal fluid pressure. Hence the maximum stress in concrete exists when the tank is empty.

13.4.2 Analysis

The analysis of circular tanks can be done using the design tables given in IS:3370 (Part IV) – 1967[3].

The net internal pressure p in the tank is given by:

$$p = p_i - p_p \tag{13.5}$$

where p_i is the internal pressure due to the contained material and p_p is the external pressure due to circumferential prestress. The internal pressure due to liquid of density γ_w varies linearly from zero at the top to a maximum value of $\gamma_\omega H$ at the bottom of the tank. p_i at a distance x from the top of the tank is given by:

$$p_i = \gamma_w x \tag{13.6}$$

[3] IS:3370 (Part IV) – 1967 (Reaffirmed 1999), *Code of Practice for Concrete Structures for the Storage of Liquids: Part IV – Design Tables*, Bureau of Indian Standards, Manak Bhavan, New Delhi.

The expression to find p_p is derived by considering the free body diagram of one half of the horizontal element of unit height when subjected to prestressing force P (Fig. 13.5).

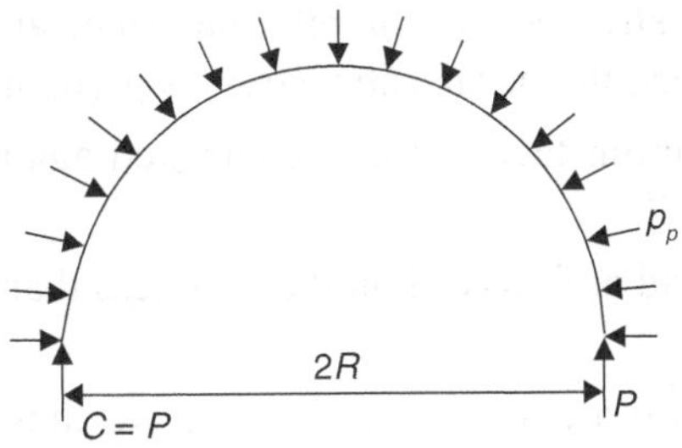

Fig. 13.5 Free body diagram of one half of the ring

Using Fig. 13.5, the force equilibrium equation in the vertical direction can be written as:

$$P = p_p R$$

$$\Rightarrow \qquad p_p = \frac{P}{R} \tag{13.7}$$

The hoop tension T and the vertical bending moment M at any height is given by:

$$T = C_T\, p_m\, R \tag{13.8}$$

$$M = C_M\, p_m\, H^2 \tag{13.9}$$

when p_m is the maximum internal pressure (net) at the base of the tank. C_T and C_M are the coefficients for finding T and M which are tabulated in IS:3370 (Part IV) for various values of $H^2/2Rt$ at different depths.

13.4.3 Design

The dimensions of circular water tank are determined based on the structural, functional, economic and aesthetic requirements. It is found out that the economic ratio of radius to height of the circular water tank is approximately equal to 2. A minimum wall thickness of 120 mm is usually provided at the top of the tank. For capacities upto 2000 m^3, the wall thickness can be kept constant for the entire depth. For large tanks, it is economical to provide tank with varying wall thickness.

The design of prestressed concrete tanks are done using the recommendations given in IS:3370 (Part III) – 1967[4]. Some of the salient recommendations given in the code are given below:

- The computed stresses in concrete and steel at all loading stages should not exceed the permissible stresses given in IS:1343.
- The liquid retaining face should be checked against cracking with a load factor of 1.2.
- The ultimate load at failure should not be less than twice the working load.
- The principal compressive stress in concrete should not exceed one-third of the cube strength.
- The average shear stress on the gross cross-section of concrete should not exceed 1/50 of the cube strength.
- When the tank is full, there should be compression in concrete at all points of at least 0.7 N/mm².
- When the tank is empty, there should not be tensile stress greater than 1 N/mm².
- Allowance should be made for the longitudinal moment induced by transverse stressing in the partially would condition. For this, the flexural stress in the vertical direction can be taken as 0.3 times the hoop compression stress due to circumferential prestressing.

For computing the resistance to cracking, the strength of concrete in direct tension f_t can be determined using the equation:

$$f_t = 0.267\sqrt{f_{cu}} \text{ N/mm}^2$$

where f_{cu} is the cube strength of concrete in N/mm². The flexural tensile strength f_{cr} may be taken as twice the strength in direct tension. f_t and f_{cr} for different cube strengths are tabulated in Table 1 of IS:3370 (Part III).

13.4.4 Detailing Requirements

A minimum cover of 35 mm should be provided to tendons on the liquid face. The recommendations given in IS:1343 for the spacing of tendons are used for the detailing of tanks.

[4] IS:3370 (Part III) – 1967 (Reaffirmed 1999), *Code of Practice for Concrete Structures for the Storage of Liquids: Part III – Prestressed Concrete Structures*, Bureau of Indian Standards, Manak Bhavan, New Delhi.

Example 13.2

Design a prestressed concrete circular water tank of radius 20 m and height 11 m. The wall of the tank is fixed at the base and free at the top. Use M55 grade concrete. The maximum allowable compressive stress in concrete at transfer is 17 N/mm^2. Use 8 mm diameter high tensile steel wires for circumferential prestressing and cables of 16 wires of 8 mm diameter for vertical prestressing. The effective stress in tendons can be taken as 1100 N/mm^2. The ultimate tensile strength of tendons is 1800 N/mm^2.

Solution:

Step 1: *Calculation of maximum hoop tension due to water pressure*

Assuming a trial thickness of wall $t = 300$ mm

Height of water tank $H = 11$ m

Radius of water tank $R = 20$ m

Form factor $\dfrac{H^2}{2Rt} = \dfrac{11^2}{2\times 20\times 0.3} = 10$

- Maximum pressure due to water p_m

  Density of water $\gamma_w = 10$ kN/m^3

$$p_m = \gamma_w H = 10 \times 11 = 110 \text{ kN/m}^2$$

- Coefficient C_T for hoop tension

  From Table 9 of IS:3370 (Part IV), for $H^2/Rt = 10$, the maximum value of C_T is 0.608. The maximum value of hoop tension occurs at a height of 0.6 $H = 0.6 \times 11 = 6.6$ m from the top of the tank.

- Maximum value of hoop tension T

$$T = -C_T p_m R$$
$$= -0.608\times 110\times 20 = -1337.6 \text{ kN/m} = -1337.6 \text{ N/mm}$$

Step 2: *Design of circumferential prestressing steel*

- Hoop stress in concrete due to water pressure

$$\sigma_c = \frac{T}{A} = -\frac{1337.6\times 1000}{1000\times 300} = -4.46 \text{ N/mm}^2$$

- Residual compressive stress in concrete at service condition f_{res}

$$f_{res} = 0.7 \text{ N/mm}^2$$

- Maximum pressure due to circumferential prestress

  The effective prestressing force should develop a hoop compressive stress equal to $4.46 + 0.7 = 5.16$ N/mm^2. Assuming that the pressure due to circumferential prestress p_p varies linearly from zero at top to a maximum at bottom (i.e. p_p will have a linear variation similar to water pressure), the maximum hoop compression due to p_p occurs at a height of 6.6 m from the top of the tank and is given by:

$$T = 0.608 \times p_{pm} \times 20 = 12.16\, p_{pm} \text{ kN/m}$$

where p_{pm} is the maximum pressure due to prestress. The required p_{pm} is given by:

$$12.16\, p_{pm} = 5.16 \times 10^3 \times 0.3 \times 1$$

$$\Rightarrow \quad p_{pm} = 127.3 \text{ kN/m}^2$$

- Effective prestressing force per unit height at base P_e (Eq. (13.7))

$$P_e = p_{pm} R = 127.3 \times 20 = 2546 \text{ kN/m}$$

- Area of prestressing steel based on residual compressive stress requirements A_{p1}

Effective prestress $f_{pe} = 1100$ N/mm²

$$P_e = A_{p1} f_e$$

$$\Rightarrow A_{p1} = \frac{P_e}{f_{pe}} = \frac{2546 \times 1000}{1100} = 2314.55 \text{ mm}^2/\text{m}$$

- Area of prestressing steel based on ultimate load factor μ_u (A_{p2})

Ultimate load factor $\mu_u = 2$

$$\mu_u = \frac{\text{Ultimate tensile force in wire at base of tank } (T_u)}{\text{Maximum hoop tensile force in tank due to water } (T)}$$

where

$$T_u = A_{p2} f_{pu}$$

f_{pu} = ultimate stress in tendon = 1800 N/mm²

T = 1337.6 N/mm

$$\Rightarrow A_{p2} = \frac{\mu_u T}{f_{pu}} = \frac{2 \times 1337.6 \times 1000}{1800} = 1486.22 \text{ mm}^2/\text{m}$$

- Area of prestressing steel A_p

A_p = maximum of A_{p1} and A_{p2} = 2314.55 mm²/m

- Required number of 8 mm diameter high tensile wires

$$n = \frac{2314.55}{\frac{\pi}{4} \times 8^2} = 46.05$$

- Spacing of wires s

$$s = \frac{1000}{n} = \frac{1000}{46.05} = 21.71\text{ mm}$$

Hence provide 8 mm wires at a pitch of 20 mm at the base.

- Effective prestressing force P_e

$$P_e = \frac{\pi}{4} \times 8^2 \times 1100 \times \frac{1000}{20} \times \frac{1}{1000} = 2764.6\text{ kN/m}$$

Step 3: *Check for hoop stress in concrete at the base in empty tank condition (i.e., at transfer stage)*

Assuming loss ratios $\beta = 0.9$ and $\eta = 0.75$

- Prestressing force at transfer stage P_o

$$P_o = \frac{\beta}{\eta} P_e = \frac{0.9}{0.75} \times 2764.6 = 3317.5\text{ kN/m} = 3317.5\text{ N/mm}$$

- Stress in concrete at transfer

$$\sigma_c = \frac{P_o}{A} = \frac{3317.5 \times 1000}{1000 \times 300} = 11.06\text{ N/mm}^2 < f_{ct} = 17\text{ N/mm}^2\text{, OK}$$

Step 4: *Check for cracking load factor μ_{cr}*

Required $\mu_{cr} = 1.2$

- Strength of concrete in direct tension f_t

  $f_t = 2$ N/mm^2 (Table 1 of IS:3370 (Part III))

- Hoop tension required for cracking T_{cr}

  Maximum hoop stress due to effective prestressing force is 5.16 N/mm^2. Hence hoop tension required for cracking T_{cr} is:

$$T_{cr} = \frac{(5.16 + 2) \times 1000 \times 300}{1000} = 2148\text{ kN/m}$$

- Cracking load factor μ_{cr}

$$\mu_{cr} = \frac{T_{cr}}{T} = \frac{2148}{1337.6} = 1.61 > 1.2\text{, OK}$$

Step 5: *Design of vertical prestressing steel*

The maximum moment in the wall occurs in the empty tank condition (transfer loading stage).

- Prestressing force at transfer loading stage P_o

  $P_o = 3317.5$ kN/m

- Maximum pressure due to P_o (p_{pm})

  $$p_{pm} = \frac{P_o}{R} = \frac{3317.5}{20} = 165.88 \text{ kN/m}^2$$

- Maximum bending moment in longitudinal direction (M)

  $M = C_M p_{pm} H^2$

  For $H^2/2Rt = 10$, $C_M = 0.0122$ (Table 10 of IS:3370 (Part IV))
  Hence,

  $M = 0.0122 \times 165.88 \times 11^2 = 244.87$ kN-m/m

- Stress concrete in longitudinal direction due to M (σ_{l1})

  $$\sigma_{l1} = \frac{M}{Z} = \frac{244.87}{\frac{1}{6} \times 1 \times 0.3^2} = 16324.67 \text{ kN/m}^2$$

  $= 16.32$ N/mm^2 ($< f_{ct} = 17$ N/mm^2, OK)

- Stress in concrete in longitudinal direction due to partial winding (σ_{l2})

  $\sigma_{l2} = 0.3 \times$ hoop stress due to P_o

  $= 0.3 \times 11.06 = 3.32$ N/mm^2

- Stress in concrete in longitudinal direction at transfer (σ_l)

  $\sigma_l =$ Maximum of σ_{l1} and σ_{l2}

  $= 16.32$ N/mm^2

- Required longitudinal stress at service condition

  $$\sigma_l = \frac{\eta}{\beta} \times 16.32 = \frac{0.75}{0.9} \times 16.32 = 13.6 \text{ N/mm}^2$$

- Residual compressive stress at service condition f_{res}

  $f_{res} = 0.7$ N/mm^2

- Required effective prestressing force in the vertical direction

  The vertical prestressing steel should be designed to resist a stress of $13.6 + 0.7 = 14.3$ N/mm^2 at service condition. Hence,

  $$P_e = \frac{14.3 \times 1000 \times 300}{1000} = 4290 \text{ kN/m}$$

- Effective prestressing force provided by a single cable P_{cable}

$$P_{cable} = 16 \times \frac{\pi}{4} \times 8^2 \times 1100 \times \frac{1}{1000} = 884.67 \text{ kN}$$

- Required number of cables per metre length n

$$n = \frac{P_e}{P_{cable}} = \frac{4290}{884.67} = 4.85$$

- Spacing of cables s

$$s = \frac{1000}{n} = \frac{1000}{4.85} = 206.19 \text{ mm}$$

Provide cables at a spacing of 200 mm c/c.

Step 6: *Design of untensioned reinforcement*

Nominal reinforcement of 0.2 percent of the gross area are to be provided in both circumferential and longitudinal direction to resist stresses due to temperature changes and shrinkage of concrete.

- Area of untensioned steel A_{st}

$$A_{st} = \frac{0.2}{100} \times 1000 \times 300 = 600 \text{ mm}^2/\text{m}$$

Hence required area of steel per face of wall = 300 mm²/m

- Number of 8 mm diameter bars n

$$n = \frac{300}{\frac{\pi}{4} \times 8^2} = 5.97$$

- Spacing of 8 mm diameter bars s

$$s = \frac{1000}{n} = \frac{1000}{5.97} = 167.5 \text{ mm}$$

Provide 8 mm diameter bars at 160 mm spacing near both faces in both circumferential and longitudinal direction.

13.5 Ring Beams

Ring beams are used to support domes. They are provided to resist the tensile forces developed due to the meridional forces in the dome. Figure 13.6 shows a typical ring beam which supports the dome kept above a circular water tank.

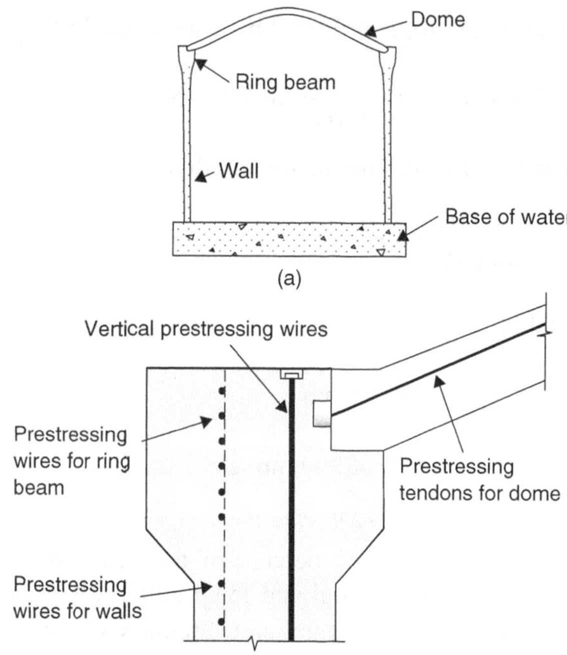

Fig. 13.6 Prestressed concrete ring beam (a) vertical section of a circular water tank and (b) tendons in the dome and ring beam junction

13.5.1 Analysis and Design

If the ring beam is continuously supported (i.e., over its entire length) on the vertical wall, then it can be assumed that the ring beam is subjected only to hoop tension.

Consider a spherical dome of radius R subjected to a total load W (Fig. 13.7). The vertical reaction per unit length (V) along the edge of the dome is:

$$V = \frac{W}{2\pi R \sin\theta} \tag{13.10}$$

where θ is equal to half of the angle subtended by the dome.

Since the dome does not carry any significant bending moments, the resultant reaction R_R will be tangent to the surface. Hence the horizontal component of the reaction per unit length H is given by:

$$H = V\cot\theta$$

$$= \frac{W\cot\theta}{2\pi R\sin\theta} \tag{13.11}$$

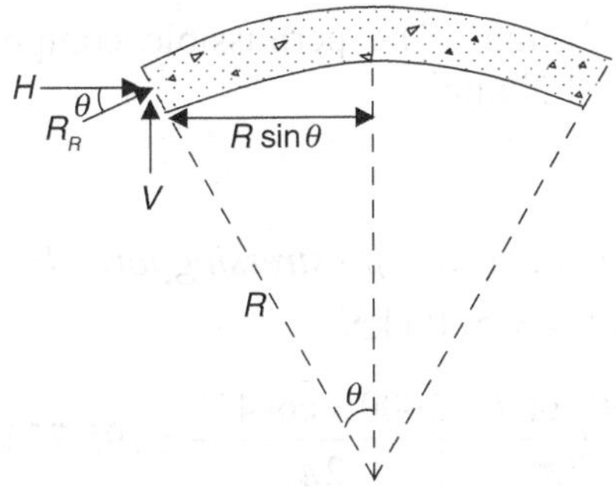

Fig. 13.7 Support reactions of the dome

The effective prestress P_e in the ring beam is determined using the assumption that the horizontal reaction H is entirely resisted by P_e (Fig. 13.8). Then,

$$P_e = HR\sin\theta = \frac{W\cot\theta}{2\pi} \tag{13.12}$$

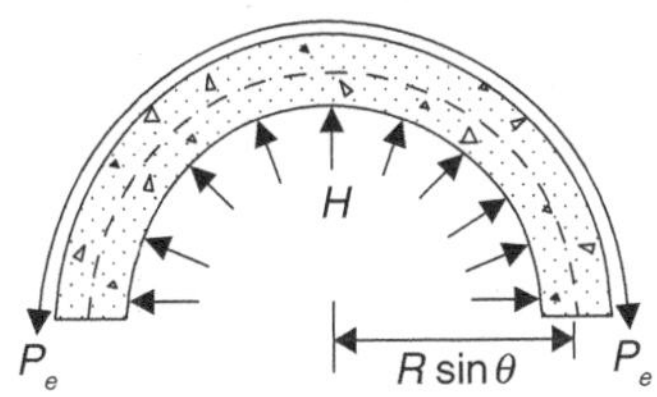

Fig. 13.8 Forces in the ring beam at service condition

The required area of tendon A_p is given by:

$$A_p = \frac{P_e}{f_{pe}}$$

After finding A_p, the prestressing force at transfer P_o is determined. The cross-sectional area of the ring beam A_c is found out using the equation:

$$A_c = \frac{P_o}{f_{ct}}$$

Example 13.3

A spherical dome shown in Fig. 13.7 carries a total load of 5000 kN. Design the edge ring beam if $\theta = 45^0$. Use 5 mm diameter high tensile wires of effective stress 1100 N/mm². The permissible compressible strength of concrete at transfer is 14 N/mm².

Solution:

Step 1: *Determination of effective prestressing force P_e*

Total design load $W = 5000$ kN

$$P_e = \frac{W \cot\theta}{2\pi} = \frac{5000 \times \cot 45^0}{2\pi} = 795.77 \text{ kN}$$

Step 2: *Design of circumferential prestressing steel*

- Area of prestressing steel A_p

  Effective prestress $f_{pe} = 1100$ N/mm²

$$A_p = \frac{P_e}{f_{pe}} = \frac{795.77 \times 10^3}{1100} = 723.43 \text{ mm}^2$$

- Number of 5 mm diameter high tensile wires n

$$n = \frac{723.43}{\frac{\pi}{4} \times 5^2} = 36.8$$

  Provide 40 wires.

Step 3: *Determination of required dimension of ring beam*

- Effective prestressing force P_e

$$P_e = 40 \times \frac{\pi}{4} \times 5^2 \times 1100 \times \frac{1}{1000} = 863.94$$

- Prestressing force at transfer P_o

  Assuming the loss ratios β and η as 0.9 and 0.75 respectively.

$$P_o = \frac{\beta}{\eta} P_e = \frac{0.9 \times 863.94}{0.75} = 1036.73 \text{ kN}$$

- Area of ring beam A

  Permissible compressive stress in concrete at transfer $f_{ct} = 14$ N/mm²

$$\text{Area of ring beam } A = \frac{P_o}{f_{ct}} = \frac{1036.73 \times 10^3}{14} = 74052 \text{ mm}^2$$

  Provide a ring beam of width 200 mm and depth 380 mm.

PROBLEMS

13.1 Design a prestressed concrete water tank of diameter 15 m and height 8 m. Assuming a flexible base, design the wall of the tank and find the spacing of 5 mm diameter high tensile steel wire for circumferential prestressing. Take f_{ct} = 14 N/mm^2 and f_{pe} = 1100 N/mm^2.

13.2 Design the circular water tank of Example 13.2, if the base of the wall is hinged.

References

Abeles, P. W., Bardhan-Roy, B. K., and Turner, F. H., *Prestressed Concrete Designers Handbook*, Second Edition, Viewpoint, Slough, 1976.

Allen, A. H., *An Introduction to Prestressed Concrete,* Cement and Concrete Association, Slough, 1981.

Bhatt, P., *Prestressed Concrete Design to Eurocodes*, Spon Press, Oxon, 2011.

Collins, M. P., and Mitchell, D., *Prestressed Concrete Structures*, Prentice-Hall, New Jersey, 1991.

Dayaratnam, P., *Prestressed Concrete Structures*, Fourth Edition, Oxford and IBH, New Delhi, 1987.

Evans, R. H., and Bennett, E. W., *Prestressed Concrete*: *Theory and Design*, Chapman and Hall, London, 1958.

Gerwick, B. C., *Construction of Prestressed Concrete Structures*, John Wiley and Sons, New York, 1993.

Gilbert, R. I., and Mickleborough, N. C., *Design of Prestressed Concrete*, Spon Press, London, 2002.

Guyon, Y., *Limit State Design of Prestressed Concrete*, Vols. 1 and 2, Applied Science, London, 1974.

Hulse, R., and Mosley, W. H., *Prestressed Concrete Design by Computer*, Macmillan, London, 1987.

Hurst, M. H., *Prestressed Concrete Design*, Second Edition, E and FN Spon, London, 1998.

Kahn, S., and Williams, M., *Post-tensioned Concrete Floors*, Butterworth-Heinemann, Oxford, 1995.

Khachaturian, N., and Gurfinkel, G., *Prestressed Concrete*, McGraw-Hill, New York, 1969.

Kong, F. K., and Evans, R. H., *Reinforced and Prestressed Concrete*, Third Edition, Van Nostrand Reinhold, London, 1987.

Leonhardt, F., *Prestressed Concrete: Design and Construction,* Wilhelm Ernst and Sohn, Berlin, 1964.

Libby, J. R., *Modern Prestressed Concrete*, Third Edition, CBS Publishers, New Delhi, 2007.

Lin, T. Y. and Burns, N. H., *Design of Prestressed Concrete Structures,* Third Edition, Wiley India, New Delhi, 2012.

Magnel, G., *Prestressed Concrete,* Third Edition, Concrete Publications, London, 1954.

Mallick, S. K., and Gupta, A. P., *Prestressed Concrete,* Second Edition, Oxford and IBH, New Delhi, 1986.

Menn, C., *Prestressed Concrete Bridges,* Birkhauser Verlag, Basel, 1990.

Naaman, A. E., *Prestressed Concrete Analysis and Design*: *Fundamentals*, Second Edition, Techno Press 3000, Michigan, 2004.

Nawy, E. G., *Prestressed Concrete*: *A Fundamental Approach*, Fifth Edition, Prentice-Hall, New Jersey, 2010.

Nilson, A. H., *Design of Prestressed Concrete*, Second Edition, John Wiley, New York, 1987.

OBrien, E., Dixon, A., and Sheils, E., *Reinforced and Prestressed Concrete Design to EC2 – The complete Process*, Second Edition, Spon Press, Oxon, 2012.

Pandit, G. S., and Gupta, S. P., *Prestressed Concrete*, CBS Publishers, New Delhi, 2008.

Pillai, S. U., and Menon, D., *Reinforced Concrete Design*, Third Edition, Tata McGraw-Hill, New Delhi, 2009.

Rajagopalan, N., *Prestressed Concrete*, Narosa Publishers, New Delhi, 2002.

Raju, N. K., *Prestressed Concrete*, Fourth Edition, Tata McGraw-Hill, New Delhi, 2007.

Ramaswamy, G. S., *Modern Prestressed Concrete Design*, Arnold-Heinemann, New Delhi, 1978.

Sawko, F. (ed.) *Developments in Prestressed Concrete*, Vols. 1 and 2, Applied Science, London, 1978.

Warner, R. F., Rangan, B. V., Hall, A. S., and Faulkes, K. A., *Concrete Structures,* Addison Wesley Longman, South Melbourne, 1999.

Wilby, C. B., *Concrete Materials and Structures*, Cambridge University Press, Cambridge, 1991.

Index